AF794607

ELECTRICITY POWER GENERATION

IEEE Press
445 Hoes Lane
Piscataway, NJ 08855

IEEE Press Editorial Board
Lajos Hanzo, *Editor in Chief*

R. Abari
J. Anderson
F. Canavero
T. G. Croda
M. El-Hawary
B. M. Hammerli
M. Lanzerotti
O. Malik
S. Nahavandi
W. Reeve
T. Samad
G. Zobrist

Kenneth Moore, *Director of IEEE Book and Information Services (BIS)*

ELECTRICITY POWER GENERATION
The Changing Dimensions

Digambar M. Tagare

IEEE PRESS SERIES ON POWER ENGINEERING

IEEE

IEEE PRESS

WILEY

A JOHN WILEY & SONS, INC., PUBLICATION

Copyright © 2011 by the Institute of Electrical and Electronics Engineers, Inc.

Published by John Wiley & Sons, Inc., Hoboken, New Jersey. All rights reserved.

Published simultaneously in Canada.

No part of this publication may be reproduced, stored in a retrieval system or transmitted in any form or by any means, electronic, mechanical, photocopying, recording, scanning or otherwise, except as permitted under Section 107 or 108 of the 1976 United States Copyright Act, without either the prior written permission of the Publisher, or authorization through payment of the appropriate per-copy fee to the Copyright Clearance Center, Inc., 222 Rosewood Drive, Danvers, MA 01923, (978) 750-8400, fax (978) 750-4470, or on the web at www.copyright.com. Requests to the Publisher for permission should be addressed to the Permissions Department, John Wiley & Sons, Inc., 111 River Street, Hoboken, NJ 07030, (201) 748-6011, fax (201) 748-6008, or online at http://www.wiley.com/go/permission.

Limit of Liability/Disclaimer of Warranty: While the publisher and author have used their best efforts in preparing this book, they make no representation or warranties with respect to the accuracy or completeness of the contents of this book and specifically disclaim any implied warranties of merchantability or fitness for a particular purpose. No warranty may be created or extended by sales representatives or written sales materials. The advice and strategies contained herein may not be suitable for your situation. You should consult with a professional where appropriate. Neither the publisher nor author shall be liable for any loss of profit or any other commercial damages, including but not limited to special, incidental, consequential, or other damages.

For general information on our other products and services please contact our Customer Care Department within the United States at (800) 762-2974, outside the United States at (317) 572-3993 or fax (317) 572-4002.

Wiley also publishes its books in a variety of electronic formats. Some content that appears in print, however, may not be available in electronic formats. For more information about Wiley products, visit our web site at www.wiley.com.

Library of Congress Cataloging-in-Publication Data:

Tagare, D. M.
Electricity power generation : the changing dimensions / Digambar M. Tagare.
p. cm.
Summary: "This book offers an analytical overview of established electric generation processes, along with the present status & improvements for meeting the strains of reconstruction. These old methods are hydro-electric, thermal & nuclear power production. The book covers climatic constraints; their affects and how they are shaping thermal production. The book also covers the main renewable energy sources, wind and PV cells and the hybrids arising out of these. It covers distributed generation which already has a large presence is now being joined by wind & PV energies. It covers their accommodation in the present system. It introduces energy stores for electricity; when they burst upon the scene in full strength are expected to revolutionize electricity production. In all the subjects covered, there are references to power marketing & how it is shaping production. There will also be a reference chapter on how the power market works"—Provided by publisher.
ISBN 978-0-470-60028-3 (hardback)
1. Electric power production. I. Title.
TK1001.T33 2010
621.31—dc22 2010022385

Printed in Singapore.

oBook ISBN: 978-0-470-87265-9
ePDF ISBN: 978-0-470-87266-6

10 9 8 7 6 5 4 3 2 1

CONTENTS

FOREWORD

It is with pleasure that I write this foreword to *Electricity Power Generation: The Changing Dimensions,* written by my esteemed friend, Mr. D. M. Tagare. Mr. Tagare and I have worked in the field of power generation and power management over the last 50 years. Mr. Tagare is known in India for his outstanding work in the field of power capacitors and filters. He was the chairman of capacitor division of Indian Electrical and Electronics Manufacturers Association and piloted growth and development of the industry for a number of years. He has published books on capacitors and reactive management that have received worldwide acclaim. This book is the result of decades of experience. It will enhance the knowledge of persons working in the power sector, both young and old. The book provides the latest knowledge on issues of electricity generation.

The electricity sector structure has changed rapidly, especially during the last 20 years. From a monopoly line-function structure, power generation and power trading have been transformed into a competitive industry. This has attracted the attention of planners, economists, managers, industrial engineers, and civil servants. Thus, there is a migration of other sector experts to the electricity sector. These experts require education and knowledge of electricity generation, not only basic, but most up-to-date, covering the latest and upcoming technologies, such as fuel cells and hydrogen. All these areas together with new and renewable energy generation are well covered in the book. Besides meeting the needs of working engineers in the sector, this book will also meet the needs of students working in the field of energy management, energy consultants, auditors, civil servants, industrial managers, economists, and planners. The book will serve as handbook and will be a useful addition to technical libraries.

Every technology over the years seeks to improve efficiencies; electricity power generation is no exception. Constant endeavor is made to achieve higher efficiencies through higher pressures and temperatures and so on. The development of power electronics, computer engineering, and information technology (IT) has added new dimensions to the electricity generation sector. Complete automation in working is achieved. Thus, the electricity engineer is required to acquire knowledge in these fields as applicable to his or her sector. The book includes updates on these subjects associated with electrical engineering. This enhances the value of the book by directly covering these areas but also by giving generous references to IEEE publications. Thus, for obtaining further detailed/intimate knowledge of the subject, an engineer has at hand relevant literature. Together with the references, this book will serve the needs of research scholars.

The most prevalent and commercial services of power generation via hydroelectricity generation, thermal power generation, and nuclear power generation are covered in Chapters 2 to 8 of the book. The significant points to be noted are:

1. Hydroelectricity generation covers pumped storage systems and oceanic-energy-based electricity systems.
2. Thermal power generation covers the requirements of rigid frequency and voltage controls, a must in modern systems. Two chapters are assigned to environmental concerns, acid rain, and carbon emissions. Rarely are such details found in a book on electricity generation.
3. In the chapter on nuclear power generation, mention is made of smaller marketable nuclear power plants developed by Japan and Russia. These plants are used to power nuclear submarines.

After extensive coverage of conventional generation plants, the author covers in Chapters 9 to 11 cover wind power generation, solar power systems, and fuel cells. Wind power generation is now an accepted method, whereas other technologies are fast developing. Persons working in the field of power generation will often be confronted with these applications, as governments in various countries are making it mandatory to get such generation up to a certain percentage of the total. Chapter 12 on hybrid systems covers the combination of the renewable systems. Chapter 17 on the hydrogen era describes a technology of future that is in the nascent stage. However, basic knowledge of the process is necessary for a practicing electrical engineer.

The author's practical approach is exhibited through Chapter 13 on cogeneration, Chapter 14 on distributed generation (DG), Chapter 15 on interconnecting distributed resources, and Chapter 16 on energy storage. These chapters provide information for field engineers, consultants on the issues, and power system engineers who face these issues in daily life. These issues, once again, are rarely incorporated in books on electricity generation. Distributed generation is coming up in a large way to improve system reliability and reduce transmission and distribution losses. However, there are interconnection problems. I am happy that the issue is fully explained here. Demand-side management is an issue in power systems that cannot be sidetracked. Power storage provides a solution.

Chapter 18 gives a brief outline of power marketing and is a must for every engineer working in a generating company. Merely increasing and optimizing generation is not adequate. This power has to be sold. Thus, without an insight in power marketing, the knowledge base of a generation executive is not complete.

Electricity reforms are taking place all over the world. The executives in the power sectors are facing constant competition. In a competitive world without sound knowledge of the subject and knowledge of the latest technological developments, the competition cannot be faced. The book will impart the necessary knowledge to senior as well as junior executives, and can be used as textbook and a reference handbook. I congratulate Mr. D. M. Tagare on this excellent publication and wish him all success. I am confident the book will receive worldwide acclaim.

P. L. NENE
Former Chairman, Madhya Pradesh Electricity Board
Bhopal, India

PREFACE

The historical growth of the electricity business has been the result of a continuously growing demand for electricity. It came about through careful nurturing under the monopolistic conditions under which it was born. This growth fell short of meeting a rising demand for electricity in quantity, quality, and price. There grew up buffeting forces, such as concerns on environmental damages associated with its production and open marketing of electricity as a commodity. The system, which had become stagnant and contented, could not attract innovative young engineers. It had to change.

These changes did come and are coming still. They came through structural changes in business, both in management and in sales. Revolutionary and possibly epoch-making changes are in the making in the outlook on sources of energy and its utilization through energy reserves. Smug reliance on buried but presently cheap fossil fuels is giving way to finding ways to utilize abundant amounts of energy in nature: water from the mountains and the seas, air, the sun's rays, and bioproducts. Research is concentrated on bottling up electricity in small or large reserves. In short, its dimensions are changing, demanding innovation. This book is about these changing dimensions and how existing producers are trying to adapt to the changes. It begins by outlining the structural changes which have and are taking place in the electricity business all over the world.

Hydroelectricity is our strongest ally during this transition. Its logical growth and renovation in operation are explained in Chapter 2. Apart from producing electricity as and when we want it, it has produced economic prosperity in the basins it serves. Pumped storage, minor electricity production based on irrigation canals, seasonal cascades, and seawater tides and waves were once experimented upon and shelved as being nonviable. They have rising dimensions now and offer scope for innovation and wealth production. Chapter 3 details these. Noncontinuous energy supply from the seas and its adoption in the existing electricity supply grids demands a look.

Thermal electricity power generation today is the backbone of electricity business. Universally available coal is the main fuel. A close look is being taken into the methods of its use. Emphasis is shifting from the size of generators to better efficiencies through higher temperatures and pressures, and, consequently, through combined cycle plant processes. With interconnections and rising transmission grid sizes, the old NERC norms meant for bringing uniformity in the rising electricity systems are giving way to new norms such as dispersed reliability reserves in specified droop characteristics of generators, very stringent system frequency controls, and so on. Power marketing has brought renovations in maintenance systems of generators in general. Chapters 4 and 5 cover these aspects of thermal generation.

Acid rain resulting from oxidation of sulfur present in fossil fuels and of nitrogen in air have a limited product volume. These can be and are contained, at a cost. Past legislation and current costs for containing acid rain, along with its measurement and reporting systems, are given in Chapter 6.

Carbon emissions, on the other hand, spread out and have to be attacked at the generating source, only by cutting down on thermal generation. This is rather a tall order, considering that thermal generation is the main electricity producer today. Containing carbon emissions in times of rising electricity requirements is undertaken as a main task by the United Nations Framework Convention on Climate Change (UNFCCC). The efforts are backed up by no less than a former president of the United States, Nobel prize winners, and governments all over the world. Limitations agreed to upon, from time to time, such as the Kyoto Protocols, are voluntary. Inducements, both for developed countries and for developing countries, are continuously worked out by UNFCCC. At the time of this writing, there are as many as 57 different schemes on the Internet for earning these inducements. Chapter 7 gives basic details. There is a good scope for self-learning and helping out on these efforts for electrical engineers.

Nuclear power generation, promises to be one of the major tools in the struggle against carbon emissions. Its association with mass destruction, the possibility of misuse of its materials, and high costs and long implementation delays are the major deterrents. Safety considerations require special categorization of its in-plant cable and wiring layouts. Safety drills are a must in these plants. Safety is also important in disposal of the waste materials. Accidents at these plants could be catastrophic, as happened at Chernobyl in Russia. Chapter 8 details all these issues. Smaller marketable nuclear power plants have been developed by Japan and Russia. These represent small plants used for power in nuclear submarines with improvements that reduce the life of waste materials. Avenues for developments in nuclear power applications are listed.

Wind power generation is a leading contender for maximum share in electricity generation by renewable energy sources. Denmark, in association with hydroelectricity sources from Sweden, claims 100% wind energy for its total energy requirements. The fluctuating and unpredictable nature of wind power creates technical as well as managerial problems in its adoption in power grids. Low-voltage ride-through and modeling parameters required by grid managers are technical problems. Maximum penetration limits and capacity factor are the managerial problems. Accurate supply quantity and timing predictions are the marketer's problems. Chapter 10, with its long list of references, covers these aspects.

Photovoltaic energy is the most vied for energy source among the renewable sources. It has a basic regularity in magnitude and timing all over the world. However, atmospheric conditions, high costs of solar cells, and large surface areas required have deterred its wider application so far. Maximum power point trackers are a universal solution for utilizing PV energy. Interfacing with grids increases the costs due to the need for power conditioners. Interfacing also raises technical problems, such as high ramping rates up or down, loss of load possibilities, and a limitation on its penetration into the grid. Presently, it is best suited as a low-volume dispersed energy generator, particularly for agricultural water pumps and residences. These aspects are covered in Chapter 11. The chapter also covers expanding application ranges for solar panels.

Fuel cells are the latest entrants on the electricity generation scene. Instead of generating thermal energy through chemical conversion like oxidation of H2 by burning, they draw out the energy electronically. Efficiencies claimed are high. Proton energy membranes (PEMs) for fuel cells operating at low temperatures (80°C) have come up fast. Present developments cast these fuel cells as low-volume dispersed electricity generators, with main applications in automobiles. They are making inroads in the electricity requirements of other customers, such as hotels, hospitals, and residential complexes, not in competition with grid supplies, but as supplementary suppliers, contributing to electricity cost reduction. Solid oxide fuel cells operating at higher temperatures have possible applications in the larger megawatt ratings. Fuel cells have scope for R&D both in basic membranes and customer-end applications. Chapter 12 deals with these issues.

Stand-alone hybrid systems are sprouting up all over the world. Chapter 12 gives detailed analytical study of six alternative hybrid stand-alone systems for a remote Mexican village. Support batteries for a PV cell array have low lives and are costly. Their use should be optimized. Fuel cells are fast replacing storage batteries in hybrids. Cases of unique midsea island-based wind energy hybrids are interesting for optimally using diesel oil. Limitations, interconnections, and controls between a hybrid and a grid become demanding as individual cases arise. Chapter 12 deals with these systems. Hybrid cars are an intermediate stage to fully electric cars. They are developing at a fast rate.

Combined generation has long been in operation and is reliably providing electricity. Equally important is its ability to utilize otherwise wasted thermal energy since their generating capacities are considerable. Existing thermal electricity generators formed strict rules for admitting this combined generation into their grids. Combined generation could influence grid parameters considerably. It was also under a sort of dual control. Its control system, vis-à-vis the grid parameters, had to be watched. It could and did operate in an islanded condition with or without a portion of the main grid. Its partnership with the grid, which was on an on-again/off- again basis, has now turned into a permanent basis, mainly due to relief it provides under environmental concerns as well as under open marketing system. To cite an example, a cooperative sugar factory in India earned revenue from sale of electricity to a state electricity board, compatible with the revenue it earned from the sale of its sugar. In another example cited, a refinery cut down its electricity bill from the grid supply. Combined generation has good scope for optimization. Chapter 13 covers combined generation.

A fairly large population of gas/oil-based piston/turbine-driven generators are dispersed across a sizeable distribution network of electric power systems. Their main purpose is to support distribution networks with active/reactive power at their weak points. They are put in by growing distribution systems as an economic alternative to other systems. They are also put in by private operators, preferably at load points. Chapter 14 shows that distributed generation (DG) power systems are not under area control and are not allowed operations that will impact system parameters such as frequency and voltage. They do alter local short-circuit levels. DGs give support to problems such as ride-through during voltage dips, small-disturbance stability, fuse ratings and relay settings, load flows, and so on. Upon a system disturbance, they must disconnect instantly through special relays like voltage vector relays and rate of change of frequency relays. Location at load centers, selling energy under peak load periods, selling contingency reserves, and so on can make them lucrative for private operators. Large-sized stand-by generators in industries are not allowed to connect into a distribution system. The Pune system, developed to help a distribution company with these standby generators, allows industries to "sell" power without actually putting current into the distribution network. A concept of distributed marketing centers is being promoted in Britain. Under this concept, local distribution generation will be given incentives to compete in the electric supply, particularly under peak-load conditions as well as under emergency conditions. It will help in breaking the barriers on wind energy penetration as well as in improving the utilization of the existing electricity systems.

Chapter 15 deals with distributed resources (DR) with higher power capacities and compatible power control systems. They connect into the upper voltage end of distribution system and are under command of the main system operation. These DRs can participate directly in power market operations. Typical examples are wind turbine generators. Interconnection systems covered here relate to system controls, system protection, steady-state control on system parameters like voltages, frequency, power factor, and so on. The interconnections impose restraints on both sides at the point of common coupling (PCC). Power quality windows specify upper and lower limits on voltages, frequencies, and harmonics. Operations recommended under islanding conditions and reconnections are given. Safety aspects are also sketched out. For exact and accurate information, it is recommended to refer to IEEE standards mentioned at the end of the reference list.

Chapter 16 on energy stores begins with the advantages of various types which are available and their applications, particularly to electrical power systems. Distribution system parameters, load density, and short-circuit capacity influence their applications. Details of a storage design to support the fluctuating nature of a PV system are given. This is followed by a detailed storage design for wind energy. Energy output from this store and its revenue earning capacity under different control nodes of storage are given. Compressed air energy storage is gaining attention. Energy stores are important for utilizing nature's abundant energy supply. The subject is an evolving challenge.

Hydrogen is looked upon as a remedy against atmospheric pollution. First, atmospheric pollution sources, their lifetimes, and rates of increases are presented. Hydrogen properties are given next. Production processes of hydrogen, their costs, and performance characteristics are given. Roadblocks to development and spread in usage are discussed. Various governments have appointed special commissions for recommendations on tackling these issues. A proposal for producing hydrogen from nuclear power is outlined. Chapter 17 covers these aspects.

Chapter 18 covers the basic structure of power marketing of electricity. The first part covers market operators, their tools, and objectives. Objectives are also laid out for transmission, still a monopoly, and wholesale markets. The second part covers capital investment prospects under the revised circumstances. Attention is given to promoting investment in the transmission sector, which is now not a market player.

Chapter 19 looks into the future with transfer of the energy source for automobiles from fuels to electricity. Thermal electricity generation will increase rather than diminish. It will not stay contained. Can future energy stores come in packs on racks in a commodity store? Electricity power production is not constrained to its old ways. It is evolving and challenging. The book tries to open up these vistas.

Many thanks go to the IEEE for its permission to dig into and borrow from their publications, on which this book is almost wholly based. I thank the IEEE Press and John Wiley & Sons, Inc. for publishing the book. I have to thank Mr. P. L Nene, ex-chairman of Madhya Pradesh Electricity Board (MPEB), for his support and Foreword. Thanks are due to Mr. Dwarkanath, ex-chief engineer, thermal generation, of MPEB, for his help. Thanks are due to my staff, particularly to Mr. Abhijit Katepallewar for his untiring help in getting this in print.

Last but not the least, thanks are due to my wife, Mira, for her enduring patience with me and support.

DIGAMBAR M. TAGARE

Pune, India
December 2010

1

ELECTRICITY HISTORY— A REVIEW AND THE ROAD AHEAD

1.1 HISTORY OF GROWTH OF THE ELECTRICITY BUSINESS

1.1.1 Societal and Organizational Changes

The start-up of electricity production and supply had an almost uniform pattern all over the world, including India. The state government issued a license covering mainly municipal areas in medium to large cities. The areas of franchises were limited to city limits. The electricity supply company had a few restrictions. They had to supply power to anybody who asked for it, and they had to abide by safety rules. The electricity quality parameters were set more or less by the machinery suppliers.

The load consisted of a few incandescent electric light bulbs per house. City street lighting was also provided by the electricity supply company.

Besides these city suppliers, there were quite a few industries that had installed their on generating plants. Textiles mills or paper mills, for example, required large quantities of steam for processing their production. Steam from the same source was used in a reciprocating steam engine to generate electricity.

The demand for electricity supply grew along with the economic growth of that particular country. In the United States and West, the growth in electric power took place at a steady and fast rate. The returns on electricity utilities were good and they required more capital to grow. This resulted in acquisitions and amalgamations. The end result

Electricity Power Generation. By Digambar M. Tagare

Copyright © 2011 the Institute of Electrical and Electronics Engineers, Inc.

was larger utilities serving larger areas. The boundaries of service areas first crossed the city limits and then state boundaries.

Supply and delivery were considered a natural monopoly in that no two licensees could operate within the same area. Monopolies had to be regulated so that they did not take advantage of consumers. Prices were regulated and standards were laid down. In return, a fair rate of return on the investments, besides normal operating expenses including depreciation, was allowed. Thus, regulation served both the utility and consumers. This regime is now termed as a regulatory regime or a (fixed) rate of return regime.

Electricity power generation, transmission, and distribution were vertically integrated. This offered several advantages for the faster growth of utilities. Irrespective of which sector earned less (transmission, for example), telescoping plans were made for all the sectors as one vertically integrated unit. The plans were based on fairly accurate forecasts of load growth over a long period (5–10 years). Again, the plans and investments were doubly vetted by the regulating agency. This gave fairly good and reliable results. It guaranteed predicted returns and maintained investor confidence [1]. All in all, this regulatory regime could be seen as the cause for growth of electricity utilities over the initial periods and even later.

1.2 INNOVATIVE TECHNOLOGY DEVELOPMENTS AND GROWTH OF CONGLOMERATES

Continually innovating technology contributed greatly to the buildup of electrical utilities. Take the size of the generators. They went up from 250 MW to 1000 MW per unit in stages. The generators were eventually cooled by hydrogen under pressure. This increased efficiencies and cut down fuel consumption to almost 10,800 BTU/kW-hr [2]. The turbines driving the generators also underwent changes. Steam pressures and temperatures were raised, leading to higher efficiencies and higher ratings. Overall, the thermal input dropped from as high as 92,500 BTU/kW-hr in 1902 to as low as 20,700 BTU/kW-hr by 1932 and to 15,000 BTU/kW-hr, by 1950 [2]. As mentioned earlier, it is now 10,800 BTU/kW-hr.

On the technology improvement side, a combined cycle process was developed wherein a gas turbine is the first link in the process. The gas exhaust from this engine is fed to a steam boiler. Steam from this boiler operates a second steam turbine. The thermal utilization factor has gone high.

On the transmission side, raising the transmission voltages resulted in lowering transmission losses. Transmission voltages went up from 132 kV to 270, 400, 500, and 750 kV AC in stages. In between, they changed over to HVDC high-voltage DC transmission systems.

1.2.1 Factors Leading to Further Growth of Conglomerates

All these developments meant higher investment costs, which only utility conglomerates with large investment capacity could afford. Over the years, wages and inflation rose on a yearly basis. Electricity utilities could keep pace with these rises and still make profits—first, due to technological evolutions and, second, due to the economies of scale resulting from larger volumes of electrical energy handled. The growth of conglomerates was thus nourished.

1.3 ECONOMIC GROWTH—GDP AND ELECTRICITY CONSUMPTION

Economic growth, as measured by gross domestic product (GDP), and electricity consumption are closely tied. This relation reflects on the growth history as well as the growth rate of electrical utilities in a country. Economic growth and electricity consumption are interdependent, in that growth of one promotes growth of the other and vice versa (see Figs. 1-1–1-3 and Table 1-1).

The percentage of energy used in the form of electricity versus the total energy in all forms of energy under use is expected to increase from 40% to 60%. This is because energy can be most efficiently utilized when in an electrical form. In the developed countries, there was rapid growth in the demand and supply of electrical energy during the period 1950–1970, then, although the growth in absolute figures kept on increasing, its yearly rate of growth somewhat tapered down. On the other hand, in non-OCB countries such as India and China, which are breaking through the economic growth barrier, the rate of growth in electrical energy exceeds the rate of growth in advanced countries. The actual growth in volumes in advanced countries is still higher than that in the developing countries because of a large established base.

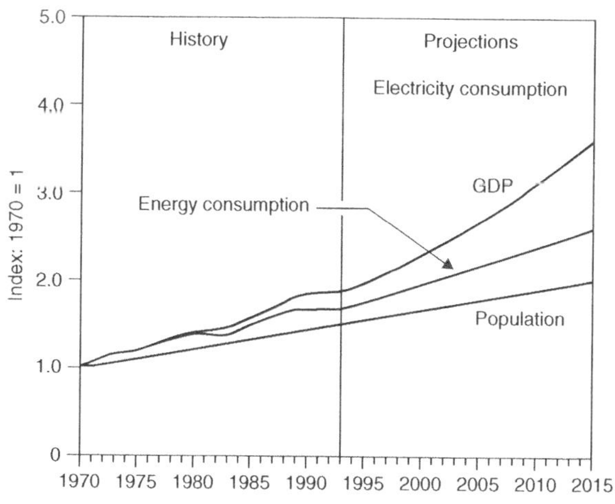

Figure 1-1. World energy, GDP, and population trends, 1970–2015. (From [3], © 1997 IEEE.)

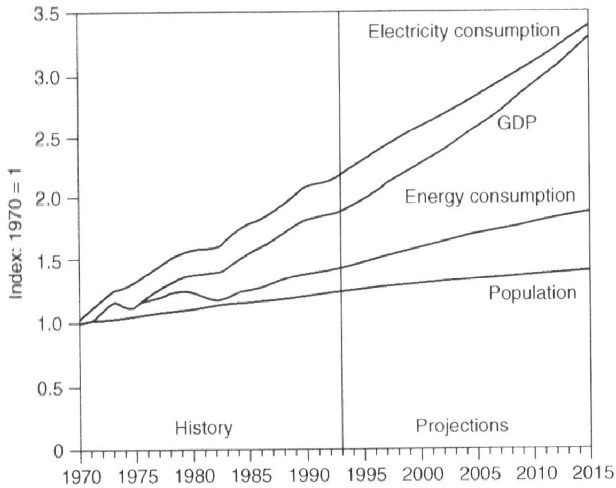

Figure 1-2. OECD energy, GDP, and population. (From [3], © 1997 IEEE.)

1.4 MONOPOLIES DEVELOP BUILT-IN DEFECTS

The increasing electricity utility sizes and the resulting market power led to several defects in the system. This led to the beginning of unbundling of the electricity utility business.

For example, government-owned electricity boards in various Indian states had an absolute monopoly with no shareholder responsibility. Their efficiencies plummeted to such an extent that total losses, including power thefts, varied from 36% to 50% of the

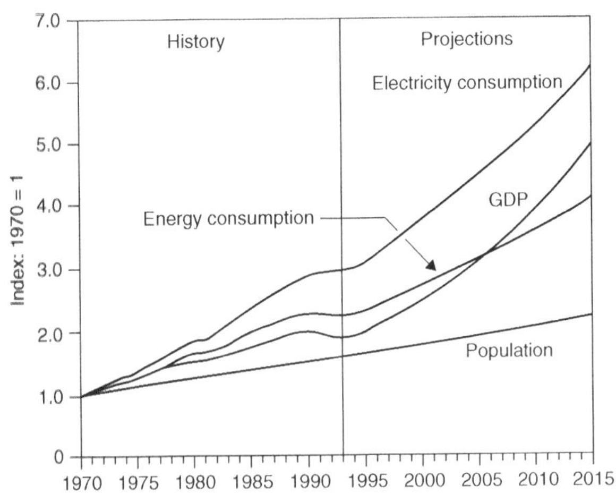

Figure 1-3. Non-OECD energy, GDP, and population. (From [3], © 1997 IEEE.)

Table 1-1. World energy consumption

Region	1970	1993	2010	2015	Annual change (%)	
					1970–1993	1993–2015
OECD	135.1	190.7	239.8	251.4	1.5	1.3
United States	67.6	87.3	104.7	108.0	1.1	1.0
Non-OECD	71.7	158.3	255.3	291.0	3.5	2.8
Non-OECD Asia	18.9	60.5	126.6	150.8	5.2	4.3
EE/FSU	39.7	60.9	74.0	79.7	1.9	1.2
Total world	206.7	349.1	495.1	542.3	2.3	2.0

Note: Totals may not equal sum of components due to rounding.
Source: [3], © 1997 IEEE.

energy produced from one state to another. These inefficiencies in turn damaged the power quality. There were unimaginably low voltages at consumer ends. Load shedding, both scheduled and unscheduled, was the order of the day.

On the other hand, in advanced countries such as the United States, the monopolies brought in immense market power that could be misused in various ways—unfair competition, for example, denial to others, access to transmission, fudging the accounts from one component of the business to another, and so on.

Monopolistic position also built in strong labor unions with forceful bargaining power. There was a severe opposition to unbundling of electricity utilities in the United Kingdom as well as in many other countries. This union power affected public relations as well as the overall efficiencies.

Recently, a new wave of open competition has challenged monopolies.

1.5 BREAKUP OF BELL SYSTEMS LEADS TO UNBUNDLING

Let us go into a bit of detail as to what happened to a big conglomerate—Bell Telephone and AT&T group in the United States. Like electric supply and services, telephone is a widespread public service utility. Bell Telephone Systems started as a local entity and grew into a nationwide network cutting across state lines. By 1990, the Bell group had four major activities:

1. Local regions catering to local telephone calls
2. Exchanges across regions catering to intrastate calls
3. Local-to-long distance calls
4. Manufacture of telephone and line equipment

They had no competition in long-distance calls. They charged high rates, with which they subsidized the local-call rates, which were kept low. The low local-call rates kept the local competition out. The competitors sought a remedy.

The dispute went on for almost 12 years, at end of which the Bell group signed a Modification of Final Judgment (MFJ), with the regulatory commission. Under this agreement, the local call business was to be handled by six independent regional Bell companies. AT&T would handle the long-distance and equipment-manufacturing groups [7]. The unbundling of Bell Telephone Systems led to a new model for public service utilities. The end results were interesting.

The local-area Bell groups now raised their rates to the competition level, which was much above what they had charged earlier. Their revenues increased and they became healthier. The AT&T group could reduce the long-distance call rates. However, their equipment group became concentrated and prospered.

The continuously evolving technologies were the biggest surprises for both the groups. With the explosive expansion of the information technology "Expressway," the demand for telephone services of all types increased beyond imagination. One wonders whether they could have handled this demand as efficiently as they are doing now had they still remained a monopolistic conglomerate!

The public benefited the most, from an increase in quality of service and falling service charges.

1.5.1 New Technologies Open Competition to Small-Scale Capital

The technological progress in developing new machines and new ideas came to the electricity business also. The new aero-gas turbine-driven generators of low MW ratings have as high efficiency as an old 1000 MW steam-driven generator with steam temperature and pressure at its maximum. The ratings per unit dropped to as low as 10 MW/unit [18]. With no paraphernalia of a coal-based large generating station, the capital costs per MW were low. The set-up time was low and they were almost mobile. These could be taken near to a load center and commissioned. Initially, these were used to serve during the peak demand periods. Compared to the almost depreciated costs of old coal-based plants with their large overheads, the total operating costs of these machines running to meet the peak demand proved to be competitive, depending on such factors as the distance of service.

These new technology developments now promoted competition between the giants and new high-efficiency mobile units. The big plants could not stand this and started selling out or merging so that higher scales of operation could give them the extra margin with which to compete. Thus, technological evolution started a trend toward unbundling of the electrical utilities.

1.5.2 Oil Cartels Deliver a Blow

Apart from the slowly developing technological challenges to continuation of the electricity utilities as they existed, there were quite a few historical events that accelerated the restructuring of the utilities. These historical events, causing a deep impact occurred, first in the United States and the West. The repercussions flowed into other countries with time.

During 1973–1974, the oil producing countries formed a cartel (OPEC) and pushed up world oil prices. There was a sudden realization that alternative energy sources had to be looked into. There was also a realization that stricter measures were necessary in the use of electrical energy as well as in its production and supply.

There was a regime revolution in Iran in 1978 that stopped the oil flow out of that country. Iran was an established and abundant supplier of oil. This cutoff of oil by Iran was another eye opener.

1.5.3 Environmental Concerns Raise Costs

Environmental concerns started occupying the center stage. All fossil- based fuels produce obnoxious gases such as NO_x and SO_x. Under environmental protection laws, the stack heights of chimneys had to be substantially raised so that the SO_2 fumes did not form local toxic pockets but spread over a wider area. Scrubbers had to be installed. These measures added to the capital costs in good measure but did not add to the power quality or to the efficiency in producing electric power.

Here again, a 10 MW aero-gas turbine was environmentally more friendly and manageable than an old 1000 MW coal-based power plant.

Nuclear power plants had come out in large numbers, since the fuel costs per kW-hr were next to nil. From the environmental point of view, these plants have a high risk, both during normal production but very much more so should an accident take place, spewing radioactive material over a large area. The environmental protection measures added substantially to the capital costs.

1.6 IMPORTANCE OF RENEWABLE ENERGY RECOGNIZED—WIND ENERGY BECOMES A CHALLENGER

Alternate renewable energy is playing a steadily rising role in discouraging continuance of large electricity utilities. Fossil-fuel-burning plants (coal and oil) emit large amounts of CO_2 into the atmosphere. The CO_2 cloud, as it grows steadily over the years, acts as a shield preventing heat radiation from the earth and increases overall earth temperatures. This has resulted in putting a cap on how much CO_2 can be discharged in air by a given plant at its capacity. Should the plant be unable to meet this requirement, it has to buy an emission allowance certificate from a "green" energy producer. A wind generator, for example, producing so many kW-hrs of green energy is entitled to a saleable emissions allowance certificate for so many equivalent tons of CO_2. This wind generator sells its electrical energy to a power grid, sells its emission allowance certificate, and additionally gets lot of fiscal benefits from the government for producing "green" energy. Of late, wind power has made significant inroads in Europe, the United States, and other countries.

Thus, with the old regulatory regime the electricity systems reached a point where the consumers started expecting larger and larger volumes of electricity with better quality and reliability. The time had come to pause and consider whether the regulatory system could be adequate for future.

1.6.1 A System Changeover is Necessary

As the regulated electricity utilities showed signs of having become too large and of being unable to meet the expectations of their customers, the thinking and approach veered toward deregulation and competition. A simple economic theory stipulates that competition will result in optimum deployment of economic resources. One begins to see the implications and results thereof in the open commodity stock markets, which have been able, despite lots of hitches, to bring optimal results both to producers and to consumers. Can we treat electrical energy as a tradable commodity and subject it to open market competition?

This seems to be the answer and the direction in which the electricity business has started to move. For a start, we will have to think of where to begin restructuring.

The answer to dismantling the present structure is neither an unmitigated yes nor no. Unlike other commodities, which can be stored if there is a sudden surplus and vice versa, electricity cannot be stored. Irrespective of the distances to be traveled or of the impedances to be overcome, it must be utilized the instant it is produced. To that extent it is an inelastic commodity, subject to violent market swings should even a temporary or brief glitch in supply occur. Baker [1] gives an example of UK prices peaking at more than 1000 pounds sterling per kW capacity and at about two pounds per kW-hr when imports from Electricité de France were interrupted by a strike in 1977. Another example is the California crisis in the electrical market and supply of electricity of 2000. Thus, we navigate in new unchartered waters with conditions that cannot be forecast.

1.7 STRUCTURAL CHANGES

1.7.1 Working of the Old Model

First, let us see how the old system worked. Figure 1-4 gives the details. The functions for each of these sectors are broadly and briefly described below.

1.7.1.1 Generation. This sector is compromised of various types of generating plants such as coal-based or oil-based thermal plants, nuclear-energy-based power plants, hydropower plants, and gas-based power plants. Different forms of energy are converted into electricity.

Functionally, it is the basis for load management—meeting varying load demands while maintaining the quality. The quality is related to frequency and voltage maintenance within narrow tolerances of declared values. Generation is also accountable for reliability.

A fall in generated frequency below the grid or network frequency will draw power from the grid so as to hold onto the frequency. A rise in frequency will work the other way. Frequency is controlled by controlling the input energy. Under low load conditions, thermal coal-based plants will have to back down so as not to exceed the turbine rpm and increase the generated frequency. Under peak load conditions, they will have

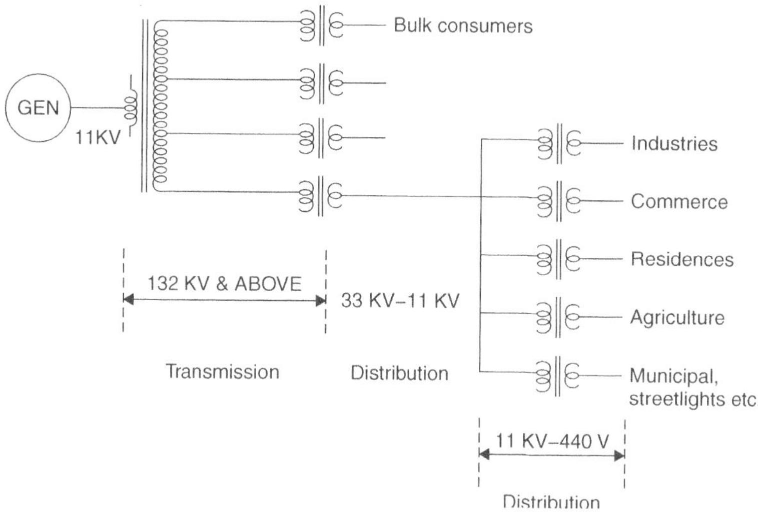

Figure 1-4. A vertically integrated electricity system—British model.

to switch on fast-responding gas/hydraulic plants to prevent the frequency from falling.

Voltages are maintained mainly with excitation control of generators as per active/reactive power requirements of the load.

For reliability, the generation, as a sectional group, has to maintain reserves, first for load management and second for network/grid contingencies.

Multiplicities of spread-out generators were directed from a load dispatch center under the old system. All the generators were also under the same ownership and management. The coordination was smooth and not overdemanding.

The jurisdiction of generation stopped at the 11/33 kV to 132/270 kV transformers.

Under the old system there were quite a few auxiliary services that were to be provided free and compulsorily by generation.

1.7.1.2 Transmission. Transmission moves electricity power from generation and distributes it over its network of transmission lines and transformers. While doing this, it has to consider the effects of its own parameters, such as capacitance, inductance, and resistance, on its functions for maintaining the quality of supply as well as for picking routes with the lowest transmission losses. It is supported where necessary by shunt or series capacitor banks and/or shunt or series reactors. It must have adequate transmission capacity for load management as well as for meeting emergencies.

1.7.1.3 Distribution and Marketing. These two functions were together in the old system. This sector started at 33 kV and went down to 400/230/110 V at the customer end. In this sector, there is some sort of continuity with the load growth, which is basically continuous and spread in different directions. The distribution losses are the highest of all the three sectors. Voltage quality problems are acute.

Since the distributor has a direct interface with the customer, quality of supply, metering, bill collection, and customer relations are of significance. The marketing sector collects the revenue from the sale of electricity.

1.7.1.4 Central/Corporate Office. It handles finances, planning, and overall coordination between the three sectors. It distributes the funds collected from marketing to the three sectors as per their requirements and working. This is the key section or the brain part of the whole system. The load dispatch center looks after efficient load and supply management as well as contingencies.

1.8 COST BREAKDOWN IN THE OLD MODEL

Figure 1-5 gives a breakdown of the cost of each of these components for the year 1995, in the old structure.

It is obvious that the first section to be opened for competition would be generation, where the scope for reducing costs is the highest. However, freeing generation alone does not take a society on the road to open market conditions. For this to happen, a customer must have a choice to buy electricity from a source, old or new, that can supply him quality electricity reliably and at a competitive price. Similarly, a generator should be able to sell his electricity to any customer, wherever he is located, at a viable and undictated price.

The connecting links are transmission and distribution, which cannot be replicated. As such, these should be available to both the supplier and user without any restraints, at reasonable worked-out costs.

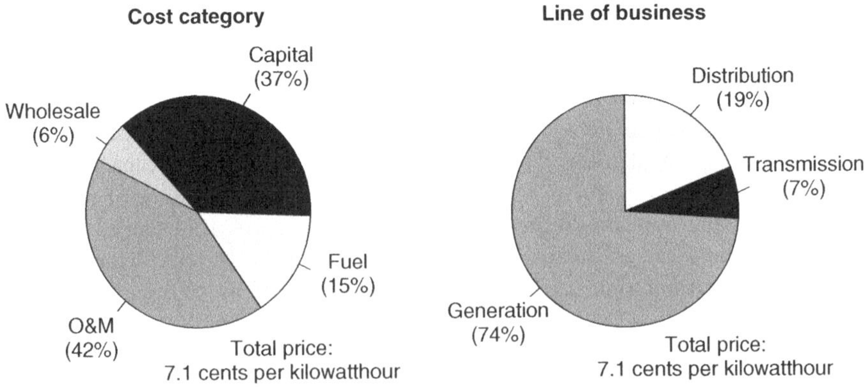

Figure 1-5. 1995 price of electricity by cost category and line of business. (From [4].)

A first step in this evolution was for the new generators to enter into power purchase agreements with existing utilities—state electricity boards, for example, or bulk-power-using customers.

1.9 STEP-BY-STEP RESTRUCTURING

1.9.1 Generation

As a first step in restructuring, the monopoly for generation was withdrawn and generation was opened up. This step has been taken in almost all the countries that were on the road to open competition in electrical business. However, freeing generation alone does not lead to open market conditions. In the first step in India, the new generators signed long-term contracts with existing state electricity boards to buy back the electric power they generated. This helped in two ways. The overall supply position improved, increasing reliability to that extent. It attracted large investments, which the state electricity boards or their backing governments could not provide.

1.9.2 Distribution

Distribution, which accounts for 19% of the cost component, looks to be a likely next candidate for opening up. However, it is not feasible to replicate complex distribution networks all over in a given area. Like transmission, it becomes a franchised monopoly structure.

What has been happening in England, India, and other countries is that the structure is segmented geographically and leased off from the state monopoly to private operators. It is still subjected to rules and regulations by the regulatory commissions as well as by the state legislative councils.

Distribution actually breaks down into two components. The first is a technical network, which receives energy from transmission and takes it to the consumer's premises in an optimum economic way. The second component, the retail part, is the interface with the consumer. This is being segregated and opened up in stages.

1.9.3 Evolution of the Free Market

Let us see how the bulk power consumers seized the opening up of access to transmission and opted out of distribution. Consumers or consumer associations above a minimum power demand are next allowed to buy directly from a generator or a generation pool, pay access and other overheads to transmission and distribution, and get their power. The minimum power bar, at which a consumer society is allowed to buy its power directly, is flexible across a number of countries in which restructuring of electricity business is going on. In the United Kingdom, for example, any load above 5 MW can now buy its requirement directly from the pool.

We carry this a step further by treating electricity as a tradable commodity and allowing traders and brokers in. They bid into an electricity pool and get allotments if their bids are competitive. The bidding can be done as in the UK system on a half-hourly basis, for the same day or for the next day in advance. It will take quite some time for this system to be adopted universally.

1.9.4 Transmission

The overriding logical step would be allowing and ensuring free access to the transmission facilities so that there will be price competition between the old generators and the newcomers, with an open choice for the consumers.

This is easier said than done. Although generation was taken out of the monopoly franchise, transmission and distribution still remained monopoly franchised zones. What is more worrisome is that the old monopoly systems owned both the sectors and they had a central office for coordinating various activities, including power flows from independent power producers. So stricter and stricter rules came into existence, opening up transmission operations and putting stricter operating norms on these.

With free access, power was now available to bulk consumers who were almost at the end of the transmission zone and independent of distribution. The bulk consumers could now negotiate and get long-term contracts on mutually advantageous terms. Thus, it is this sector of consumers that benefited the most initially. Retail consumers did not benefit much at the start.

Transmission has to keep pace with the growing demand for its services. The demand and the service capacity should grow to keep in step with the growth rate for the electricity supply. However, under the free access policy the number of users has gone up. The numbers of connections, cross connections, the flow directions, and so on, are now not uniform and predictable, as under the old system. For the same quantum of energy handled, the demand for servicing capacity for transmission has gone up or will go up substantially. This becomes a prescription for congestion.

1.10 THE NEW DECISION AUTHORITIES

There is now a pool authority that occupies the same important place as the load dispatch center did in the older system. It is not a free authority, but riddled with lots of rules and restrictions. On the financial side, it allocates quota against bids. It also plays with the costing system so that price considerations take care of congestion in the system, peak demand, and so on. On the technical side, it passes on flow allotments to the grid controlling authority, which now has to work out economically optimum dispatches.

The grid authority and the pool authority are no longer working under the authority of the older utilities, although these utilities might still own the transmission facilities.

On direct producer-to-bulk-consumer contracts, the contracting parties deal directly with financial matters. The pool authorities add access fees, transmission charges, a reliability charge that takes care of cost of marginal reserves compulsorily held by the producers, charges against standard costs, and its own management costs.

The retailers, including distributors, collect funds against bills to consumers, which still are under regulatory control. In turn, they pay into the pool for energy contracted for and purchased. The pool then distributes these funds to generators (per their offers), transmitters, and distributors.

1.11 OPEN POWER MARKETING NOW RERESTRUCTURING ELECTRICITY POWER SYSTEM

In its drive to carry power quality and cost benefits uniformly to more and more segments of a society, the electricity power system (EPS) is evolving into wide-area electrical networks. For example, the PJM system in the United States covers seven states, and Union for the Coordination of Transmission of Electricity (UCTE) serves almost all of Europe. They are supported by a tremendous growth in metering and controls with the latest software and microprocessor systems.

Under the old NERC requirements for a generating utility on contingency reserves, a reserve equal to the loss of capacity by the largest component in their system, either in generation or in transmission, had to be maintained. Load balancing and maintaining the frequency was a compulsory ancillary service.

In large systems today, both load balancing involving frequency control and contingency protection are taken over by the transmission system operator (TSO). With three tiers of frequency control, frequencies have to be maintained within ±200 mHz. The frequency characteristics of the system have gone as high as 20,750 MW/Hz for UCTE and even higher for PJM. The reserves for generators are spread uniformly by the droop characteristic of the generator. Power markets are setting up new norms.

The regulator in a power marketing system sets a cap on peak power prices to prevent unfair marketing power by unscrupulous competitors. As these peak prices and caps go up, and also with simultaneous technology improvements, the renewable energies are getting competitive. Under the old Kyoto proposals, a country was supposed to produce 10% of its increasing energy requirements by the year 2012. Wind energy has reached 20% already. Power marketing and the wide-area spread of EPS is producing new norms and promoting new directions of growth.

It looks like restructuring from an old vertically integrated system is likely to become an event in the history of EPS evolution very soon.

REFERENCES

1. G. C. Baker, The Wave of Deregulation: Operational and Design Challenges, *Power Engineering Review*, November 1999, pp. 15–16.
2. Department of Energy/Energy Information Administration, DOE/EIA-0562(00), The Changing Structure of the Electric Power Industry 2000: An Update. Energy Information Administration, Washington D. C., 2000.

3. Editorial, World Energy Consumption, *Power Engineering Review,* August 1997, pp. 15–16.
4. Department of Energy/Energy Information Administration, DOE/EIA-0383, Annual Energy Outlook 1997, Energy Information Administration, Washington D.C., December 1996.

BIBLOGRAPHY

Addison, E., Going Global, *Power Engineering Review,* September 1994, pp. 7–8.

Bell, R. E., Bell Breakup Plus Five. Mixed Reviews, *IEEE Spectrum,* December 1998, pp. 26–31.

Brooks, E. R., Utility Perspective (U.S. Regulatory Evaluation), *Power Engineering Review,* November 1995, pp. 9–10.

Concordia, C., Electric Power Systems: Past, Present, and Future, *Power Engineering Review,* February 1999, pp. 7–8.

Debadishi, E. S., Developing Countries—Restructuring with Benefits from Competition, *Power Engineering Review,* June 2004, pp. 16–23.

Draper, E. L., Assessment of Deregulation and Competition, *Power Engineering Review,* July 1999, pp. 16–18.

Fiedler, K. J., Preparation for Retail Competition, *Power Engineering Review,* July 1999, pp. 8–11.

Gellings, G. W., Power Delivery System of the Future, *Power Engineering Review,* November 2002, pp. 7–2

Hayes, B., Utility Restructuring. *Power Engineering Review,* April 1992, pp. 17–18.

Renz, B. A., Technology's Role in Our Changing Industry, Power Engineering Review, April 1998, pp. 11–13.

Shearman, J., Emerging Responses: How Utilities are Copying in New Environments. *Power Engineering Review,* November 1994, pp. 13–17.

Steinmeir, A. G., State and Federal Regulatory Thrust, *Power Engineering Review,* April 1992, pp. 19–22.

Temin, P., A Parable of Network Disintegration: The Bell System Breakup, *Power Engineering Review,* November 1994, pp. 10–12.

Terzic, B., Distribution Companies of the Future, *Power Engineering Review,* November 2002, pp. 13–16.

Townsend, Liberalization and Internationalization of Electric Supply Industries, *Power Engineering Review,* September 1994, pp. 9–11.

Voropai, N.I. and E. J. Handschin, Liberalization and Modernization of Power Systems, *Power Engineering Review,* February 2001, pp. 16–19.

Yeager, K. E., Competitiveness through Technology and Cooperation, *Power Engineering Review,* September 1997, pp. 6–8.

Zorpette, G., Loosening the Bonds of Oil, *IEEE Spectrum,* March 1991, pp. 34–38.

Zorpette, G., Power and Energy—Clean Air Act Amended. Gulf Crisis Convulses Oil Markets, *IEEE Spectrum,* Jan 1991, pp. 61–64.

Zorpette, G., Power and Energy—Merger Mania, *IEEE Spectrum,* June 1990, pp. 28–30.

2

RISKS, OPERATION, AND MAINTENANCE OF HYDROELECTRIC GENERATORS

2.1 THE PRESENT SCENARIO

In Chapter 1 we showed that the demand for electricity is rising with living standards. It reflects both a growing economy and increasing industrial production. This is illustrated in Tables 2-1 and 2-2.

It may be noted that proportion of electricity generation by hydro power to total energy produced varied from 7.247% to 7.11% for the years 2000 to 2006. One would have expected it to rise further, considering the benefits it carries, as stated earlier. In what follows, we will study its steady progress as well as the risks associated with a hydroelectric project.

Hydro power has been one of the oldest suppliers of electricity. It is a clean, economically feasible, renewable energy option. It carries multiple social benefits such as irrigation, water supply, flood control, and improved navigation.

2.2 TYPES AND SIZES OF HYDROELECTRICITY PROJECTS

We will classify hydroelectricity (HE) projects into the following categories:

- Large lake-backed hydroelectricity projects, operating year-round. Generally mega sized.
- Pumped storage systems
- Small Hydro Projects (SHPs)

Electricity Power Generation. By Digambar M. Tagare

Copyright © 2011 the Institute of Electrical and Electronics Engineers, Inc.

Table 2-1. Actual production of electrical energy from various energy sources over the period 1995 trough 2005—United States (thousand megawatt-hours)

Period Total (all sectors)	Coal [1]	Petroleum [2]	Natural gas	Other gases [3]	Nuclear	Hydroelectric, conventional [4]	Other renewables [5]	Hydroelectric, pumped storage [6]	Other [7]	Total
1995	1,709,426	74,554	496,058	13,870	673,402	310,833	73,965	–2,725	4,104	3,353,487
1996	1,795,196	81,411	455,056	14,356	674,729	347,162	75,796	–3,088	3,571	3,444,188
1997	1,845,016	92,555	479,399	13,351	628,644	356,453	77,183	–4,040	3,612	3,492,172
1998	1,873,516	128,800	531,257	13,492	673,702	323,336	77,088	–4,467	3,571	3,620,295
1999	1,881,087	118,061	556,396	14,126	728,254	319,536	79,423	–6,097	4,024	3,694,810
2000	1,966,265	111,221	601,038	13,955	753,893	275,573	80,906	–5,539	4,794	3,802,105
2001	1,903,956	124,880	639,129	9,039	768,826	216,961	70,769[R]	–8,823	11,906[R]	3,736,644
2002	1,933,130	94,567	691,006	11,463	780,064	264,329	79,109[R]	–8,743	13,527[R]	3,858,452
2003	1,973,737	119,406	649,908	15,600	763,733	275,806	79,487[R]	–8,535	14,045[R]	3,883,185
2004	1,978,620	120,771[R]	708,854[R]	16,766	788,528	268,417	82,604[R]	–8,488	14,483[R]	3,970,555
2005	2,013,179	122,522	757,974	16,317	781,986	270,321[R]	87,213[R]	–6,558	12,468[R]	4,055,423[R]
2006	1,990,926	64,364	813,044	16,060	787,219	289,246	96,423	–6,558	13,977	4,064,702

Notes:

[1] Anthracite, bituminous coal, subbituminous coal, lignite, waste coal, and synthetic coal.

[2] Distillate fuel oil (all diesel and No. 1, No. 2, and No. 4 fuel oils), residual fuel oil (No. 5 and No. 6 fuel oils and bunker C fuel oil), jet fuel, kerosene, petroleum coke.

[3] Blast furnace gas, propane gas, and other manufactured and waste gases derived from fossil fuels.

[4] Conventional hydroelectric power, excluding pumped storage facilities.

[5] Wood, black liquor, other wood waste, biogenic municipal solid waste, landfill gas, sludge waste, agriculture byproducts, other biomass, geothermal, solar thermal, photovoltaic energy, and wind.

[6] The quantity of output from a hydroelectric pumped storage facility represents production minus energy used for pumping.

[7] Nonbiogenic municipal solid waste, batteries, chemicals, hydrogen, pitch, purchased steam, sulfur, tire-derived fuels, and miscellaneous technologies.

Value is less than half of the smallest unit of measure.

R = Revised.

Note: Totals may not equal sum of components because of independent rounding.

Sources: Energy Information Administration, Form EIA-906, "Power Plant Report;" Form EIA-920, "Combined Heat and Power Plant Report"; and predecessor forms.

Table 2-2. Global electricity installed capacity in various countries from 2001 to 2005 (millions of kilowatts)

Region/Country	2001	2002	2003	2004	2005
North America					
Canada	67.230	66.882	69.029	70.197	70.680
Mexico	9.634	9.636	9.628	9.643	10.565
United States	78.916	79.356	78.694	77.641	77.541
Others					
Total	155.780	155.873	157.351	157.481	158.786
Central and South America					
Argentina	9.602	9.612	9.762	9.780	9.898
Brazil	61.063	62.523	65.311	67.792	68.999
Chile	4.128	4.122	4.155	4.279	4.279
Colombia	8.065	8.721	9.077	8.893	8.915
Paraguay	7.390	7.410	7.410	7.410	7.410
Peru	2.860	2.965	2.996	3.032	3.056
Uruguay	1.534	1.534	1.538	1.538	1.538
Venezuela	13.215	13.207	12.491	12.491	13.864
Others					
Total	114.186	116.433	119.427	122.171	124.996
Europe					
Austria	8.092	8.096	8.118	8.128	8.170
Croatia	2.076	2.076	2.063	2.063	2.076
Former Serbia and Montenegro	2.910	2.910	2.910	2.910	2.910
France	20.821	20.847	21.004	20.899	20.984
Germany	4.328	4.831	4.937	4.058	4.073
Greece	2.373	2.377	2.379	2.380	2.400
Italy	13.389	13.456	13.557	13.703	16.728
Norway	26.766	26.319	26.262	26.758	26.088
Portugal	3.929	3.963	3.990	3.991	4.315
Romania	6.120	6.122	6.242	6.248	6.279
Spain	12.672	12.744	15.550	15.525	15.600
Sweden	16.506	16.523	16.187	16.098	16.302
Switzerland	13.240	13.285	13.295	13.310	13.315
Turkey	11.175	11.673	12.241	12.579	12.645
Others					
Total	160.279	161.241	165.448	165.584	169.478
Eurasia					
Georgia	2.662	2.662	2.662	2.700	2.700
Kazakhstan	2.200	2.200	2.230	2.247	2.247
Kyrgyzstan	2.949	2.969	2.969	2.910	2.910
Russia	43.900	44.700	44.828	45.221	45.531
Tajikistan	4.054	4.054	4.054	4.054	4.528
Ukraine	4.700	4.731	4.758	4.351	4.785
Others	5.363	5.617	5.623	5.658	6.050
Total	65.828	66.933	67.124	67.141	68.451

(continued)

Table 2-2. *Continued*

Region/Country	2001	2002	2003	2004	2005
Middle East					
Iran	2.637	2.803	3.028	4.430	4.833
Others					
Total	5.036	5.504	5.441	6.949	7.452
Africa					
Congo (Kinshasa)	2.440	2.515	2.535	2.469	2.410
Egypt	2.810	2.678	2.745	2.745	2.745
Mozambique	2.180	2.184	2.136	2.136	2.136
Others					
Total	20.700	20.945	21.359	21.295	21.636
Asia and Oceania					
Australia	7.711	7.711	7.770	7.788	7.781
China	79.352	83.006	86.075	94.896	105.242
India	25.140	26.260	26.910	29.569	31.001
Indonesia	4.393	4.383	4.383	4.535	4.566
Japan	22.019	21.621	21.697	22.006	22.048
Korea, North	5.000	5.000	5.000	5.000	5.000
Malaysia	2.054	2.118	2.106	2.115	2.095
New Zealand	5.193	5.260	5.320	5.345	5.345
Pakistan	5.010	5.039	5.046	6.491	6.497
Philippines	2.304	2.520	2.518	2.867	3.217
Taiwan	4.422	4.422	4.511	4.511	4.512
Vietnam	3.292	4.128	4.155	4.155	4.155
Others					
Total	174.107	179.893	184.153	198.412	211.065
World Total	695.916	706.822	720.303	739.032	761.863

Source: EIA, Annual International Energy 2005.

The hydro projects are also classified as per size:

- Mega sized
- Small sized, further divided into small, mini, and micro sizes.

2.3 ADVANTAGES OF HYDROELECTRICITY

Hydroelectricity offers the following advantages:

- In the long run, hydroelectricity is the cheapest energy one can produce. Energy costs are next to nil, compared to ever increasing costs of fossil fuels. With projected life periods extending to over 50 years, capital costs are also low. Old hydro projects greatly help present energy producers on their balance sheets and regulated tariffs.

- Operationally, this should be a favorite of load controllers, in that there is hardly any wasted time between "start" and "synchronizing." Typically for HE generators, this interval is in minutes. For steam generators, this is very high, extending up to 12 hours in some cases.
- The project offers bulk energy storage facilities—practical, feasible, and in operation under pumped storage projects. Energy storage systems are presentely the focus of electricity research and development, hydrogen-based storage of electricity, for example.
- There is no scope for international cartels by energy suppliers.
- Environmentally, it produces almost no atmospheric pollution.
- Accompanying advantages in irrigation transform the face of the downstream region with increased agricultural production and, indirectly, with growth of supporting industries and cities.
- Flood control will be a great boon in the Asian region. In the past, floods have taken lives and incurred huge property and animal losses. Hydroelectricity projects have in-built flood control, an added advantage.

The majority of hydroelectricity projects have dam and canal building as a part of composite projects providing both power for industry (mainly) and irrigation for agriculture. As a result, not only do investments go up, but a lot of other problems arise that affect the consideration, planning, and execution of projects. Relative to other forms of power generation, progress here is very slow. Let us explore the factors involved.

2.4 SLOW PROGRESS OF HYDROELECTRICITY PROJECTS

Factors responsible for a slow progress are:

1. Land acquisition, evacuees, and resettlement of evacuees
2. Archeological problems—shifting of ancient sites
3. Environmental aspects
4. Added features of hydroelectricity projects:
   - Loss of a natural public drainage system
   - Improvement or establishment of new navigation scheme
   - HVDC transmission system for bulk transfer of power

2.4.1 Land Acquisition, Evacuees, and Resettlement

With creation of dams, large areas—forest or semi-agricultural lands—have to be acquired. Historically, this has been done under a legal process by the government with acquisition authority in that area. This process is slow, running into years.

The HE projects generally are in semimountainous, undeveloped regions. The populations evacuated usually had low living standards. Their lands had low valuations. They were paid predetermined compensation and were given some residential blocks on alternate sites. However, the compensation money evaporated in no time and the evacuees became migrants and destitute. Plan-wise, acquisition and resettlement costs form a big chunk of an HE project. This becomes a disincentive for private capital investment.

This is a universal problem, common to all the countries in the world. A number of United Nations conferences have discussed it. It has been agreed in principle that land evacuees should be long-term beneficiaries of the prosperity resulting from a hydroelectric project.

2.4.2 Archeological Problems

Very naturally, ancient civilizations grew largely along the banks of rivers, both upstream and downstream. Ancient pilgrimage cities also grew along rivers. There is a problem when heritage sites also come within the scope of areas to be sunk permanently underwater. The sites have to be dug out and rearranged at other favorable sites.

The problem is exacerbated when archeological artifacts are discovered. Hundreds or thousands of years of sedimentation may have laid deep layers of earth on these artifacts. Archeologists spend a great deal of time and money in locating these and digging them out.

2.4.3 Environmental Problems

Endangered species must be protected. Old river flows and cascades, as well as water tables, support a variety of flora and fauna and a variety of fish. Some of these rare varieties are likely to be wiped off the earth's surface for good. Environmentalists wage a tenacious legal and public campaign against this and force compromise and accommodation. All this eats up time.

Old water tables in the northern Asia have supported migratory birds for centuries. The water tables are in the vicinity of rivers. As such, the drawing down of these water tables has reduced the number of annual migratory birds previously visiting these areas (see Appendix 3-1 in Chapter 3).

2.4.4 Added Features of Hydroelectric Projects

Loss of Natural Sewage Clearance. Village sewage is cleared by rivers in small and medium localities. Instead of a river flow, the riverbanks will now be covered by the stagnant waters of a lake. This causes health and water problems in upstream areas. A significant amount of money has been spent in providing modern sewage disposal facilities in such areas.

Additional Navigational Facilities. Under HE projects, river water flows have increased in size and are almost perennial. This has improved river navigation considerably, both in speed and in volume. However, barrages and dykes now have to be added, significantly increasing the total project costs.

Additional and Reienforced Transmission Facilities. These are not strictly a part of HE projects. However, without developing these megasized power transmission facilities, HE projects by themselves become meaningless. This is best illustrated by the Brazilian example, where there are a number of HE projects in the north region that are not connected to the national grid. As a result, Brazil imports electricity from its neighbors [1].

2.5 FACTORS PROPELLING THE PHASED PROGRESS OF THE HYDROELECTRIC INDUSTRY

Hydroelectricity had a slow but phased growth in the past. It is interesting to study the phases of this growth and the reasons cited for the slowdown.

Sadden [2] divided the phases into three periods. Although the following applies to developing countries, with a time frame shifted forward by 15–20 years, it applies to developed countries as well.

2.5.1 Phase 1 (1900–1920)—Technocentric Phase

Steady but slow growth of electricity supported by private investment began with the start of electricity generation, by about 1900. It continued until the end of World War I (1914–1918). Living standards improved slowly and so did the demand for electricity. The demand was met by private sector utilities and industries (for their internal use, mainly). The growth was technocentric and projects were led by engineers. The small capital requirements could be met by project developers. This was a monopoly regime. Tariffs mostly were governed by demand and supply. The private sector utilities assumed the commercial risk and the project risk.

2.5.1 Phase 2 (1920–1980)—Capital-Directed Phase

This phase began after the end of World War I. Industrial growth accelerated and the economy in general began building up. The demand for electric supply grew fast, demanding larger electricity projects. Capital requirements went up. Mergers and acquisitions and also floating of equity capital resulted in vertically integrated, larger monopolies for raising the necessary capital. On hydropower projects, which were infrastructure development projects, governments stepped in with financing. Financing at low interest rates and longer repayment periods were also available from international bodies like the World Bank and the Asian Development Bank. Tariffs were regulated by the governments in such a way that commercial risk was underwritten. The vertically integrated utilities had experience and overall control for the large-sized projects that materialized during this phase. To that extent, project risk was well met.

Depending on a country's political maturity and economic situation, foreign direct investment also entered the picture.

The growth in this phase was more economy-driven and under the control of finance experts, and relatively less under technical experts.

2.5.3 Phase 3 (1980 Onward)—Sociotechnical Phase, Infrastructure Nature

By 1980, the demand for electricity had grown substantially. This led to considerations of megasized hydroprojects. These projects came with additional features, as listed earlier. In other words, these projects turned into infrastructure projects. They were undertaken, supported, and executed by entities created by governments, and headed by management experts. The basic technologies and machinery supplies had become standardized. Engineers played only a supporting role. International agencies as well as (in some countries) foreign investors contributed substantially.

Medium and small hydro projects continued to be financed by private venture capital. Due to the prolonged period until the start of the return on investment, debts and leveraging of debts assumed a higher position than share capital. Marketing came under the influence of supply and demand conditions as well as under regulators.

2.6 HYDRO PROJECTS FALL SHORT OF ATTRACTING PRIVATE INVESTMENT

During this period of free and fast economy, private capital turned to investments with a high and quick return. This has led to debt leveraging. It also has required security of projects during debt service. There is a lot of competition for attracting private capital investment. Hydropower projects, with their long gestation period and initial negative cash flow features, did not qualify. Yet, the demand pressure kept on generating hydro projects at a reasonably fast rate.

The restructuring of the electricity business and power marketing through electricity exchanges is a new and very serious risk to hydropower projects. Electricity tariffs for 8–10 years ahead cannot be projected. There can be no long-term commitments, except on very broad terms, which may not be accepted by financers.

All the issues listed earlier require political and social acceptance. A private investor might find this difficult.

2.7 DAM BUILDING PROGRESS OVER A CENTURY

Figure 2-1 shows the extent of dam building worldwide during the above three phases.

2.7.1 Principal Risks Associated with Development of Hydro Projects

Table 2-3 lists these risks.

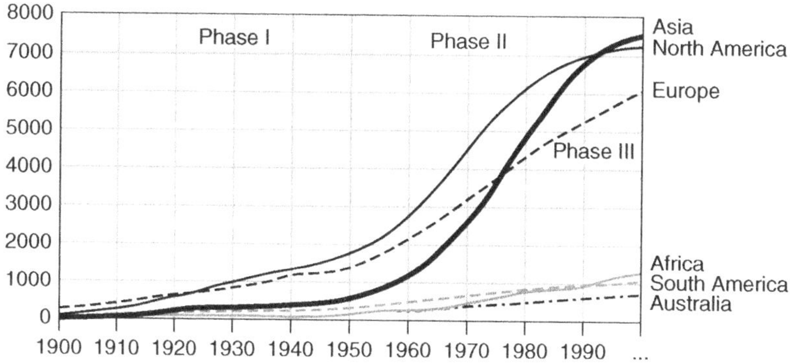

Figure 2-1. Extent of dam building worldwide, 1900–2000. (*Source:* ICOLD, 1998; excludes time trend of dams in China. From [2], © 2002 IEEE).

Table 2-3. Hydro development risks

Political/Economic	Government rules and regulations Inflation Tax rate variations Economic force majeure
Commercial	Demand Power purchaser credit Power purchaser longevity Interest Refinancing Capital and credit availability
Currency	Exchange rates Repatriation
Technical (geology and hydrology)	Location-specific
Environmental	Inundation and loss of land base Impacts on terrestrial and aquatic species Approvals procedures
Social	Resettlement Public attitudes about development Project area impacts and compensation
Return	Long term and could go negative, as happened in the case of Brazil
Construction Time	Schedule delays and associated costs

Source: [2]. © 2002 IEEE.

2.7.2 India Has a High Proportion of Hydroelectricity

Figure 2-2 below gives quite a favorable picture of the capacity position of hydroelectricity in India. Note that at 27.5% of the total installed capacity in India by March 31, 2007, hydroelectricity is as an important player.

In India, electricity business is a totally government affair. India has a large agricultural economy and agricultural population. Popular governments naturally gave a high preference to investment in hydro projects that irrigated large land areas. As a result, the capacity of hydroelectric projects grew substantially. It may be noted that small hydro projects below 5 MW capacity are not included in the figure.

2.8 DESIRABLE CONFIGURATION FOR HYDRO PROJECTS TO ATTRACT PRIVATE INVESTMENT

Sadden [3] lists the following requirements for attracting hydro projects:

1. Location with high heads, naturally requiring less reservoir area and reducing environmental impacts and approving procedures
2. Run-of-the river projects
3. Surface-based configuration of pump house and other structures (no tunnels and caverns)
4. Short development cycle and debt repayment

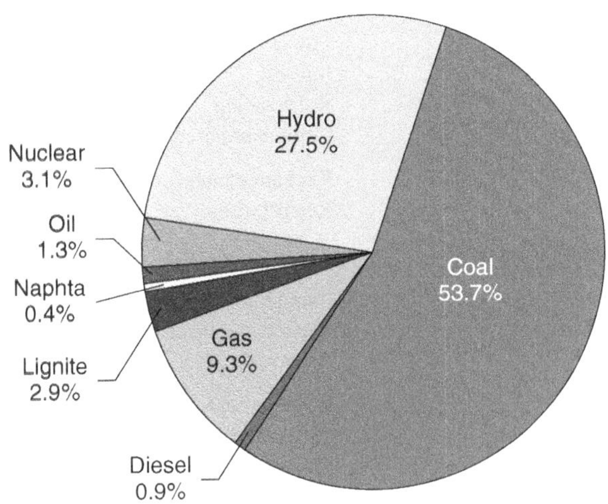

Figure 2-2. Breakdown of generation capacity covered by the database. The total corresponds to 124,486 MW as of March 31, 2007. (From Annual Report 2005-06, Ministry of Non-conventional Energy Sources.)

2.8.1 Challenges

A large demand for electricity exists worldwide. Hydro potential to the extent of about 20% of this requirement also exists. The rewards are low cost of production and environmental acceptability. There are inherent risks, generally associated with dam construction. The challenge lies in selecting an appropriate type, estimating and meeting the risks involved, and arranging the resources necessary for implementation.

Hydropower has its own prominent place in the energy suppliers' constellation.

2.9 OPERATION OF A HYDROELECTRIC PLANT

2.9.1 Typical Layout

Figure 2-3 below gives a general idea of a hydroelectric power house. For more information, please refer to Chapter 3, Appendix 3-1—Itaipu.

Some points to note—

- Penstock pipes bring in water from the reservoir to the turbine through a control valve. The pipe thickness is designed to withstand the static head of water plus likely water hammer pressure, which occurs when the turbine is suddenly switched off.
- Tailrace. The rated flow of water should not cause a back pressure. The tailrace should have the same slope as that of the outfall channel.

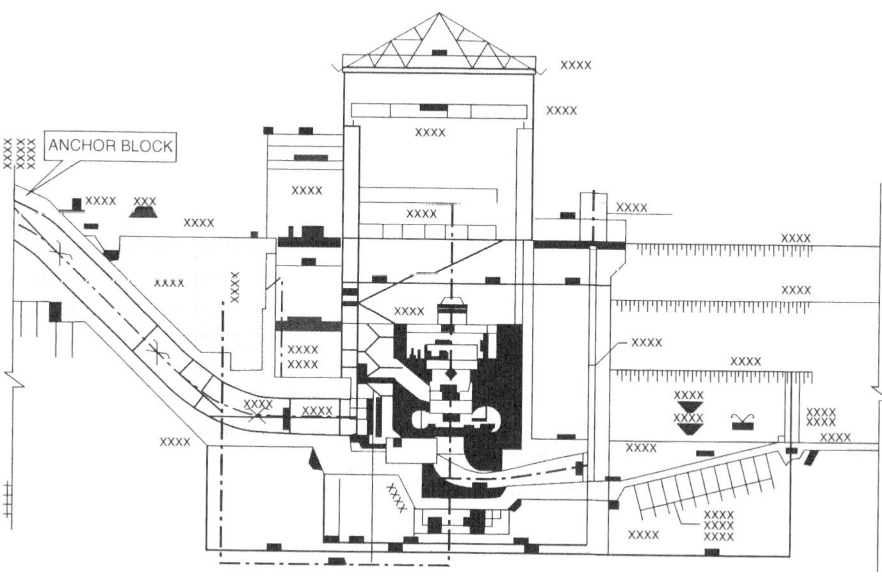

Figure 2-3. A typical layout of a hydroelectric plant. (From [3], © 1982 IEEE.)

2.9.2 Capability Curve for a Hydrogenerator

This curve is an essential tool with which the plant operator keeps watch on the stable operation of a generator. The load dispatcher might ask for any combination of reactive power and active power. These combinations might vary unpredictably. It is for the plant operator to hold them within the scope of capacity curves. Figure 2-4 below shows a typical capability curve for a 25 MVA hydrogenerator.

2.9.3 Efficiency of a Hydro Unit

For a given water flow, we must produce maximum possible power by efficiently using the natural resource. Operationally, we can put this as using minimum water flow to meet a given load requirement. This water flow is related to water head available for the hydro plant. The available head is defined as the difference between the free surfaces of water upstream and downstream of a hydro plant. This head determines the water inflow, flow through the turbine, discharge flow, and so on.

In the hydro plant system, we have energy losses due to resistances to the water flow at the grids, within the canals and conducts, at the regulator, at bifurcation, and so on. These losses are proportional to the square of flow velocity. To these losses, we add the head loss due to tailrace effect, which is due to the accumulation of water downstream of the hydro plant.

Operation of a turbine alternator (T/A), plant is determined by the head available and the flow through it. Figure 2-5 gives the efficiency and the designed and maximum flow for a T/A.

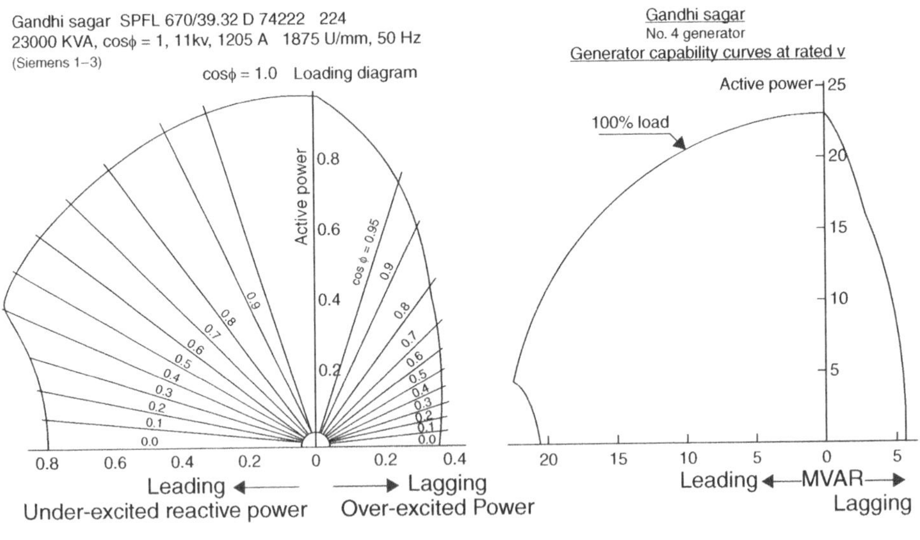

Figure 2-4. Capability curve of a hydrogenerator. (Courtesy of A. R. Dwarkanath.)

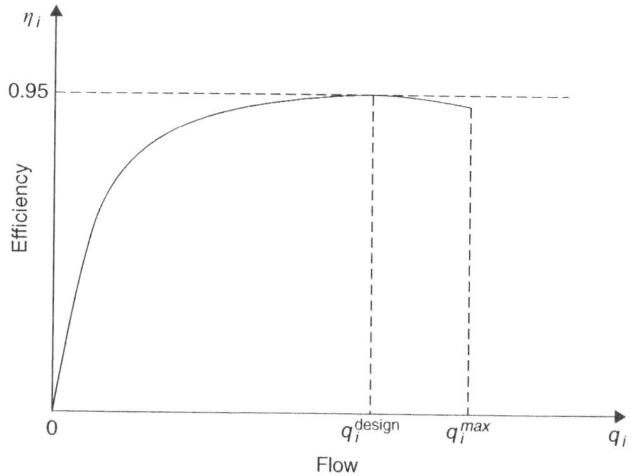

Figure 2-5. Example of an efficiency curve for a turbine. (From [4], © 2004 IEEE.)

Figure 2-6 gives the hydraulic power of a hydrogenerator for given flows at various available heads. The following may be noted:

1. The turbine-alternators are rotating at a near fixed speed corresponding to synchronization with the grid frequency. As such, the friction and other fixed losses are constant.
2. There are flow ranges within which we get maximum output power for given water heads.

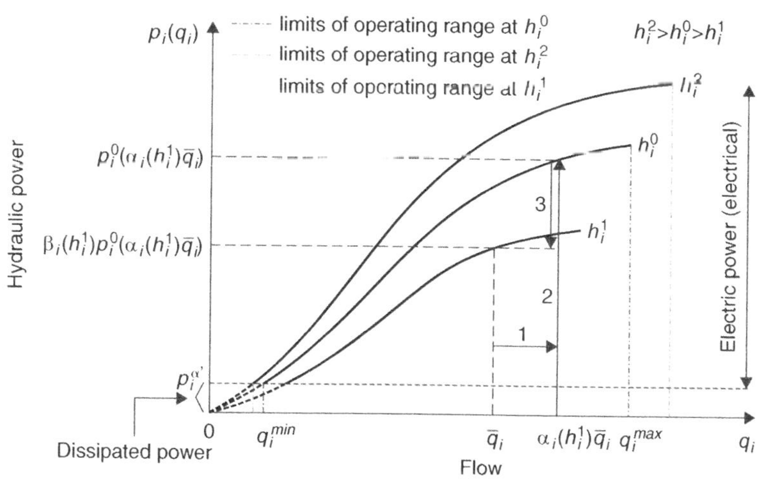

Figure 2-6. Similar transformation of power/flow curves. (From [4], © 2004 IEEE.)

3. There are restricted zones outside these ranges where cavitations on one hand and vibrations on the other can occur.

The same figure when reversed can be used to determine the flow for a desired power output.

2.10 UNIT ALLOCATION WITHIN A LARGE HE PLANT

A certain load has to be met at a given time when the available head in a canal is known. The unit with best characteristics for this head and desired output can be selected [4]. Unit allocations for an assigned load demand are made accordingly within a multigenerator hydro plant. Please note that characteristics are unit specific.

2.11 SPEED CONTROL OF A WATER TURBINE

This has special importance since, as one of the major partners in the overall electrical power systems, synchronous frequency generated and maintained by the hydro system plays a dominant role. The turbine speeds have to be controlled scrupulously within narrow variation limits, under all types of load conditions. This has to be coupled with maximized energy production at economic water flow conditions as well.

The handicaps are the huge inertia of the rotating mass, including water. This mass has an inherently large response time. Along with speed control, stability has also to be ensured.

Basically, a governor controls the water jet impinging across a turbine blade. A large number of microprocessor-controlled systems have been presented in papers on this subject [5–11].

An equally important role is played by the gate control upstream. It controls the quantity of water flow as pointed out earlier.

2.11.1 Governor for Water Turbine Generators (WTGs)

Functions Involved. A governor controlling a water turbine (WT) has to look after the following operations in the sequences for putting a water turbine in operation:

1. Startup process	0–t1 (see Figure 2-7)
2. Revving it up to synchronizing speed	t1–t2
3. No-load period	t2–t3 and t3–t4
4. Picking up load	t4 onward

To this we can add time to disconnect the load gradually or in a sudden unplanned cut-off. A microprocessor-based governor, with the help of a servo motor mechanism, controls these operations mainly through control of the gate opening.

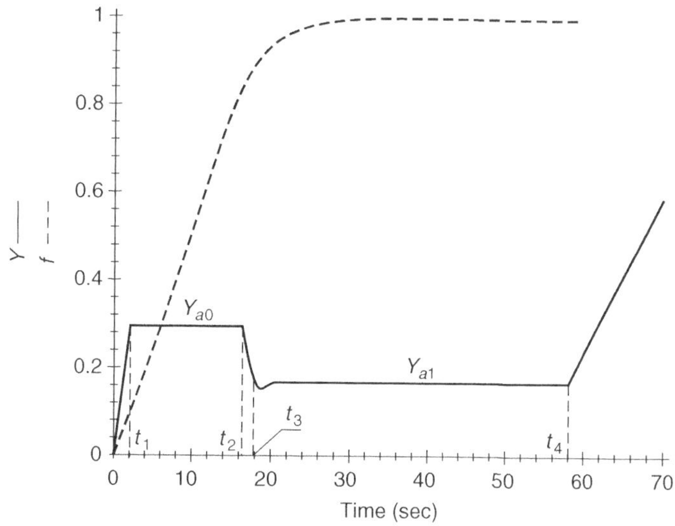

Figure 2-7. Start-Up Process. Y is the gate opening and f is the present frequency. (From [12], © 1990 IEEE.)

Control of Gate opening. Please refer to Figure 2-8. Several factors are involved in gate opening control:

1. Economy in water use. Water at high potential is valuable. Water flow from 0 to t4 does not produce a saleable product, that is, electricity for dispatch. This has to be minimized.
2. Gate opening controls the flow rate. The flow rate along with the head of water controls the performance for a given type of turbine. Figure 2-7 gives the extent of gate opening for various operations.

2.12 STARTUP PROCESS FOR A WTG

The following conditions pertain to the startup of a WTG:

1. Start-up from idle should be fast from 0 to t1.
2. Running the turbine within an acceptable speed band for synchronizing with the grid frequency requires a steady flow. At the end of this period, voltage, phases, and frequencies come within acceptable limits for synchronizing, which is now achieved.
3. The idling period for the no-load condition does not require as much energy as under 2 above. Thus, a low gate opening is maintained until it is time to start picking up the load.

4. While delivering load, gate opening, flow rate, head of water, and type of turbine are interlinked as discussed earlier.

Table 2-4 gives an idea of how rigidly turbine speeds are controlled.

2.13 SPEED CONTROLS ARE RIGID

Some national specifications are shown in Table 2-5.

2.14 SPEED INCREASE DUE TO SUDDEN LOAD CUTOFF

Upon sudden disconnect of the load, speed tends to increase. A sharp gate shutoff increases the pressure in the penstock. Both speed and pressure increments must satisfy established regulation guarantees. Figure 12-8 shows the performance under such conditions.

Typical field test reports are given in item 7 of Table 2-4. This represents a practical idea.

2.15 FREQUENCY AND HARMONIC BEHAVIOR AFTER A SUDDEN LOAD REJECTION

Two hydraulic systems in Sweden were experimented upon for behavior upon sudden load rejection. These were (1) an asynchronous generators (ASG) and, (2) a synchronous generator (SG). ASG's prevail when KVA ratings are less than 1000. They draw

Table 2-4. Field test result

	Test item *a*	Designed indices	Test results*
1	Speed fluctuation for no-load operation	$\leq\pm0.15\%$	±0.09–$\pm0.13\%$
2	Speed deadband i_x	$\leq0.02\%$	0.009–0.015%
3	Dead time of main servomotor	≤0.2 s for rejection of $\geq10\%$ load	0.12–0.18s for rejection of 10% or 25% load
4	Time from start-up to synchronization	Determined according to custom	55–90 s
5	Settling time of output power	Determined according to customer specifications	10M–15 s for 0–100% or 100–0%
6	No-load disturbance of ±4 Hz	Ensure to select optimal parameters	See Fig. 3 T_{r1} = 4–10 s with zero or small overshoot
7	100% load rejection	Ensure regulation guarantee and specifications	See Fig. 4 T_{r2} = (16–22)T_w one $\geq3\%$ overshoot

Source: [12].

Table 2-5. Specifications of some countries for speed controls

Item n	U.S.	F.R.G.	U.S.S.R.
Speed fluctuation for no-load operation	$\leq\pm 0.15\%$		
Speed deadband i_x	$\leq 0.02\%$	$i_x/2 \leq 0.02\%$	
Dead time of main servomotor T_x	≤ 0.2 s for rejection of $\geq 10\%$ load		≤ 0.2 s for max. output of governor

Source: [12].

magnetizing current from the grid and have poor power factors. To improve these power factors, shunt capacitors, roughly at 50% of the KVA rating of the turbine alternators, are used. Upon sudden load rejection, both these types of generators try to speed up. To determine this speedup, their instant and falling rpm (frequency) were measured.

The ASG selected was rated at 440 KVA at 400 V, PF 0.84 rpm @ 760. Capacitor rating was 150 KVAR. ASG was loaded to 20% of its rated value. At 1.79 s, the load was tripped. The capacitors were left on. As the load is tripped, generator current drops to zero. Internal voltage drop of the generator due to its supply of load current becomes nil and generated or induced voltage becomes equal to the terminal voltage, which is high. Apart from an abrupt change in magnitude, there is a sudden change in phase also. This latter change causes an instant spike in frequency. The rate of change of fre-

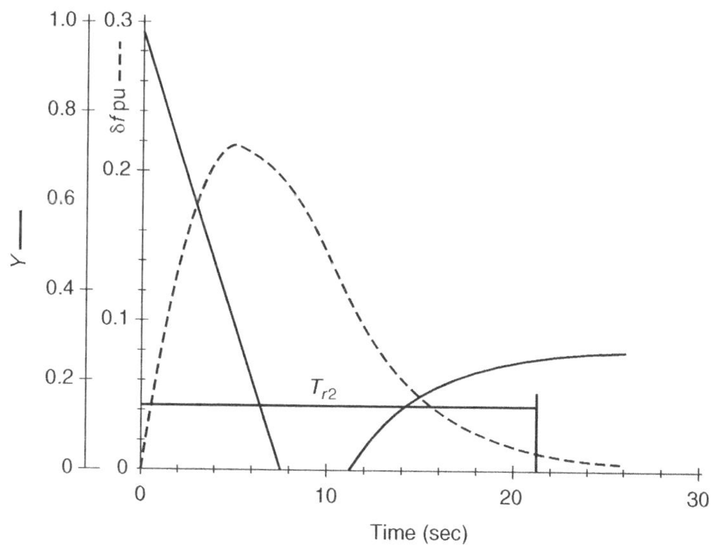

Figure 2-8. Response with 100% load rejection. Y = The gate opening. δf = Change in the present frequency from the set frequency. (From [12], © 1990 IEEE.)

quency after the load rejection depends on the inertia constant of the generator and imbalance between generator and load.

Figure 2-9 shows the sudden spike in frequency followed by a continuous rise later for ASG.

Another SG unit selected was rated at 880 KVA. Its load rejection was done at 2.34 s and the exciter was blocked at 11 s.

Figure 2-10 shows the principal frequency response for the SG and Figure 2-11 shows the estimated SG voltage output.

There is a principle frequency until the controller shuts off the exciter current for the SG. With this shut off, the generator voltage falls.

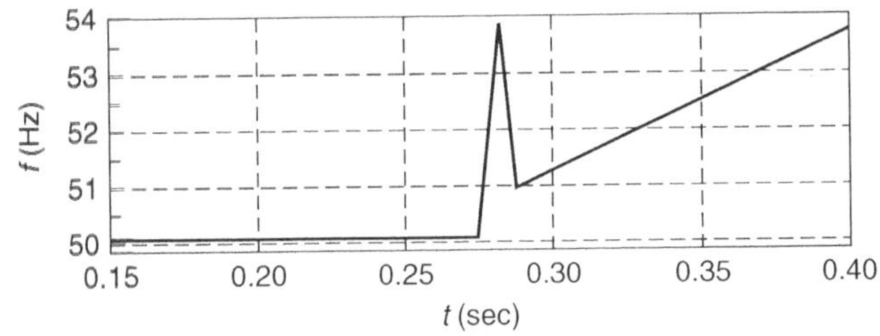

Figure 2-9. Principal frequency response for ASG. (From [13], © 2003 IEEE.)

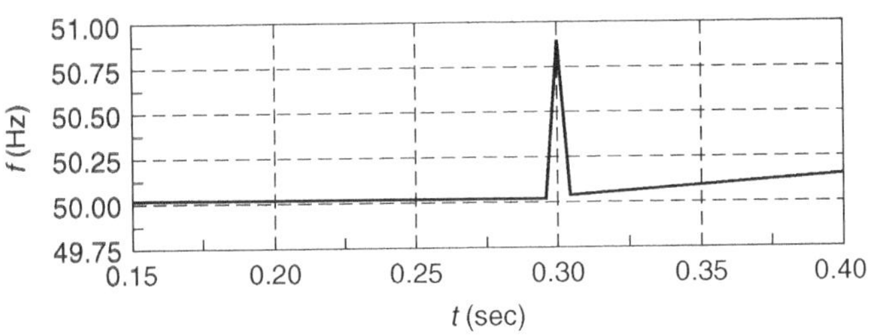

Figure 2-10. Principal frequency response for SG. (From [13], © 2003 IEEE.)

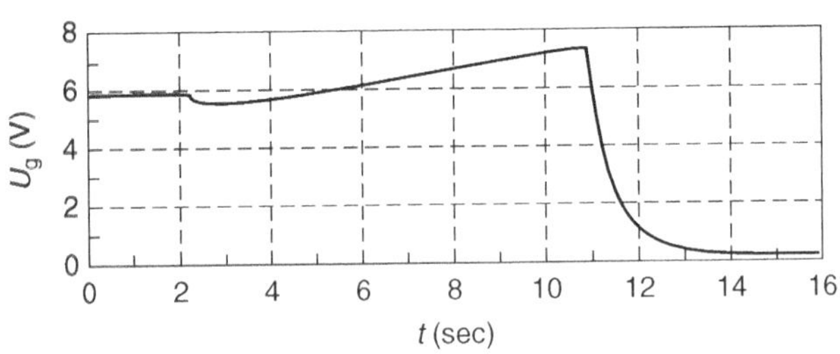

Figure 2-11. Estimated SG output voltage amplitude. (From [13], © 2003 IEEE.)

If the inertia constant is high, the rpm change is not steep, that is, the frequency slope is not steep. If the generator load imbalance is high, the slope is steep.

The ASG, on the other hand, is running at a frequency above synchronous. This drops on abrupt load disconnect. The generator tries to run as a motor and slip frequency keeps on increasing. However, this is stopped by the protection system through mains cutoff.

2.15.1 Voltage Behavior After a Load Cutoff

Consider a case in which power-factor-correction capacitors are connected across a hydro generator that it is cut off from the grid. The generator and capacitor will run in a boost-up voltage mode. In an alternate to the bus capacitor system, a rotor is fed the excitation current from a cyclo-convertor at variable frequency. This controls the TG frequency in synchronization with the grid the frequency. By making the grid frequency an independent reference parameter for the cyclo-converter, it can get instantly switched off, cutting out the rotor field excitation control.

On the other hand, the excitation control can be adjusted with a proper time lag to produce the leading KVAR required by the system and do away with power-factor-correction capacitors altogether. However, this is not very advisable. Turbine Alternators are normally run at unity power-factor output with outside capacitor help. In case of an emergency demanding maximum reactive power output to prevent a grid collapse, full amperage capacity of the generator can be utilized for maximum reactive power production under the limits of generator design, as depicted by the capability curve.

2.16 EFFECT OF PENSTOCK PRESSURE PULSATIONS

During certain load levels on generators coupled to low-head-type turbines, there is pulsating water flow in the draft tube. This causes pulsating pressure in the penstock. This in turn causes pulsation of the electric power output. These are interrelated to the geometry of spiral case, tube, and so on. The power system stabilizer, which is expected to bring stability to the power output, generally takes power output as a reference parameter. Should pulsation occur in the "rough zone," MVAR and terminal voltages will pulsate. Again, there could be interarea power pulsation at its own specific frequency between the generator and connected segment of the grid. Should the penstock pulsation frequency coincide with these (0.3 Hz as a typical case), then the turbine shaft will come under severe stress. To avoid this, signals from a pressure transducer fitted on the penstock are fed to the power system stabilizer as a counterfeedback [14].

2.17 AC EXCITATION OF ROTOR FIELD

A novel excitation control reported by the Kansai Electric Co. of Japan uses AC excitation in a fixed-speed position and changes power input (pumping mode) or power output (generator mode) with generator control. In the AC excitation control of the rotor field, the excitation frequency from a cyclo-converter is governed by two reference parameters, the grid frequency and the active power. Thus, if the rotor slows down, the excitation frequency changes so as to keep the stator pole and the opposite rotor pole locked, as in a synchronous machine. The adjustment of the rotor speed also corresponds with the active power input/output.

This has several advantages in a pumped water storage scheme. When the generator is acting as a motor, an adjustable speed pump has far better efficiency than a fixed-speed pump. In the output mode, if the load variation is fast (quick) the sluggishness of the governor control introduces instability. By contrast, the cyclo-converter response is fast enough to cope with the speed of change in load. Similarly, the penstock pulsations do not expose the turbine shafts to pulsating torques.

Rotor field excitation through an inverter based on a cyclo-converter has another advantage. Its inertia constant is practically nil.

2.18 UNIT COMMITMENT FROM HYDROELECTRIC GENERATORS, INCLUDING PUMPED STORAGE SYSTEMS

Hydrothermal agreements cover both hydro and thermal unit commitments. Neutral considerations enter into this agreement. It may be remarked that hydro generators are in a monopolistic situation even in today's power marketing systems.

On the intake side, particularly for pumped storage hydro, they buy power when the seller has no customer offering better prices in competition. This applies to nighttime generation by thermal and excess/slack-period generation by wind.

On the selling side, they have advantages rather than constraints on the environmental front. Their input energies can be stored and held back. To that extent, they mostly sell when prices are advantageous. Their unit commitments are self-imposed, arising out of their committed sales.

They are under constraints from the regulators under security requirements for the system.

Unit commitments for thermal generation are critical for the whole system. They are dealt with in Chapter 5.

All of these operational features are taken care of by a control system relying heavily on software programs. These are discussed in the next chapter.

2.19 ICMMS OF HYDROELECTRIC GENERATING UNITS

Intelligent control maintenance management systems (ICMMS [15]) of modern automation technology have three basic components: controls, maintenance, and management. Controls govern quality and performance. The field of controls, particularly automatic controls, has been highly researched and developed.

Maintenance aims at high reliability and availability. This field has been taken for granted for a long time, with very scanty research being done. Scheduled maintenance systems still rule this aspect. Of late, more maintenance-free equipment is being introduced.

Maintenance management systems include technical management and efficiency in all aspects of management, such as economic benefit and social benefit. This requires the highest level of attention from expert personnel.

2.20 CONTROLS AND COMMUNICATIONS IN HYDRO SYSTEMS

With the megasizes involved in hydroelectric generators, the control system rests primarily in the load dispatch center and secondarily at the generating stations [16].

The system breaks into operating controls (meeting local demand per bus) and security controls for deciding generator mix and regulating capacity. Mainly, this involves:

- Automatic start–stop of a unit
- Automatic record of disturbance "trend"
- Automatic generation control
- Economic dispatch with maximum energy production based on gross hydro energy output versus water discharge through turbines
- Unit commitment
- Automatic voltage control
- Reservoir management program
- Interface with regional load dispatch office on generating unit parameters, switchyard backup data, and set points for MW, MVAR, and so on
- Short-term and long-term reporting, and history building on energy reports, hydraulic reports, and so on
- Satellite clock synchronization

2.21 GENERAL MAINTENANCE

Maintenance in a hydraulic system includes:

1. Scheduled and breakdown maintenances. Scheduled maintenance cannot always prevent breakdown maintenance. Breakdown maintenance is very costly, involving with loss of production, emergency buying of materials, and employing expert man power, and so on.
2. Advanced maintenance. In advanced maintenance systems, one has an optimized asset management with nil costs in breakdown maintenance and with scheduled maintenance being replaced by predictive maintenance.

3. Predictive reactive maintenance. Predictive maintenance is based on condition monitoring and forecasting, information processing, and fault detection, all of these being coupled by an artificial intelligence technology.

2.22 LIMITATIONS OF SCHEDULED AND BREAKDOWN MAINTENANCE

Figure 2-12 gives the effect of scheduled maintenance. The bathtub curves as shown are generalized, assuming that a particular class of apparatus is manufactured, installed, and operated, again in a generalized common way. These types of curves will vary in detail from one class of unit to another.

For example, failure rates might show up as the apparatus components settle in immediately after being commissioned. Once these have been fine-tuned, after the initial runs, failure rates are intermittent. Then the apparatus works without noticeable trouble over a period. As the wear advances, we approach the rising end of the bathtub curve and failure rates rise. Generally, zone 1 and, to an earlier extent, zone 2 do not draw much maintenance attention. If a regular maintenance, including replacement of parts that call for attention, is taken care of, the bathtub slides further to the right, extending asset life. Attention now is concentrated on how this could be done intelligently.

2.23 REACTIVE MAINTENANCE—KEY ELEMENTS

Reactive maintenance is based on three key elements:

1. Monitoring and forecasting
2. Diagnosis and prognosis
3. Maintenance decision making

In Monitoring and forecasting, one looks for the key behavioral parameters of the system and its components and establishes indices for the same. We monitor these and analyze the trend. In the latest developments, we carry this online, that is, without disturbing the working of the system.

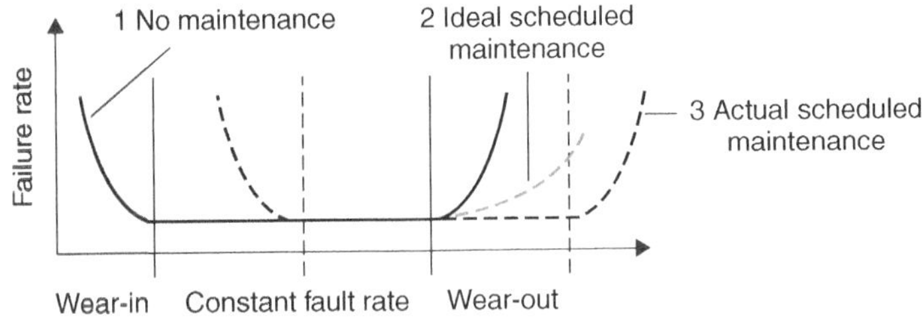

Figure 2-12. Effects of scheduled maintenance. (From [15], © 2004 IEEE.)

Next comes the crucial part, collection of historical/statistical data with respect to these indices. We can then forecast the health of the system, including life left until near-breakdown or next breakdown. Thus, online monitoring along with the database is the foundation of predicative maintenance.

In diagnosis and prognosis, all failures and degradations are diagnosed, including cause and location of failures as well as degradations. It includes effects of these on other parts of the system. Knowledge of the degree of degradations can help us to arrive at an approximate time to failure.

Maintenance decision making involves answering the question, "Is maintenance necessary and if so when?" Note that under scheduled maintenance one expends lots of valuable man-hours, as well as machine hours (if the maintenance cannot be done without stopping the machine), all of which could possibly be avoided if we knew that the existing health of the machine could be taken for granted for a further predictable period. The diagnosis will also tell us precisely what materials will have to be replaced and how this maintenance can be adopted and executed.

2.24 KEY COMPONENTS OF AN ICMMS—CASE OF A HYDROELECTRIC SYSTEM

Some of the key components of an ICMMS for a hydroelectric generator are:

- Control station with a controller for the water turbine generator. Data is monitored on the speed of the turbine, that is, its frequency, output of controller, actual stroke of main servomotor, and so on.
- Technical management station. This essentially oversees the water turbine governing system. It does a performance assessment and monitors the operational arrangement of the unit. It watches the dynamic motion and failure status (if and when failure occurs) of the electro-hydraulic servomechanism.
- Unit simulating platform. This is where the components of the systems are modeled. Here, dynamic responses of processes can be simulated and studied in advance. These include startup, synchronization, frequency or speed regulation, load increasing or decreasing, load rejection, and shutdown etc.

2.25 INTELLIGENT ELECTROHYDRAULIC SERVOMECHANISM

In the old system, communication between a controller and the servomechanism was done with analog signals. This is now done with digital signals, which gives us a smooth and coordinated way to collect data and control the system. The kind of failures and degradations on the watch list are trembling coil breakages, zero-point creepages, mechanical backlashes, and so on. These signals are sent to the maintenance system for further processing.

Please refer to Chapter 3, Appendix 3-2 for details.

2.26 ONLINE MONITORING AND FORECASTING

2.26.1 Partial Discharges (PDs) in the Stator Coils of Alternators

The old method to monitor PDs was to apply a steady high AC voltage across individual stator coils and measure the magnitude of PDs against these voltages. A record tells us whether these are rising over a period and, if so, at what rate. Maintenance is indicated when the PD level crosses a reference datum. This method has several disadvantages, including taking the alternator out of service for periodic testing. This type of testing work can be planned during off-load periods, when hydro electric sets are kept idle.

In the online measurement technique, PD detection coils are fitted on individual stator coils. The rotor excitation voltages are ramped up in stages, up to 100% of the rated value. The alternator is loaded, as it would be normally, say to 80% of the rated value. PD measurements are taken across these detector coils.

This method has the advantage that the PDs are measured under dynamic working conditions. The stator coils are subjected to vibrations, overhang stresses, thermal conditions, and so on, as they would be under normal working conditions. There are disadvantages also. The PD signals are contaminated by noise signals, which must be differentiated and filtered out. The PD has to travel across a conductor in a slot. This passage is a ladder type with capacitances to slots in parallel and conductor inductances in series. The signals could be attenuated to such an extent that it would be difficult to locate exactly the PD source point.

The PDs arise out of defects in the insulation. These defects arise due to aging deterioration, moisture pollution, inadequate design, and so on. The slot insulation defects might arise out of voids, de-lamination, and so on. Generally, this insulation consists of mica flakes bound with resins and has a high strength against PDs. PDs will arise due to degradation of the coil or corona shielding, poor design of the stress-grading system for the overhang position, and so on. These localized defects could lead to breakdown, even if the overall insulation as measured with a megger is found to be satisfactory.

PDs generated due to possible effects as above could be classified into two groups:

1. Self-contained PDs whose growth rate is negligible. These can stay in the insulation without harm for years.
2. PDs with accelerating growth have to be watched for. These are accelerated by internal electrical discharges and by the heat generated during these discharges. Monitoring and analyzing the PDs is not a simple, straightforward proposal. As pointed out, the collected signals also have noise levels. These signals have to be filtered to weed out the unwanted portions. With the insulations normally used, a passband between 30 kHz to 500 kHz is acceptable, and rest of the signals are filtered out. There are clusters of recorded data that are segregated into shapes of homogenous classes. Clusters denoting noise levels are again weeded out. The rest of the clusters are divided into subpatterns and these subpatterns become relevant to one of the specific topologies. From here, one next associates the source of the macro category of defects with the subpattern [17].

PD magnitudes thus identified are filed. Comparison of these with the earlier magnitudes tells us the growth in magnitude and, just as importantly, the rate of growth. Both of these tell us the safely acceptable interval to the next survey and, alternatively, how soon corrective maintenance is necessary. With these techniques, hydrogenerators designed for a historical/statistical life of 20 years have served well beyond 20 years. Just as importantly they have helped us to not have to peform breakdown maintenance as we near the end of a 20 year life. One may compare this with the old scheduled maintenance method of measuring PDs and insulation at fixed intervals.

2.26.2 Air Gap Monitoring of Vertical Hydraulic Generators

Over a period of working years, the rotor becomes eccentric with respect to the stator and starts rubbing against it. In a worst-case scenario, a rotor scoured the stator to such an extent that the stator had to be completely replaced. This caused a production loss for 2.5 months and cost as much as 50% of the cost of a new generator [17]. An online air gap monitoring has become a must for predictive maintenance.

The stator assemblies have large masses of stamping, copper, insulation, and so on. To some extent, these become flexible and shift slightly. The forces causing this shift are magnetic (field effect), vibratory, structural (temperature changes), and geotechnical. In one example cited, the control on a solid-state exciter current generator failed and the excitation current shot up to such an extent that the rotor pulled in the stator [17]. On the rotor side, the centrifugal forces can distort and shift some components such as the rotor rim. This shift causes considerable vibrations. These are also detectable with the mapping of the air gap.

Depending on the MW ratings, air gaps vary from 25 mm up to 50 mm. The monitoring accuracy expected is within ±0.5 mm. Air gaps have to be measured along their circumference at a number of places. The plot gives us the concentricity of the rotor with respect to the stator. Signals are processed mathematically and pole surface roundness is accounted for.

Sensors are placed on the stator. These sensors monitor capacitive current in the air gap when a constant voltage at a high frequency (455 KHz) is applied across it. Preferred position of the sensors in a vertical alternator is at the top. The stator assembly and the frame are less rigid at the top than at the bottom, so maximum relative movement is expected there.

Air gap monitoring can also be done with optical sensors. However, these accumulate dirt, grease, and so on and become undependable. Although accuracy in an optical system is higher than in a capacitive measurement system, the latter is preferred.

2.27 SUBSYNCHRONOUS RESONANCE (SSR) AND TWISTING OF ROTOR SHAFTS

Hydroelectric systems are tied into high-voltage transmission systems. The majority of these HE installations are away from load centers. The interconnecting links are long and need a voltage regulation as per load. This is provided by installing high-voltage series capacitors of designed impedances. They have an advantage of giving automatic voltage regulation. The protection systems for the series capacitors were not initially very robust and reliable, but now these systems are being used more extensively.

These series capacitors, depending on the other system parameters, including the inductive impedance of a turbo alternator, can resonate at low or subsynchronous frequencies. Similarly to what we explained earlier on penstock pressure pulsations, large energies travel back and forth. Under their impact, turbine rotor shafts twist and shear. This happens more when the electrical subsynchronous resonance frequency coincides with the mechanical resonance system of the turbo alternator [6].

As pointed out earlier, repair and replacement costs, including loss of reserves, are tremendous, so monitoring must take care of SSR.

REFERENCES

1. J. Martini, S. C. Martini, and L. B. Reis, Brazilian Energy Crises, *IEEE Transactions on Power Engineering Review,* April 2002, pp. 21–24.
2. B. Sadden, Hydropower Development in Southern and Southeastern Asia, *IEEE Transactions on Power Engineering Review,* Power Engineering Review, IEEE Volume 22, Issue 3, March 2002, pp. 5–9.
3. B. Sadden, Aspects of Hydrodevelopment in South and Southeast Asia, *IEEE Transactions on Power Engineering Review,* Volume 1, 2001.
4. M. Breton, S. Hachem, and A. Hammadia, Accounting for Losses in the Optimization of Production of Hydroplants, *IEEE Transactions on Energy Conversion,* June 2004, pp. 346–351.
5. L. Ye, Control/main strategy, power tolerant mode, *IEEE Transactions on Power Engineering,* August 2001, pp. 345.
6. B. Strah, B., *IEEE Transactions on Energy Conversion,* June 2005, pp. 424-434.
7. R. E. Doan, *IEEE Transactions on Energy Conversion,* June 2004, pp. 449-455.
8. D. Schneiter, *IEEE Transactions on Energy Conversion,* June 1995, pp. 342-353.
9. O. P. Malik, *IEEE Transactions on Energy Conversion,* June 1995, pp. 354-359.
10. J. Jiang, *IEEE Transactions on Energy Conversion,* March 1995, 188-194.
11. D. M. Tagere, Reactive Power Management, Tata-McGraw Hill Publishing, 2004.
12. Y. Luqing, W. Shouping, L. Zhaohui, O. P. Malik, and G. S. Hope, Field Tests and Operation of a Duplicate Multiprocessor-Based Governor for Water Turbine and its Further Development, *IEEE Transactions on Energy Conversion,* Volume 5, Issue 2, June 1990, pp. 225–231.
13. V. V. Terzija and M. Akke, Synchronous and Asynchronous Generators Frequency and Harmonics Behavior After a Sudden Load Rejection, *IEEE Transactions on Power Systems,* May 2003, pp. 730–736.
14. K. E. Bollinger, L. D. Nettleton, and J. H. Gurney, Reducing the Effect of Penstock Pressure Pulsations on Hydro Electric Plant Power System Stabilizer Signals, *IEEE Transactions on Energy Conversion,* Volume 8, Issue 4, December 1993, pp. 628–631.

15. J. F. Lyes, T. E. Goodeve, and H. Sedding, Parameters Required to Maximize a Thermoset Hydro-Generator Stator Winding Life. Part II-Monitoring, Maintenance, *IEEE Transactions on Energy Conversion,* September 1994, pp. 628–635.

16. J. W. L. Simpson, R. C. Tychsen, Q. Su, T. R. Blackburn, and R. E. James, Evaluation of Partial Discharge Detection Techniques on Hydro-Generators in the Australian Snowy Mountains Scheme. Tumut 1 Case Study, *IEEE Transactions on Energy Conversion,* March 1995, pp. 18–24.

17. J. Borghetto, A. Cavallini, A. Contin, G. C. Montanari, M. de Nigris, G. Pasini, and R. Passaglia, Partial Discharge Inference by an Advanced System. Analysis of Online Measurements Performed on Hydrogenerator, *IEEE Transactions on Energy Conversion,* June 2004, pp. 333–339.

BIBLIOGRAPHY

Andersson, S. and S. O. Lindstrom, Retrofit of Trangslet Hydro Power Station, *IEEE Transactions on Energy Conversion,* March 1992, pp. 29–34.

Baptistella, L. F. B. and J. C. Geromel, Decomposition Approach to Problem of Unit Commitment Schedule for Hydrothermal Systems, Control Theory and Applications, IEE Proceedings, Volume 127, Issue 6, November 1980, pp. 250–258.

Binato, S. and M. V. F. Pereira, Decentralized Planning of Hydroelectric Power Systems, *IEEE Transactions on Power Systems,* February 1995, pp. 492–498.

Borghetti, A., A. Frangioni, F. Lacalandra, and C. A. Nucci, Lagrangian Heuristics Based on Disaggregated Bundle Methods for Hydrothermal Unit Commitment, *IEEE Transactions on Power Systems,* Volume 18, Issue 1, Feb. 2003, pp. 313–323.

Chaa-An, L., R. B. Johnson, A. J. Svoboda, T. Chung-Li, and E. Hsu, A Robust Unit Commitment Algorithm for Hydro-Thermal Optimization, *IEEE Transactions on Power Systems,* Volume 13, Issue 3, August 1998, pp. 1051–1056.

Chao-An, L., A. J. Svoboda, T. Chung-Li, R. B. Johnson, and E. Hsu, Hydro Unit Commitment In Hydro-Thermal Optimization, *IEEE Transactions on Power Systems,* Volume 12, Issue 2, May 1997, pp. 764–769.

Chuang, F., Y. Luqing. L. Yongqian, Y. Ren, I. Benoit, C. Yuanchu, and Z. Yuming, Predictive Maintenance in Intelligent-Control-Maintenance-Management System for Hydroelectric Generating Unit, *IEEE Transactions on Energy Conversion,* March 2004, pp. 179–186.

Da Silva, E. L., M. Th. Schilling, and M. C. Rafael, Generation Maintenance Scheduling Considering Transmission Constraints, *IEEE Transactions on Power Systems,* May 2000, pp. 838–843.

De Moraes, J., J. Rodriguez Villalba, V. F. Salatko, Electrical and Related Design Aspects of Itaipu Hydroeletric Project, *IEEE Transactions on Power Apparatus and Systems,* Volume PAS-101, Issue: 5, May 1982, pp. 1012–1020.

Delfino, B., G. B. Denegri, A. Morini, E. Cima Bonini, R. Marconato, and P.Scarpellini, Black-Start and Restoration of a Part of the Italian HV Network: Modelling and Simulation of a Field Test, *IEEE Transactions on Power Systems,* August 1996, pp. 1371–1379.

Doan, R. E. and K. Natarajan, Modeling and Control Design for Governing Hydroelectric Turbines with Leaky Wicket Gates, *IEEE Transactions on Energy Conversation,* June 2004, pp. 449–455.

El-Serafi, A. M. and S. O. Faried, Effect of Adaptive Reclosing on Turbine-Generator Shaft Torsional Torques, *IEEE Transactions on Power Systems,* June 1994, pp. 1730–1736.

Gonzalez, C. and J. Juan, Leveling Reliability in Systems with Large Hydro Resources, *IEEE Transactions on Power Systems,* February 1999, pp. 23–28.

Gonzalez, C., J. Juan, J. Mira, F. J. Prieto, and M. J. Sanchez, Reliability Analysis for Systems with Large Hydro Resources in a Deregulated Electric Power Market, *IEEE Transactions on Power Systems,* February 2005, pp. 90–95.

Harrison, G. P., H. W. Whittington, and A. R. Wallace, Climate Change Impacts on Financial Risk in Hydropower Projects, *IEEE Transactions on Power Systems,* November 2003, pp. 1324–1330.

Jiang, J., Design of an Optimal Robust Governor for Hydraulic Turbine Generating Units, *IEEE Transactions on Energy Conversation,* March 1995, pp. 188–194.

Kheirmand, A., M. Leijon, and S. M. Gubanski, Advances in Online Monitoring and Localization of Partial Discharges in Large Rotating Machines, *IEEE Transactions on Energy Conversion,* March 2004, pp. 53–59.

Kundu, D., Technical Requirements to Connect Parallel Generators to the Ontario Hydro Distribution Electricity System, *IEEE Transactions on Energy Conversion,* March 1992, pp. 8–14.

Lindenmeyer, D., A. Moshref, M. C. Schaeffer, and A. Benge, Simulation of the Start-up of a Hydro Power Plant for the Emergency Power Supply of a Nuclear Power Station, *IEEE Transactions on Power Systems,* February 2001, pp. 163-169.

Luqing, Y., W. Shengtie, B. Fengshan, O. P. Malik, and Z. Yuming, Control/Maintenance Strategy Fault Tolerant Mode and Reliability Analysis for Hydro Power Stations, *IEEE Transactions on Power Systems,* August 2001, pp. 340–345.

Lyles, J. F., T. E. Goodeve, and H. Sedding, Parameters Required to Maximize a Thermoset Hydro-Generator Stator Winding Life. Part I: Design, Manufacture, Installation, *IEEE Transactions on Energy Conversion,* September 1994, pp. 620– 627.

Lyra, C. and L. R. M. Ferreira, A Multiobjective Approach to the Short-Term Scheduling of a Hydroelectric Power System, *IEEE Transactions on Power Systems,* November 1995, pp. 1750–1755.

Malik, O. P. and Y. Zeng, Design of a Robust Adaptive Controller for a Water Turbine Governing System, *IEEE Transactions on Energy Conversation,* June 1995, pp. 354–359.

Mo, B., A. Gjelsvik, and A. Grundt, Integrated Risk Management of Hydro Power Scheduling and Contract Management, *IEEE Transactions on Power Systems,* May 2001, pp. 216–221.

Nabona, N., J. Castro, and J. A. Gonzalez, Optimum Long-Term Hydrothermal Coordination with Fuel Limits, *IEEE Transactions on Power Systems,* May 1995, pp. 1054–1062.

Pollock, G. P. and J. F. Lyles, Vertical Hydraulic Generators Experience with Dynamic Air Gap Monitoring, *IEEE Transactions on Energy Conversion,* December 1992, pp. 660–668.

Ponrajah, R.A., J. Witherspoon, and F. D. Galiana, Systems to Optimize Conversion Efficiencies at Ontario Hydro's Hydroelectric Plants, *IEEE Transactions on Power Systems,* August 1998, pp. 1044–1050.

Prowse, D. C. H., P. Koskela, T. A. Grove, and L. R. Larson, Experience with Joint AGC Regulation, *IEEE Transactions on Power Systems,* November 1994, pp. 1974–1979.

Radinskaia, E. and F. D. Galiana, Generation Scheduling and the Switching Curve Law, *IEEE Transactions on Power Systems,* May 2000, pp. 546–551.

Roy, S., Optimal Planning of Generating Units Over Micro-Hydro Resources Within a Catchment Area, *IEEE Transactions on Energy Conversion,* March 2005, pp. 231–236.

Samuelsson, O. and M. Akke, On-Off Control of an Active Load for Power System Damping—Theory and Field Test, *IEEE Transactions on Power Systems,* May 1999, pp. 608–613.

Schewe, P. F., Electric Idyll, *IEEE Transactions on Power Systems,* October 2006, pp. 40–47.

Schniter, P. and L. Wozniak, Efficiency Based Optimal Control of Kaplan Hydrogenerators, *IEEE Transactions on Energy Conversation,* June 1995, pp. 348–353.

Shawwash, Z. K., T. K. Siu, and S. O. D. Russell, The B.C. Hydro Short Term Hydro Scheduling Optimization Model, *IEEE Transactions on Power Systems,* August 2000, pp. 1125–1131.

Siu, T. K., G. A. Nash, and Z. K. Shawwash, A Practical Hydro, Dynamic Unit Commitment and Loading Model, *IEEE Transactions on Power Systems,* Volume 16, Issue 2, May 2001, pp. 301–306.

Souza Jr., O.H., N. Barbieri, and A. H. M. Santos, Study of Hydraulic Transients in Hydropower Plants Through Simulation of Nonlinear Model of Penstock and Hydraulic Turbine Model, *IEEE Transactions on Power Systems,* November 1999, pp. 1269–1272.

Strah, B., O. Kuljaca, and Z. Vukic, Speed and Active Power Control of Hydro Turbine Unit, *IEEE Transactions on Energy Conversion,* June 2005, pp. 424–434.

Tong, S. K. and S. M. Shahidehpour, Hydrothermal Unit Commitment with Probabilistic Constraints Using Segmentation Method, *IEEE Transactions on Power Systems,* Volume 5, Issue 1, February 1990, pp. 276–282.

Tuohy, A., P. Meibom, E. Denny, and M. O'Malley, Unit Commitment for Systems with Significant Wind Penetration, *IEEE Transactions on Power Systems,* Volume 24, Issue 2, May 2009, pp. 592–601.

Wang, C. and S. M. Shahidehpour, Ramp-Rate Limits in Unit Commitment and Economic Dispatch Incorporating Rotor Fatigue Effect, *IEEE Transactions on Power Systems,* Volume 9, Issue 3, August 1994, pp. 1539–1545.

Woschnagg, E., Turbogenerator Field Winding Shorted Turn Detection by AC Flux Measurement, *IEEE Transactions on Energy Conversion,* June 1994, pp. 427–431.

Xichang, X. Y. and Z. Quanren, Practical Implementation of the SCADA+AGC/ED System of the Hunan Power Pool in the Central China Power Network, *IEEE Transactions on Energy Conversion,* June 1994, pp. 250–255.

3

HYDROELECTRIC GENERATION—PUMPED STORAGE, MINOR HYDROELECTRIC, AND OCEANIC-BASED SYSTEMS

3.1 WATER AS AN ENERGY SUPPLIER AND AN ENERGY STORE

Water raised to a height above ground level has potential energy stored in it. We can draw on it as an energy supplier as illustrated in the earlier chapters. During and after use of water as an energy supplier, there are no byproducts and energy withdrawal is a clean, eco-friendly operation. With rising demand for eco-friendly production of electricity, attention has shifted to smaller sources of hydro electricity, such as irrigation canals, seasonal water streams, and run-of-the river schemes. A moving mass of water can also impart and deliver energy in a electromechanical system. The main forms that are attracting attention again are tidal energy and waveform energy from sea water. There are also discussions going on about drawing electrical energy from oceanic thermal differences in water temperatures.

Let us turn to water as an energy store, wherein we use electricity generated during off-peak periods to pump water to a higher level and produce energy during peak periods by letting it go down again. Typical round-trip efficiency between energy output and energy input is 80%. This energy store is of an immense importance to most of the renewable energy sources whose energy output is mostly unpredictable and variable.

At the economic forum of major countries in the world, an agreement was reached to bring down CO_2 emission levels to 85% of the 1990 levels by the year 2050. This means we have to scale back thermal power generation, which accounts for around

Electricity Power Generation. By Digambar M. TagAre
Copyright © 2011 the Institute of Electrical and Electronics Engineers, Inc.

56–60% of supply universally, to a negligible percentage. The main contenders to replace this source will be hydroelectricity in all its forms, nuclear electricity, and renewable energy, of which wind is the major contender, followed by solar.

The renewable sources, including hydro-based, are mostly scattered. This brings distributed energy sources and their accommodation in the grid to the forefront, as well as the need for distributed energy storage. The main contender for this today is hydrogen, with the drawback that its production is not eco-friendly. Its return-trip efficiency is low, about 32%.

We can safely forecast that water will take the lead as an energy supplier in all its forms and as an energy storage medium even on a distributed scale.

Water is first a medium for storing energy and then a supplier. Let us study some aspects of this in detail.

3.2 PUMPED WATER STORAGE SYSTEM FOR ELECTRICITY GENERATION

Imagine that there are two lakes, one upstream and one downstream of a river or stream. During low-demand periods on the electrical power system, output of the base-load generators is utilized to pump water from the lower lake to the upper lake. During high-demand periods, this water drives the same electrical motors/pumps as turbine-generators, which generate electricity and deliver it to the grid.

Commercially, electricity is used for pumping up when the time-of-use (TOU) rates are the lowest. Electricity is produced under peak demand periods when TOU rates are the highest. This makes the whole process attractive. Besides, delivery of power is at one's command. The following description gives the details.

A typical case is described in Wright and Brinson [1], involving a hydropower system located at Salem, South Carolina in the United States.

In the pumping mode, the system pumps up water from a 7565 acre lake downstream to a 313 acre lake upstream. It generates electricity in the reverse mode, with the motor acting as a generator. The generators are located in a cavern, some 165 m underground.

The generators are vertical, synchronous machines, with an attached blower and with a pony motor to rev up the motor to near synchronous speed before connecting it to the grid at the grid frequency. Each unit is rated at 313/360 MVA, 60/80°C rise, 0.9 PF, 19 kV, 3 Ph, 60 Hz, 300 RPM. Net head available is 367 m. When in a pumping mode, it is rated at 392,613 HP, with an overload capacity of 15%. A microprocessor-controlled exciter provides excitation current. A pony motor mounted on the generator shaft is rated at 18,000 HP. It is an induction type. It brings up the assembly to 300 RPM in about 7 minutes. The speed and current of this motor are controlled with a water rheostat. After the synchronization of the generator with the grid, the pony motor trips off. During the normal running, the friction and windage losses of the pony motor are quite low. For starting up the pumping operations, the air from the pump/turbine is vented out and water is allowed to rise.

The cavern type of installation presents two specific problems. Space and heat losses do not permit the step-up transformers to be located next to the generators, which is desirable in view of large currents involved at 19 kV. They are located above ground and vertical buses carry this current from the generators (there are four of these) to the step-up transformers (19 kV:565 kV). These bus ducts and the generator room have to be air-conditioned. Secondly, the rock has a high resistivity. Grounding presents a formidable problem with ground resistivity of the rock at 240,000 ohm-cms.

Relief is provided against complete failure and blackout possibilities with a small 750 kW diesel generator. There are also two auxiliary services.

3.3 OPERATION OF A PUMPED STORAGE SYSTEM

Normally, pumped storage systems should be operated for an optimized economical advantage for mainly shaving off the peaks from the daily load curves. The marginal costs of electricity generation are the highest during peak periods. The pumping up consumes costly energy from the thermal stations. The overall efficiency of a pumped storage systems averages to around 67%. Lee [15] has calculated that if a storage battery were to be substituted for storage of energy for use in peak periods, its efficiency would be higher at 70–80%. Use of thermal power from the grid for pumping up the water deducts certified emission reductions (CERs) that otherwise would be fully available for a nonpolluting hydropower system.

The pumping up operation in pumped storage systems should be carried on for an optimally minimum time. This gives longer equipment life by reducing wear and tear. Also, prolonged pumping times increase the power consumption in the pumping up mode. In a typical calculation, Lee [2] gives an economical disparity of $20.7 million per annum between one system with pumping up period of 7.7 hr and generation periods of 5.8 hrs, compared to another that has both of these daily operations restricted to 4 hr each.

Constraints on these operations are:

- A minimum water level that must be maintained in the upper lake, remembering that turbine efficiencies are tied to water heads across its blades.
- Daily load variations, for example, practically no demand on weekends.

See Figure 3-1 below.

3.4 PUMPED STORAGE SYSTEMS HAVE LIMITED SCOPE

Pumped storage systems have comparatively low cyclic efficiencies as pointed out above. Besides, their geographical requirements are not easily met. They require two water lakes in place of one, meaning twice as many problems in land acquisition and resettlement compared to a straightforward HE project. Both the lakes need to be adjacent and must have an economically viable water level difference. This makes pumped storage less attractive than small hydroelectric projects.

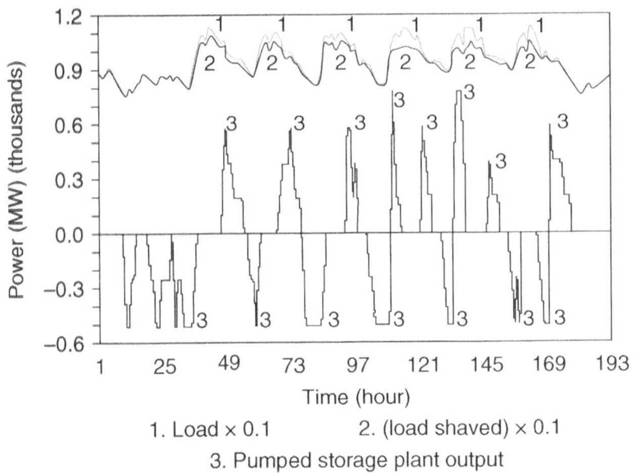

Figure 3-1. Pumped storage plant generation. (From [2], © 1992 IEEE.)

It may also be noted that pumped storage systems do not add to the total electricity production. They consume more electricity than they produce, since there are losses in both directions. Refer to Table 2-1. It shows that their yearly contribution to the total production is negative. The main value of a pumped storage system lies in its capacity as an energy storage medium that can be operated for maximum advantage. This capacity is now in demand.

3.5 PUMPED STORAGE SYSTEMS AND WIND ENERGY

When viewed from the points of using costly and environmentally objectionable thermal electricity, coupled with specific land requirements, it would appear that pumped storage systems for electricity production do not have a very bright future, but this is not quite so. It looks like pumped storage systems are likely to be favorites not too far in future.

Pumped storage systems can store huge amounts of energy and return it back, in electrical form, as needed. Wind energy has large trappable energy potential. It requires a large amount of energy storage to enable it to grow because of the variable nature of winds. Wind energy's biggest advantage is that it incurs no fuel costs—it is free energy.

The following status-quo reports are self explanatory:

1. Renewable energy use in producing electricity is getting a universal push. European Union countries have set a target of meeting 25% of their total annual electricity requirements from wind energy by the year 2012.
2. Denmark already claims to produce 100% of it electricity requirements from wind energy. It is tied into an international grid with Sweden and Norway. Both these countries have huge hydroelectricity production capacity. They act as energy stores for Denmark.

3. Portugal claims to have reached the 25% target for a number of days in 2008. It has integrated its grid with that of Spain and has access to hydroelectricity installations in both countries. It is targeting to reach a level of 40% and is supposedly looking for pumped storage systems.
4. The Azores Islands in Portugal claim to have achieved 63% of target with renewables of all types, including wind, hydro, and solar, available there.
5. Germany is reportedly planning to replace old wind generators with large-sized new generators.

3.6 SMALL HYDROELECTRIC PLANTS (SHPs)

Small dam-based hydroelectric plants have been in operation for quite some time. With growing demand and production of electricity, they did not attract much attention because of their small scale of operations. However, the picture is changing fast in their favor now.

Small HE plants have the advantage of not requiring large tracts for water storage. Land requisition problems are down to almost nil. Capital costs for machinery and equipment are comparatively low. Both these factors result in a very small, temptingly low, completion period.

There is a rising, universally accepted environmental concern now. Larger electricity generation projects, both hydro and thermal, have a sizeable price tag attached on account of social and environmental concerns, so much so that it is estimated that, presently, this cost matches the cost of generating all of the electricity power. Not only are small hydro free of these costs, but they gain advantages from it, such as liberal subsidies on capital costs and winning gains on CDMs (clean development mechanisms) through certified emission reduction certificates, earned due to displacement of carbon as a source of energy in electricity generation.

Initially, depending on the size, one starts experimentally with the lowest cost equipment, say a turbo generator. The turbo generator technology improves in the meantime and gives you a higher efficiency machine later. It is affordable to replace the existing one, since the sizes are small.

Needless to say, manpower required, operating and maintenance expenses, and so on are relatively low.

3.7 TYPES OF SHP PROJECTS—SIZES

In India, small HE projects are defined as per their sizes:

Small hydro—2 to 25 MW
Mini hydro—0.1 to 2 MW
Micro hydro—up to 0.01 MW

Hydro schemes below 25 MW are eligible for CDM benefits.

Slovenia sets a limit of 2 MVA for small hydro plants with a limit of 50 kVA for privately owned SHPs. All SHPs have to tie into the national grid [3].

Worldwide, the definition for small hydropower plants varies by country (Table 3-1). Classification of hydro schemes in India is shown in Table 3-2.

3.8 LOCATION-WISE DESIGNATIONS OF SHPs

There are three types of SHPs with regard to location:

1. Dam-based SHPs
2. Canal-based SHPs
3. Run-of-the river SHPs

3.9 COMPONENTS OF AN SHP

Components of an SHP include efficient desilters, self-cleaning water intakes, trash racks, spillways, and tail races in the construction of canal-type SHPs. The routine work consists of excavation, concrete lining, steel work, and building a diversion canal. A pump house to accommodate the turbo generator, controlling and synchronizing equipment, and, of late, connection to a SCADA network comes next.

Table 3-1. Range for small hydroplants

Country	MW
UK	5
UNIDO	10
Sweden	15
Colombia	20
Australia	20
India	25
China	25
Philippines	50
New Zealand	50

Source: [3, p. 85].

Table 3-2. Micro, mini, and small hydroplants

Class	Station capacity in kW
Micro	Up to 100
Mini	101 to 2000
Small	2000 to 25,000

Source: [3].

The most cost-determining component is the turbo generator. These come in a wide variety, particularly for the low-head-canal-based SHPs. Turbine types are summarized in Table 3-3. Figures 3-3 to 3-5 give sketches of these.

3.10 TYPICAL LAYOUTS OF SHPs

3.10.1 The Generator

There are two types of generators commonly used: an induction generator and a synchronous generator. However, the synchronous generator is seldom used in SHPs.

Induction Generator. The induction generator is the lowest cost and most easily and reliably adaptable one. If one revs it up to near its synchronous speed, determined by the number of poles in the windings and the grid frequency, and then switches it on, it locks up with the grid frequency and voltage and stays locked up. However, it has its own disadvantages. It needs an external reactive power supply to activate its magnetic field. Unlike a synchronous generator, it cannot generate its own reactive power. This means that one has to install power capacitors in its tie-up with the grid. This way, the grid has control on both voltage and frequency. A synchronous generator, on the other hand, has independent control on both. It can generate and inject reactive power and support grid voltage at its point of coupling. An induction generator cannot do that.

Problems with Induction Generators. Connecting external capacitors across an induction generator creates problems of its own:

1. Assume a condition wherein the grid gets disconnected. There is no direct load across the generator. As a result, the generator feeds the capacitors, which in

Table 3-3. Turbine performance characteristics

Sr. No.	Turbine type	Rated head Hr (m)	Capacity (MW)
1	Vertical fixed propeller	2-20	0.25-15 and above
2	Vertical Kaplan (adjustable blade propeller)	2-20	1-15 and above
3	Vertical Francis	8-20	0.25-15
4	Horizontal Francis	8-20 and over	0.25-15
5	Tubular (with adjustable blades and fixed gates)	2-18	0.25-15
6	Tubular (fixed blade runner with wicket gates)	2-18	0.25-15
7	Bulb	2-20	1-15
8	Rim	2-9	1-8
9	Right angle drive propeller	2-18	0.25-2
10	Open	2-11	0.25-2
11	Closed flume	2-20	0.25-3
12	Cross flow	8-20	0.25-2

Source: [4].

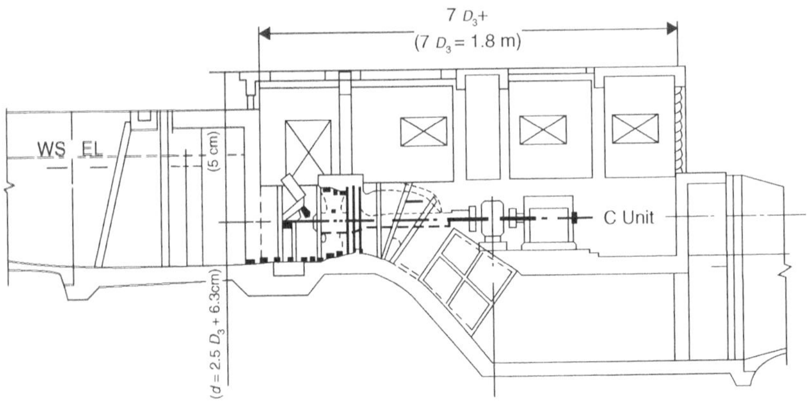

Figure 3-2. Typical layout of power house with tubular turbine. (From [4], © IEEMA 2008.)

turn boosts up the magnetic flux. A high voltage builds up and breaks down the insulation. The capacitor or generator coils might be the casualty. It is essential that when there is no output current from the generator into the external circuit, the capacitor be disconnected.

2. As a strict grid discipline, when the grid is disconnected, the independent generator should be disconnected from its point of connection. It should not work in an islanded condition. High-speed detection relays and controls are essential. Please refer to Chapter 4 on this point.
3. The rotating mass of the induction generator needs a big heave-in to pull itself into synchronism with the grid frequency. This requires an instantaneous heavy

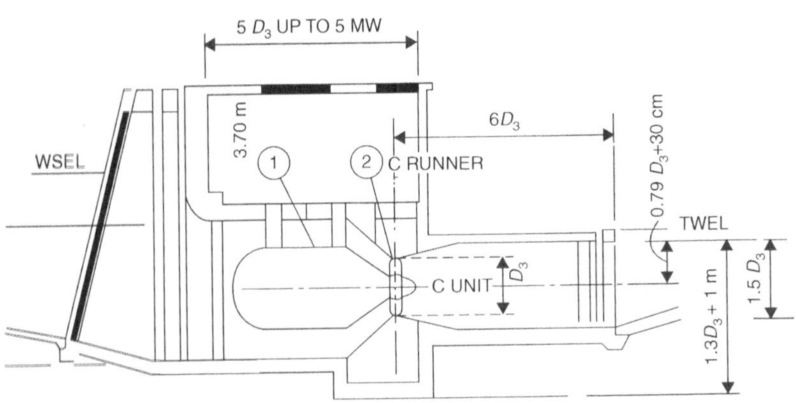

Figure 3-3. Typical layout of power house with bulb turbine. (From [4], © IEEMA 2008.)

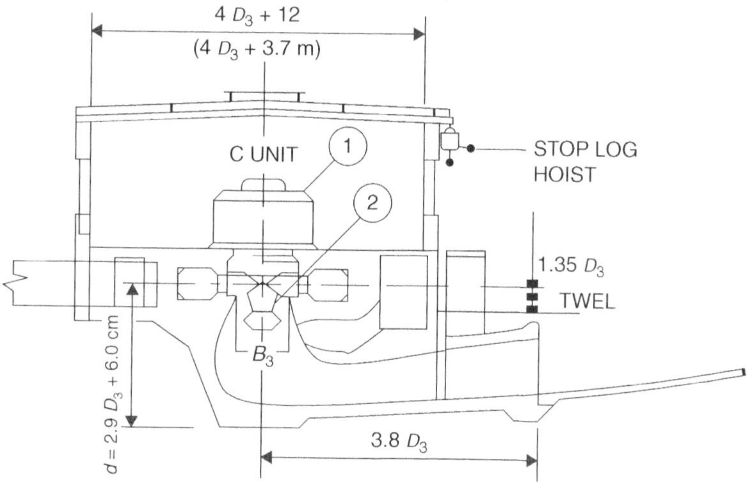

Figure 3-4. Typical layout of power house with propeller turbine. (From [4], © IEEMA 2008.)

power. If the attained speed is too far behind the synchronous speed, this will cause an objectionable dip in the grid voltage. This point must be watched.

4. There have been objections on grounds of harmonic distortions in the grid supply on account of induction generators operating with permanently connected capacitors. This has been investigated by Vorsie in Slovenia [3]. He has found total current harmonic distortion, in a typical case, varying from 37.6% on full load to 56.6% against no load on the induction generator. However, the distortion will be sized by the short-circuit level of the grid at the point of connection.

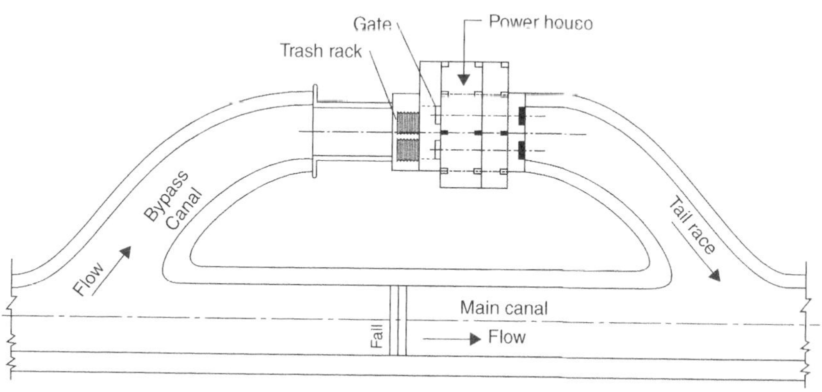

Figure 3-5. Typical layout of canal-based small hydropower scheme. (From [4], © IEEMA 2008.)

5. Privately owned and utilized very small SHPs can act as independent standby supplies, similar to diesel-engine-driven generators. In this case, there is no guarantee of the quality of supply and they cannot be connected into the grid.

3.10.2 Dam-Based SHPs

The Ministry of National Bureaus of India has surveyed and identified as many as 4861 sites with a total power potential of 12,841 MW. These sites are thrown open for private development. Sizeable subsidies are offered, mostly based on the power rating and the terrain. The majority of them are along mountainsides.

3.10.3 Canal-Based SHPs

With emphasis on agricultural development, lots of dams and reservoirs have been built. Canals are an integral part of these projects. It is realized that there is considerable potential in developing canal-based SHPs.

Some 5400 sites with a potential of 14,300 MWs have been identified and thrown open for development by private parties by the Indian Government. Here again, subsidies based on kW rating, are offered. For some 593 SHPs already built and in operation, some quite useful data has been collected.

3.11 PROJECT COSTS OF AN SHP

For an average project, the cost breakdown is given as:

E and M	54%
Toward civil work	34%
Toward other expenses	11.5%

A verified cost formula is:

$$\text{Cost for the project} = K \cdot P^x \cdot H^y$$

where K is a constant, x and y are other constants, P is the power rating in kW, and, H is the available head in meters.

Typical values in rupees (Rs) are [23]:

$$K = 390855$$

$$x = 0.2050$$

$$y = 0.1483$$

Again, a typical cost of production has been worked out at 1.50 to 3.00 Rs per unit. The CDM benefit available works out to 0.40 to 0.45 Rs per unit.

A central testing laboratory has been established to test and certify the units for SHPs.

With rising prices of oil and trading of electricity, coupled with environmentally free and readily available supply of hydropower at affordable and subsidized costs, small hydro projects have a great potential for development, not only in India, but all over the world.

3.12 DRAWING ELECTRICITY FROM THE OCEAN

3.12.1 Nature of Energy Available from the Oceans

Energy available from the ocean consists of tidal energy, wave energy, and thermal (difference) energy.

Tidal Energy. Tides are formed by three attraction forces on the mass of water in an ocean: the gravitational attraction by the earth and the pull-out attraction exerted by the sun and the moon. The latter two attractions are roughly proportional to

$$(\text{mass of celestial body}) \times (\text{mass of water under influence})/(\text{distance in between})^2$$

When the sun and moon are coaxial in position, the pull is the highest. When they are at right angles, it is the lowest. On full moon day and on the first day of the new moon, the pull-out forces are the highest. The pull-out forces cause the rise and fall of the tides. Rotational forces of the earth give motion to the waves. Thus, a tidal wave has two kinds of energies associated with it: the potential energy associated with the rise of tide and height of the wave, and the kinetic energy associated with their motions.

These tides have a predictable pattern. In some parts of the world, the tide will rise twice in 24 hours (diurnal tides). In some other parts, they will rise once in 24 hours (semidiurnal tides). The output curves will display a 14 day cycle. The bursts of energy that can be drawn, typically, start three hours ahead of the peak of the tide and spread to over 4 to 6 hours past the peak. The peak itself will advance by one hour each day and come back to the original position after 14 days. However, average energy that can be drawn during a 14 day cycle does not change much from one cycle to the next. The energies available for conversion into electricity are as great as those from wind. They are predictable in time, varying in magnitude and fixed in timing for withdrawal, which may not fit in with the time of their requirement in the electricity grid.

The most common requirement to enable efficient utilization of electricity from the ocean is that of electricity storage. Tidal energy plant designs in the earlier days started with a selection of an estuary opening into the sea in a location known for high tides. A barrage with sluice valves and generators was built. Water was taken in during high tide, stored, and let out during ebb tide. The energy output was proportional to the water storage area and the heights to which it could be stored. Utilization efficiency was improved by running the generators twice—during water-in and during water-out periods. It was further improved by using pumps to compliment the water storage.

Basically, these installations are akin to pumped water installations across mountainsides, as described earlier, except that tidal energy is used to pump up and store the water instead of grid energy and also except for the drawback of no control on electricity generation time, which is, however, fixed in schedule.

Tables 3-4 and 3-5 give data on existing and conceived tidal power plants throughout the world.

Table 3-5 below gives data on major ideal energy projects.

3.12.1 Le Rance Tidal Power Plant

This installation was the first tidal power plant [5]. It was installed in 1966 in France. It is of considerable size, with 240 MWs power capacity. Being an early venture, capital costs were high. A coffer dam ahead of the long barrage added substantially to the costs. These days, coffer dams are dispensed with.

Useful guidance data on effects on the ecosystem of the Le Rance plant have been collected. The barrage increased the average turbidity in the water, which was harmful to some types of algae, but it promoted plant growth as well as fish populations. Special audio devices are used to keep fish away from turbine blades.

One side effect of this large estuary-based project was a tremendous increase in tourists, for whom many sorts of water sports and activities became available.

Table 3-4. Existing tidal water plants

Location	Mean tidal range (m)	Basin surface area (km^2)	Power (MW)	Annual production (GW-hr)	Inauguration date
Le Rance (France)	7.85	17	240	600	1966
Kislaya Guba (USSR)	2.4	2	0.4	—	1968
Jiangxia (China)	5	2	3.2	11	1980/1986
Annapolis (Canada)	6.4	6	20	50	1985

Source: From [5], © IEEE 1993.

Table 3-5. Major conceived tidal electricity projects

Location	Mean tidal range (m)	Basin surface area (km^2)	Power (m)	Annual production (GWh)
Severn (UK)	7.0	520	8640	17000
Shepody (Canada)	10	115	1800	4800
Cumberland (Canada)	10.9	90	1400	3400
Cobequid (Canada)	12.4	240	538	1400
Gazolim (South Korea)	4.8	100	480	530
Kutch (India)	5.00	170	900	1600
San José (Argentina)	5.9	750	500	9500

Source: From [5], © IEEE 1993.

In general, these basin-type tidal water projects require a lot of capital. Their gestation periods are large—10–12 years in some cases. They are mostly undertaken under government finance and initiative.

3.13 UNDERWATER TURBINE AND COLUMN-MOUNTED GENERATOR

A new innovation to harnessing tidal energy is the underwater turbine with column-mounted generator. The tidal waves run on the seabed much as they run on the seawater surface. Just like an offshore wind farm with aerial turbogenerators on tall poles, we can dream of water turbine generators on the seabed. These are feasible and experimental models have been set up.

One has to locate offshore sites in areas with tide velocities at 2.25 to 2.5 m/sec at depths of 20–30 m below sea levels. The units have large turbine blades driven by the wa-

Figure 3-6. Underwater turbine and column-mounted generator. (From [6], © IEEE 2003.)

ter; with mechanical connections, these drive a generator on a platform above the water (Figure 3-6).

With generators rated at 500 kW each, one can have a tiny offshore installation rated at 2 MW with just four of these installations. The blades can be adjusted from the top so that energy can be driven by tides flowing in as well as flowing out. This innovation can replace capital-intensive basin-type tidal power plants in the future.

3.14 WAVE ENERGY

Tindal defines wave energy as the energy imparted by a passing wind across open water [8]. Half of the energy is in the form of elevated energy in the height of the wave and half as kinetic energy in the motion of the wave. Let P = power in the wave, L = length of the wave, T = period of vibration or $1/T$ = frequency, H = height of the wave, Z = amplitude of the wave, and W = watts. Then [10],

$$\frac{dP}{dL} = cTH^2,\ c \approx 976\ \mathrm{Wm^{-3}s^{-1}}$$

This power is vast. It has been estimated for various countries in Europe as shown in Table 3-6.

Two methods have been experimented upon to extract this power.

The Islay wave energy conversion system is a "Wells turbine" type wave energy system—an oscillating water column system—in which a gully chamber has an oscillating water column that compasses air into a turbine called a Wells turbine. This, in turn, drives an induction generator and produces electricity (Figure 3-7).

This system has an advantage that electricity delivery, because of induction generator, can be fed directly into an electrical distribution system. Overall efficiencies are claimed to be between 34% to 38.7%.

In the Archimedes wave spring system, a large buoy floats up and down on the waves (Figure 3-8). A rope with a spring at ts end is tied to this buoy. It actuates a 3 Ph linear

Table 3-6. European wave energy potential

	Techical potential (TW-hr/year)	
Country	Near shore	Offshore
Denmark	2–3	5–8
France	3–5	12–18
Germany	0.3–0.5	0.9–1.4
Greece	1–2	4–7
Ireland	7–11	21–32
Italy	3–5	9–16
Portugal	4–6	12–18
Spain	3–5	10–16
UK	14–21	43–64

Source: [7], © IEEE 2005.

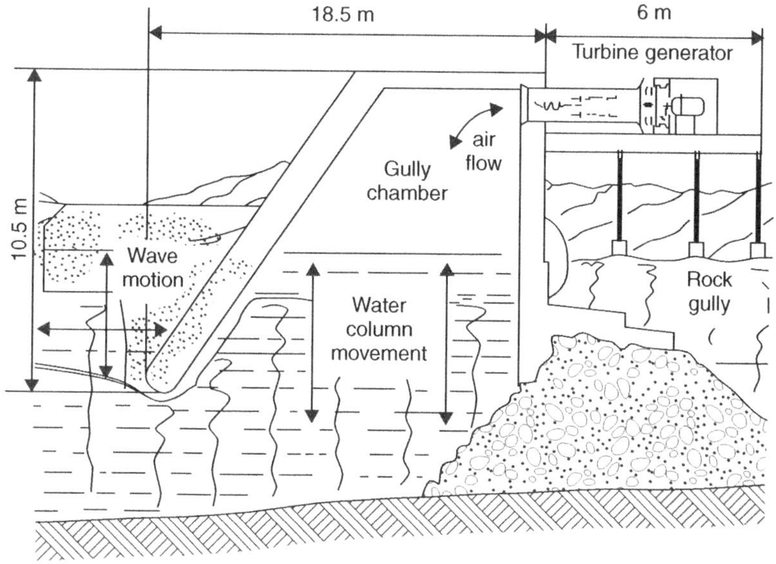

Figure 3-7. Illustration of the Islay wave energy conversion system. (From [8], © IEEE 1996.)

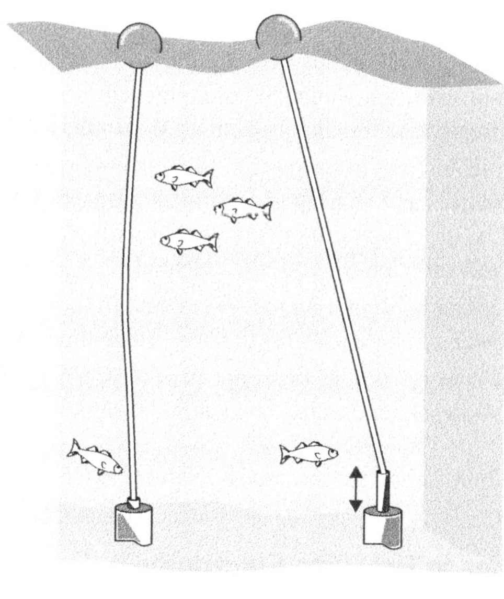

Figure 3-8. Outline of buoy, rope, and linear generator for direct conversion of wave energy to electricity. (From [7], © IEEE 2005.)

permanent magnet generator fixed in place on the ocean floor. In another system under development, a linear permanent magnet generator, an armature wire travels up and down across a series of magnets. Direct AC voltage is generated, presently at a variable frequency. It is proposed to develop a system wherein the generator can feed directly into the grid system at the grid frequency. These generators were formerly in limbo. Now there seem to be good possibilities for this application. Unlike other systems for drawing electric energy from the ocean, this system has the potential to produce generators of adoptable economic size to produce affordable electricity.

APPENDIX 3-1 WORLD'S LARGEST HYDROELECTRIC PROJECTS

Itaipu Hydro Project

The Itaipu hydro project is rated at 12,600 MW of hydroelectric power and is one of the largest hydroelectric power plants in the World. It was constructed during 1982–1988 as a joint venture between Brazil and Paraguay. It is located on Parana River on the border between Brazil and Paraguay.

The Brazilian electricity system operates at 60 Hz, whereas that in Paraguay operates at 50 Hz. Power is shared equally between the two countries, with the result that the entire setup consists of two duplicate but identical interconnected systems. There is a frequency converter tied between the two sections. Figure 3-9 shows a cross section through the power house.

The following points are noteworthy:

- A continuous antiflooding gallery around the turbines has a capacity to accommodate 14,000 m^3 of flood water. Should a mishap occur, it provides sufficient time to take corrective action and prevent a catastrophe.
- The equipment levels rise in proportion to their importance. Turbines are at the lowest level, controls and step-up transformers are next, 500 kV GIS and cables are above, and so on. Transformers are cooled with water from the penstock. A safety control and valve on the water is essential.
- Auxiliary power can be drawn on the 50 Hz side from the 220 kV grid. Similarly, auxiliary power can be drawn for the 60 Hz side through the 750 kV grid.
- For emergency energy supply, there are two 16 MVA synchronous generators and two 4 MVA diesel generators, rated at 50 Hz and 60 Hz for each unit.

Detailed specification for the project are shown in Table 3-7.

Signs of the Times in Brazilian Electricity

Reconstruction of the energy business in Brazil started in 1995. Brazil has a number of hydroelectricity stations. In 1997 the electricity supplied by hydro was 94% of all electricity supplies. By 1999, it came down to 87%. For 2001, the installed capacity in Brazil breaks down as 59,000 MW in hydro, 4200 MW in conventional thermal, and

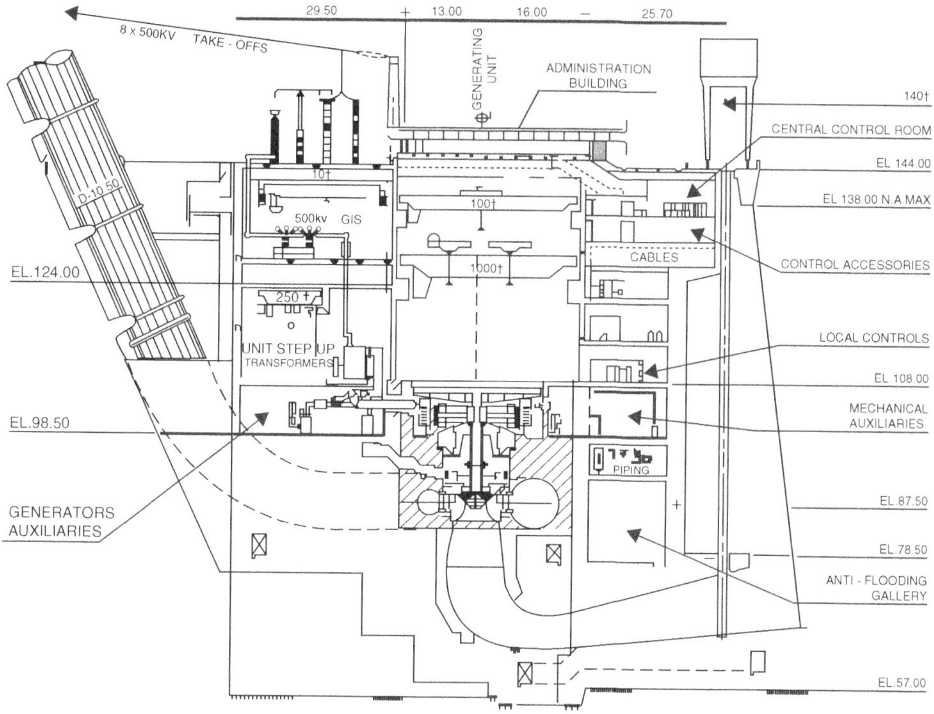

Figure 3-9. Cross section through the power house, Itaipu hydro project. (From 10], © IEEE 1982.)

1900 MW in nuclear. Total capacity is 65,700 MW.

After reconstruction, the government stopped investing in hydro generation and in transmission. This is attributed mainly to the effects of reconstruction. With cheaper gas available and faster periods of construction, emphasis shifted to gas thermal generation and hydro went into limbo. There are no private takers for hydro.

Today, Brazil buys electricity from Paraguay and Argentina. Electricity consumers have gone on a voluntary cut in consumption of 20%. Continuous drought conditions since 1997 have exuberated the conditions [9].

APPENDIX 3-2 REMOTE CONTROL OF THE HYDROELECTRIC SYSTEM AT GURI

The Guri project in Brazil (Figure 3-10) is the second largest hydroelectric project in that country [4]. Its unique control system covers 2 power houses, 3 switchyards, and 9 spillway gates. Power house No. 1 has 10 generating units totaling 3000 MWs. Power house No. 2 has 10 generating units totaling 7000 MWs. Besides electricity control, the control system is required to control some nine spillway gates so that water flow to downstream power houses can also be controlled.

Table 3-7. Project and equipment data of the Itaipu hydroelectricity project

River average flow, m^3/s	8460	
Reservoir normal elevation, m	220	
Reservoir useful storage, m^3	19×10^9	
Spillway capacity, m^3/s	62.2×10^3	
Dam type	Hollow gravity	
Plant rated capacity, MW	12,600	
Number of units	18	
Equipment Data		
Turbines		
Type	Francis	
Rated output, MW	715	
Design head, m	118.4	
Rated head, m	112.9	
Design speed, r/min	91.6	
Specific speed	246	
	50 Hz	60 Hz
Operational speeds, r/min	90.9	92.3
Runner diameter, mm	8746	8746
Runner weight, t	300	300
Generators		
Type	DIN W42	
Terminal voltage, kV	18 + 5%	
Cooling: stator winding rotor	water, radial air	
	50 Hz	60 Hz
Rated output, MVA	823.6	737
Inertia constant (kW sec/kVA)	4.4	5.3
Reactances, direct axis		
Synchronous (*Xdu*), per unit	0.90	0.90
Transient (*X′du*), per unit	0.30	0.30
Subtransient (*X″du*), per unit	0.20	1.18
Ratio *X″q*/*X″d*	1.10	1.17
Rotor diameter, m	16.00	16.00
Rotor weight, t	1700	1700
Stator weight, t	722	710
Number of poles	66	78
Stator winding		
Type	wave	
Number of parallels	6	
Number of coils	504	
Water connections	168	

Table 3-7. *Continued*

	50 Hz	60 Hz
Excitation system		
Positive current converter	50 Hz	60 Hz
Rated current, A	9000	7100
Base field voltage, V	230	219
Positive ceiling, V	1380	2250
Response time, S	0.05	0.05
Negative ceiling, V	1240	2025
Negative current converter		
Rated current, A	2400	1350
Ceiling voltages, V	1380	2250
Excitation and unit service transformers		
Construction	cast coil	
Temperature rise	900°C	
Insulation	class F	
BIL	125 kV/50 kV	
Unit transformers		
	50 Hz	60 Hz
Rated output, MVA	3 × 275	3 × 256
Voltage, kV	$18–525/\sqrt{3}$	
Taps on high-voltage winding	+ 2 × 2.5%	
Connections	delta–wye	
Cooling	FOW	
Construction	Core	
Temperature rise	650°C	
BIL, high-voltage winding	1550°C	
High-voltage bushings	oil/SF6	
Main auxiliary transformers		
Rated output, MVA	3 × 15/5	
High voltage, kV	$525/\sqrt{3}$ + 2 × 2.5%	
Low and tertiary, kV	13.8 and 13.8	
Connections	wye–wye–delta	
Cooling	air–oil	
Switching	tap transfer	
BIL, kV	1550 – 110	
500 kV gas-insulated switchgear (GIS)		
Assemblies		
BIL, kV	1550	
Switching impulse, kV	1240	
50 or 60 Hz, 1 min., kV	740	
Field tests:		
Oscillating, switching surge, kV	940	
AC, 1 min., kV	570	

(*continued*)

Table 3-7. *Continued*

500 kV gas-insulated switchgear (GIS) (*cont.*)	
Assemblies (*cont.*)	
Rated gas pressure, kPa	600
Conductor and enclosure	aluminum
Rated current, A rms	4000
Circuit breakers	
Interrupting capacity, kA	63
Number of interrupters	4 per pole
Operating mechanism	hydraulic
Disconnecting switches	
Operation	group
Operation mechanism	motor
Grounding switches	
Operation mechanism	manual and spring
Type	high speed
Line entrance	motor
Lightning arresters, kV	444

The chain of commands is:

- Regional dispatch center—EDELCIA
- Local load dispatch center—ELDC or LDC
- Guri station

There is primarily control from the regional dispatch center, which establishes plant dispatch for power, plant output, voltage requirement, and river flow. The regional dispatch center (EDELCA) sends set points for the above controls to Guri and Guri operates two main programs: automatic generator control and automatic voltage control. These programs can also operate under remote control.

The main features of the computer control systems are as follows. There are two computer systems: one at the load dispatch center (ELDC) and other at Guri.

The control system consists of the following elements:

1. Preventive controls for generator control, control to meet local bus demand, and security control to evaluate generation mix and regulating capacity
2. SCADA
3. Automatic generator start–stop
4. Automatic disturbance trend recorder with a capacity of 40 samples for six fast-response and six low-response analog inputs and 10 samples for four status inputs. Time between readings: 0.1 s, 0.5 s, and 5 s.
5. Automatic generation control (AGC)

Figure 3-10. Geographical location of the Guri hydroelectric system. (From [10], © IEEE 1992.)

6. Economic dispatch—maximum gross hydroelectric output versus water flow through turbines
7. Unit commitment
8. Automatic voltage control (AVC)
9. River flow control—reservoir management
10. Logging and historical data management—daily, weekly, and monthly energy reports, hydraulic reports
11. Automatic satellite clock synchronization

Remote Terminal Units (RTUs)

There are 20 RTUs for 20 generators, two common service RTU units, and two switchyard RTUs. Each RTU consists of a monitor, a controller, and a dispatch receiver.

Operation of Generator RTU

Main functions:

1. Three levels of controls:

   Normal—When the control loop is via remote station (LDC)

   Back up—When the RTU loses contact with LDC station, the loop is through local master station controller.

Standalone—Local control through a man–machine interface

Redundancy—Two adjoining RTUs have common communication lines. Should one RTU lose its line, it gets on to the next generator's RTUs line. Beyond this all the generators RTU have a common (emergency) communication line.

2. Sequence of events recording, with a time lag of 2 ms
3. Pulse accumulators—Kwh, etc.
4. Automatic disturbance trend recording
5. Local process control loops—unit speed, MW, kV, etc.
6. Process schedulers (startup, shutdown, etc.)
7. Real-time database
8. Local man–machine interface

Common Services RTUs

Main features of the common services RTUs are:

1. Substation monitoring logic, check hardwired relay logic, etc.
2. Real-time database for all the generating unit RTUs

Switchyard RTUs

The only function of the switchyard RTUs is monitoring.

Automatic Generation Control (AGC) and Automatic Voltage Control (AVC)

Guri computers operate AGC and AVC within set points received from EDELCA. However, there is provision for LDC also to operate these controls.

Working of the Guri Control System

The master station communicates with the 24 RTUs as mentioned earlier. The main program consists of AGC and AVC as described above. Fundamentally, generator outputs are adjusted to meet the station needs. They also act as a communication link to the LDC.

The AGC has following the functions:

1. Monitoring functions to
   - Determine electrical connectivity of online generating units
   - Validate critical variables (i.e., frequency deviation, time deviation, MW, Mvar, kV, ELDO interface)
   - Monitor operating limits (i.e., kV, MW, Hz)

- Verify that a response time for generating unit is followed
- Monitor status changes

2. Control functions
   - Ramping of set points (MW, Hz, kV, and Mvar)
   - Capture and filter bus control error

Smooth control is obtained after combining proportional and integral terms with a second order filter, which has following threshold levels for filtering:

15 MW: dead band (no control)
250 MW: normal mode
500 MW: power engineering review, permissive mode
600 MW: emergency mode

All the signals need not be communicated from the generator RTUs to the station control and upward. They can be retrieved as and when needed. These redundant signals are MW, MVAR, kV and generator KA. This redundancy has reduced communication line crowding and improved the performance of AGC and AVC.

The above system was pioneered in 1992 in Brazil and has been followed by others.

REFERENCES

1. L. S. Wright and E. M. Brinson, The Design, Application, and Testing of Electrical Power Systems at the Bad Creek Pumped Storage Project, *IEEE Transactions on Energy Conversion*, March 1992, pp. 15–19.
2. T.-Y. Lee and N. Chen, N., The Effect of Pumped Storage and Battery Energy Storage Systems on Hydrothermal Generation Coordination, *IEEE Transactions on Energy Conversion*, Volume 7, Issue 4, December 1992, pp. 631–637.
3. J. S. Martini, Electrical India, March 2007.
4. S. K. Singhal, "Analysis for quick estimation of canal based small hydro power schemes," IEEMA Journal (India), September 2008, pp. 76–80.
5. J. P. Frau, Tidal Energy: Promising Projects: La Rance, a Successful Industrial-Scale Experiment, *IEEE Transactions on Energy Conversion*, Volume 8, Issue 3, September 1993, pp. 552–558.
6. C. Lang, Harnessing Tidal Energy Takes New Turn, *IEEE Spectrum*, Volume 40, Issue, 9, September 2003, pp. 13.
7. M. Leijon, H. Bernhoff, O. Agren, J. Isberg, J. Sundberg, M. Berg, K. E. Karlsson, and A. Wolfbrandt, Multiphysics Simulation of Wave Energy to Electric Energy Conversion by Permanent Magnet Linear Generator, *IEEE Transactions on Energy Conversion*, Volume 20, Issue 1, March 2005, pp. 219–224.
8. C. E. Tindall and X. Mingzhou, Optimizing a Wells-Turbine-Type Wave Energy System, *IEEE Transactions on Energy Conversion*, Volume 11, Issue 3, September 1996, pp. 631–635.

9. J. Martini, S. C. Martini, and L. B. Reis, Brazilian Energy Crises, *IEEE Power Engineering Review,* April 2002, pp. 21–24.

10. F. Castro, M. Pescina, and G. Llort, Reliability Improvements of the Guri Hydroelectric Power Plant Computer Control System AGC and AVC, *IEEE Transactions on Energy Conversion,* September 1992, pp. 447–452.

BIBLIOGRAPHY

Aoki, K., M. Itoh, T. Satoh, K. Nara, M. Kanezashi, Optimal Long-Term Unit Commitment in Large Scale Systems Including Fuel Constrained Thermal and Pumped-Storage Hydro, *IEEE Transactions on Power Systems,* Volume 4, Issue 3, August 1989, pp. 1065–1073.

Churchill River Hydro Developments, *IEEE Power Engineering Review* (This article was abstracted from the Churchill River Power Projects Website, http://www.ptm.ca/churchill), August 1998, pp. 21–23.

de Moraes, J. , J. Rodriguez Villalba, and V. F. Salatko, Electrical and Related Design Aspects of Itaipu Hydroeletric Project, *IEEE Power Engineering Review,* Trans-Power Apparatus and System, May 1982, pp. 1012–1020.

Fostiak, R. J. and H. R. Davis, Electrical Features of the Rocky Mountain Pumped Storage Project, *IEEE Transactions on Energy Conversion,* March 1994, pp. 206–213.

Hammons, T. J., F. Taher, A. B. Gulstone, B. K. Blyden, R. Johnston, E. Isekemanga, K. Paluku, A. C. Calitz, N. N. Simanga, African Electricity Infrastructure, Interconnections, and Changes, *IEEE Power Engineering Review,* January 1997, pp. 6–16.

Hammons, T. J., P. H. Corredor, A. M. Fonseca, A. C. G. Melo, H. Rudnick, M. Calmet, and J. Guerra, Competitive Generation Agreements in Latin American Systems with Significant Hydro Generation, *IEEE Power Engineering Review,* September 1999, pp. 4–4.

Harrison, G. P. and A. R. Wallace, Sensitivity of Wave Energy to Climate Change, *IEEE Transactions on Energy Conversion,* Volume 20, Issue 4, December 2005, pp. 870–877.

Nanahara, T. and A. Takimoto, A Study on Required Reservoir Size for Pumped Hydro Storage, *IEEE Transactions on Power Systems,* February 1994, pp. 318–323.

Natarajan, K., Robust PID Controller Design for Hydroturbines, *IEEE Transactions on Energy Conversion,* September 2005, pp. 661–667.

Osburn, G. D. and P. L. Atwater, Design and Testing of a Back-to-Back Starting System for Pumped Storage Hydrogenerators at Mt. Elbert Powerplant, *IEEE Transactions on Energy Conversion,* June 1992, pp. 280–288.

Polinder, H., B. C. Mecrow, A. G. Jack, P. G. Dickinson, and M. A. Mueller, Conventional and TFPM Linear Generators for Direct-Drive Wave Energy Conversion, *IEEE Transactions on Energy Conversion,* Volume 20, Issue 2, June 2005, pp. 260–267.

Schewe, P. F., Electric Idyll, *IEEE Transactions on Power Systems,* October 2006, pp. 40–47.

Turanli, H. M., Preparing for the Next Generation at Manitoba Hydro, *IEEE Power Engineering Review,* March 2002, pp. 19–23.

Veltrop, J. A., Future of Dams, *IEEE Power Engineering Review,* March 2002, pp. 12–15.

4

THERMAL POWER GENERATION—STEAM GENERATORS

4.1 THERMAL ELECTRICITY GENERATION HAS THE LARGEST SHARE—THE PRESENT SCENARIO

Thermal electricity generation has traveled a long course—from steam operated reciprocating engines to steam turbines, from diesel engines to gas turbines and from there to combined-cycle power plants (CCPPs). The thermal efficiencies and ratings per unit have steadily climbed. CCPPs are expected to be the predominant generators in the near future.

Today, 60% of world's electricity comes from thermal generation. The megascale operations in thermal electricity have been attacked on many fronts. Fossil fuels—their main energy suppliers—have become vulnerable and objectionable. The electricity business is demonopolized, that is, restructured. The open marketing course demands all-around optimization. Technological innovations are bringing in varied types of competitions. However, it has been and is expected to continue to be the largest supplier of electricity. It has always been based on private capital raised by utilities and others, except during the monopoly period when, in India, the states intervened and supplied public capital. In the future, it will be based on private finance mainly. It will not attract state finance or international infrastructure finance.

Under the restructured scenario, many of the earlier advantages are lost. Both, project risks and commercial risks have now to be faced under different conditions and considerations.

Copyright © 2011 the Institute of Electrical and Electronics Engineers, Inc.

Technological innovations have continued. Both engineers and financiers have equally important parts to play, except that software engineers hold an advantage over "basic only" electrical engineers.

4.2 PLANNING OF THERMAL STATIONS—RISKS AND CHALLENGES

After all is said and done, there is still a need and a challenge for setting up large-sized thermal stations. Some of the risks and challenges involved in this follow.

4.2.1 Project Risks

The issues involved in decision making are:

- Resource planning: Competing resources are hydro, nuclear, renewable and dispersed generation
- Alternatives to generation: buy, depend on demand-side management instead of increasing generation, nonutility generation
- Purchase considerations: real-time heat rates versus replacement heat rates, emission allowances; CO_2 certificates
- Restraints: dynamic restraints—emission controls, voltage controls (reactive power supply), load balancing (unit commitments), transmission availability, time of supply rates, caps by regulations, security reserves

These conditions could influence the decision-making process:

- The supply exceeds the demand. The predicted demand does not materialize. Too many generators jumped in.
- The supply is short since the demand went up unpredictably. It is too late to realize that the project was underestimated.
- A condition in between, wherein the supply and demand match evenly. The evenness is controlled by market forces. The risk proves to be well taken.
- A new tie line in transmission comes up, making it easy for the next-door generator to dump its extra capacity into our system.
- There will be distributed generators at nodal points.
- Large customers will opt for in-house generation as developments in technologies make smaller generators equally efficient and economically viable.

The commercial risks are:

1. A predicted price might not materialize.
2. There will be unpredicted competition.

3. There will be price caps from regulators. In the worst-case scenario, the regulator might order a refund on past collections from customers.

There are challenges in raising financing. Major portion of financing will come from lending institutions. Under the new circumstances, these institutions will have to be extra careful. Expectations from these institutions are listed below. One experienced utility [1] states that there should be:

1. Long-term contractual revenue resources with a high degree of certainty that covers, with sufficient margins, both project costs and variable costs; in other words, a power purchase agreement (PPA) with long-duration customers
2. An experienced and reputable turnkey construction contractor to cover schedule, price, and performance risk (no overruns with delays)
3. Proven technologies with many examples in successful operation
4. Long-term, stable fuel supply
5. Stable legal, regulatory, environmental, and financial teamwork
6. Strong and experienced project sponsors

4.2.2 Fuels for Thermal Generation

Commonly used fuels are—coal, oil, and natural gas. Over the last decade, natural gas has shown rapid growth. This is followed by the renewable energies, which are expected to grow at a much faster rate henceforth. Nuclear power is also gaining attention and is expected to grow rapidly.

4.3 COST BREAKDOWN AND CONSUMPTION PATTERN OF ELECTRICITY

Figure 4-1 shows the energy consumption pattern of electricity and Table 4-1 shows the consumption pattern.

4.4 MAIN ENERGY SUPPLIERS

4.4.1 Coal

Coal leads the race in the amount of electricity produced and expected to be produced in the near future. Coal reserves are abundantly spread almost all over the world. Unlike oil and gas products controlled by a few OPEC countries that have vast reserves of these, there is no cartel center on coal prices. Coal prices are relatively stable and competitive. Thus, when one looks for economy of scale, coal leads the way with single generating stations offering up to 10,000 to 20,000 MW. If one is planning "big" and over a long time period, then coal is still a contender.

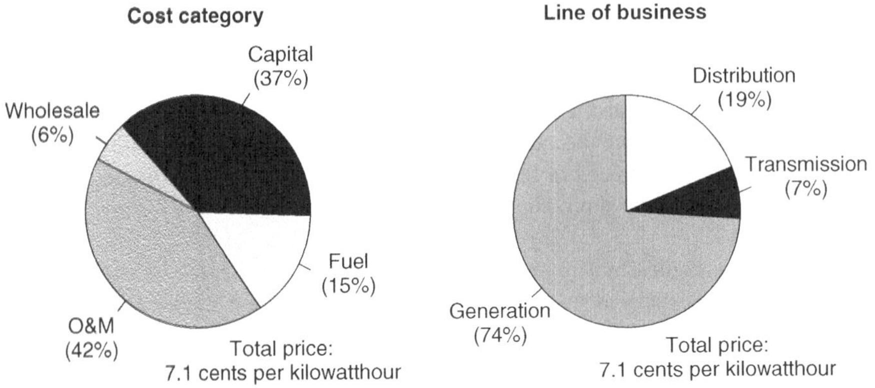

Figure 4-1. EIA Report DOE/EIA-0614. (*Source:* Energy Information Administraton, *Annual Energy Outlook 1997,* DOE/EIA-0383(97), Washington, D.C., December 1996.)

Coal has its own problems. It produces harmful gases, such as SO_x and NO_x, and it produces solid particulates in the exhaust. Recovery of energy per kilogram of weight varies from coal produced from one mine to another. Technical requirements (size of land, availability of water, rail connection, etc.) dictate the siting of a coal-based thermal station, rather than a location near an electrical load center. Technologies are being developed to meet these shortcomings.

CO_2 produced by coal burning is indirectly harmful. CO_2 emissions are getting bigger with increasing thermal production. There is a strong movement toward putting restrictions on CO_2 emissions.

No doubt, coal offers challenges in its use but these are being met by advanced clean coal technologies [2]. Some of these technologies are

- Pulverized coal is being used more and more.
- Rankin cycle, the oldest method in use, which is still popular
- Kalina method, in which ammonia (NHz) is mixed with water
- Pressurized, fluidized-bed combustion method
- combined cycle method

Table 4-1. Percentage change in consumption of different energies

	1993	1995	2000	2005
Oil	100 base	Nil	110.68	145.04
Natural gas	100 base	104.2	125.5	212.23
Coal	100 base	105.4	117.7	160.5
Nuclear	100 base	104.07	110.5	97.7
Renewable	100 base	106.68	121.0	176.9

Source: EIA Report DOE/EIA-0614.

- Integrated coal-gasification-based CCPP method
- Integrated coal gasification combined cycle

For reducing NO_x at the coal burning stage, the following are tried:

- Staged combustion with improved low NO_x
- NO_x contents increase with higher combustion temperatures. Achieve higher temperature in stages. Note that higher input temperatures to turbines give higher turbine efficiencies.
- Selective catalytic reduction

For Reducing SO/NO_x in the fuel, use

- Active carbon
- Limestone injection
- Moving-bed type reactor with synthesized absorbent
- Electric discharge with ammonia injection in fuel gases
- Wet scrubbing of fuel gases with limestone gypsum

Advances in metallurgy have enabled equipment manufacturers to push up the boiler temperature and pressures, thus improving steam turbine efficiencies. This has helped coal to keep up its pace.

Coal ash handling is another environmental problem. Building construction and material manufacturers and cement manufacturers are encouraged to put their plants next to coal-based thermal stations.

4.4.2 Natural Gas

Natural gas has been a forerunner in the competition of energy suppliers for thermal electricity generation. The relative amount of CO_2 as well as SO_x and NO in its emissions are comparatively low. Additionally, it is cleaner than coal, with no ash or solid particulates in its exhaust. It can be economically and conveniently transported from the gas wells to electricity load centers through pipelines. Chemically, it transforms favorably into other products that have a mass demand, such as chemical fertilizers. With fuel cells on the rise, natural gas promises to become a favorite of residential and commercial consumers of electricity.

Electricity generation based on natural gas using the latest technologies gives higher thermal-to-electricity generation ratios than those based on coal or liquid fuel technologies. Gas-based generators are compact in size, requiring smaller sites, and have low initial capital requirements.

The gas turbine or combustion turbine technology gives a fast starting, low overall capital cost project ideal for meeting peak demands quickly.

Gas fields are being discovered all over the world. Huge gas fields have been discovered in the bay of Bengal and on the east coast of India, as well as in the middle and northwest part of the country.

The technical and environmental factors are in favor of using more and more natural gas as a feed for thermal generating stations. As a consequence, its percentage change in increasing use is the highest of all the feed energy materials.

However, natural gas is tied with fossil fuels in prices. These prices are volatile and difficult to project. The share of electricity production through natural gas is projected at 45.1 quadrillion BTUs by 2015, as against 77.2 for coal and 497 million BTUs for renewable energy sources [3].

4.4.3 Mineral Oils

When mineral oils appeared on the scene as energy suppliers, they instantly became the favorites of electricity producers. Oils could be transported though pipelines over long distances from oil wells to thermal stations. Lighter distillates of crude oil, such as petroleum, kerosene, and diesel oil, have been ruling over transportation segment of energy consumers. These fetch good market prices. Distillates from the residual portion of crude oils produce furnace oil, which has been used as a thermal energy supplier. When furnace oil came first on the scene, it was a top contender for use in electrical energy because of its ease of handling and storage, as well as because of lower capital costs compared to those of coal-burning thermal stations. However, concentrations of its resources in a few countries gave it unpredictable price volatility. One such upheaval in 1978 led to reconsiderations of optimizing electricity consumption against rising demand for it. Norms for efficiency in electricity use and reduction in demand through demand-side management came to the forefront. These steps helped the U.S. electrical industry to hold down additions to electricity generating capacity for well over a decade.

Environmentally, oil is as bad as coal as far as gases such as NO_x and SO_x and carbon dioxide are concerned. Environmental control measures increased the capital costs of oil-based thermal plants at a later stage. Price volatility and uncertainty has somewhat pulled down share of oil as an energy source for electricity to less than 10% over the period of 1995–2005. It is not expected to alter much.

4.4.4 Nuclear Power

Nuclear power is dealt with in detail in the next chapter.

4.5 WORKINGS OF A COAL-FIRED STEAM GENERATOR UNIT

Figure 4-2 shows the elements of a typical coal-fired unit.

4.5.1 Coal Flow

A belt conveyor brings in and drops coal into a coal bunker. This coal passes to a pulverizer. From there it is blown up to a burner in the boiler, where it is burned. Water is heated into steam at the required temperature and pressure in this boiler.

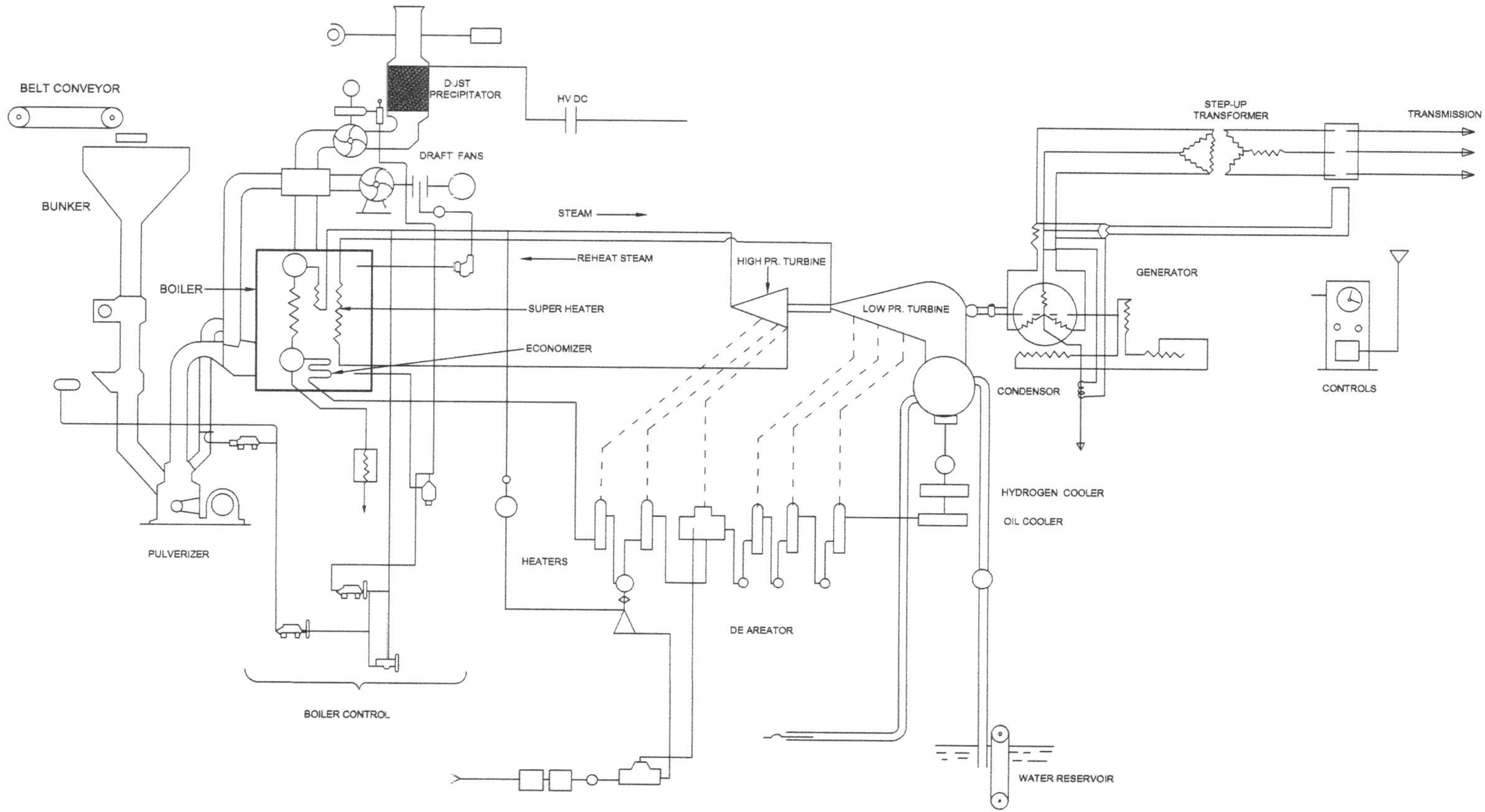

Figure 4-2. Elements of a typical coal-fired plant.

The flue gases pass through an air heater, where heat is extracted to warm up water being fed into the boiler for raising steam. The flue gases, with the help of an induced-draft fan, exit to the atmosphere through a dust precipitator. Various measuring and control systems in this system are shown in the diagram.

Steam is raised at high temperature and pressure. It passes onto and drives a high-pressure turbine followed by a low-pressure turbine. Both turbines drive a generator on a common shaft. Generator output, which is at medium voltage, is stepped up to the grid voltage and is delivered into the grid.

Steam output from the turbines goes to a condenser. Heat is extracted on the way. Feed water is made up from a feed water tank and from this condensate, and fed into the burner.

Beginning around the 1950s and 1960s, higher efficiencies and larger scales of operation led to increasing the operating pressures and temperatures of steam turbines. A pressure of 3208.20 psia is called the critical pressure for steam. The general scheme is shown in Figure 4-2, divided into two types: a subcritical system operating at around 2400 psia and a super critical system operating above the critical pressure.

4.6 TYPES OF BOILERS

There are two types of boilers. The old boilers, which have fairly long lives and still operate today, are drum type, in which heat is absorbed in furnace walls and in a drum. The steam is heated again externally in the old system through a superheater. It then passes to the steam turbines.

In the newer boilers, heating is achieved through an economizer, furnace walls, and superheater, working in tandem. This is termed as "once-through" system.

Figure 4-3 shows the types of steam generation systems described above.

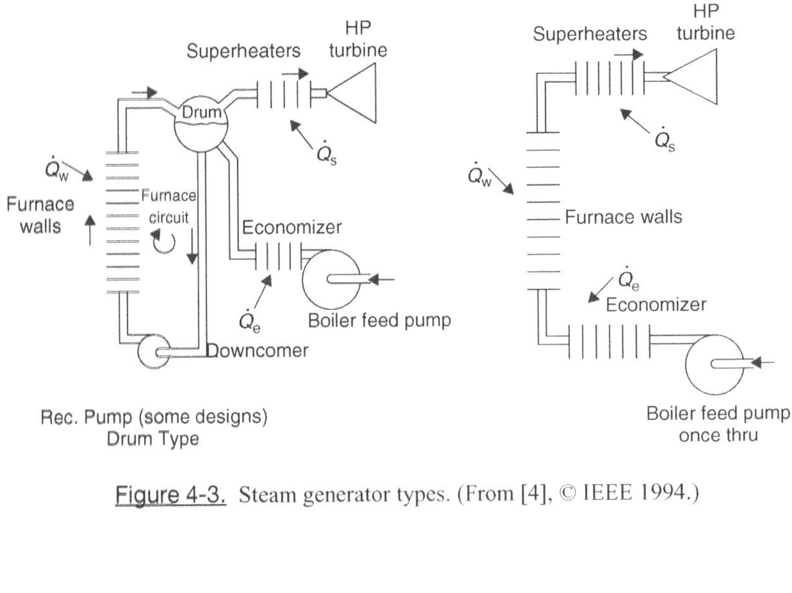

Figure 4-3. Steam generator types. (From [4], © IEEE 1994.)

The fuel flow and draft air flows control the rate of heat generation. This rate is determined by the electrical load on the generator driven by the steam turbines.

The once-through system operates at high pressures throughout. Even the back pressures on the feed water pumps are high. Coupled with high steam temperatures, this requires thicker and high-strength construction materials. These thicker sections, once raised to higher pressure and temperature conditions and left stabilized there, can have predesigned long lives. If the load variations are steep and frequent, they will develop metal stress fatigue, which decreases their life and can cause early or sudden failures. The megasized new classes of generators are thus best suited for base-load service. The older classes of steam generators operating at lowers pressures and temperature have relatively thinner metal cross sections and are more amenable for intermediate-load servicing.

4.6.1 A Modern 1000 MW Boiler

As the size of a steam generator plant increases, it becomes, under today's conditions, necessary to concentrate on higher efficiencies and lower and controlled emissions of all types, including CO_2 emissions. Figure 4-4 shows a sectional sketch of a modern 1000 MW boiler.

The following points are noteworthy:

- Coal pulverization is very fine. This helps in two ways: combustion is complete and there are no coal particles in exit flues or gases. The burner operates at an optimum rate of air flow. Excess air flow carries away the heat energy.
- An advanced "PM" burner gives low NO_x combustion. The exit flue is scrubbed to further remove escaping NO_x emissions.

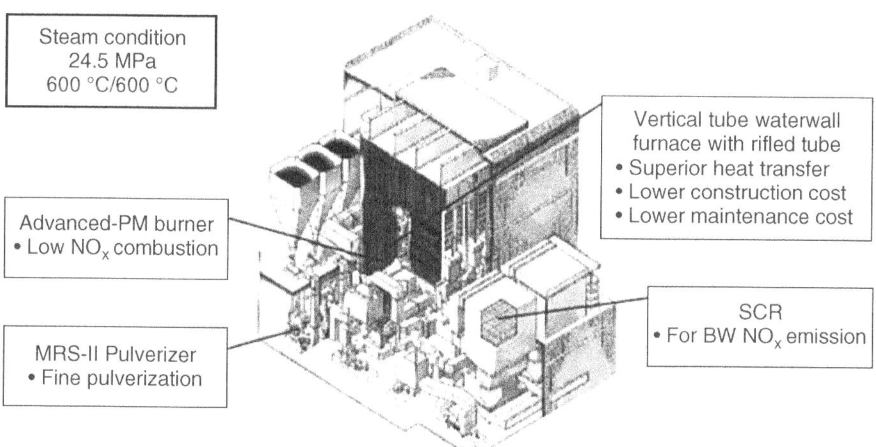

Figure 4-4. Example of a modern 1000 MW boiler (Chugoku EPCO Misumi number 1). (From [5], © IEEE 2001.)

- The operating parameters—pressure and temperature—for steam are at high pressure (24.5 MPa) and temperature (600°C). A combined-cycle operation incorporating a gas turbine and two-stage steam turbines further improve this efficiency.

4.6.2 Vertical Water-Wall Furnace with Rifled Tubes

This boiler has a sophisticated construction. It has a vertical-tube water wall and rifled tubes as shown in Figure 4-5. This construction gives superior heat transfer and lower construction and maintenance costs. The unit operates with gas.

4.6.3 Integrated Coal Gasification Combined Cycle Furnace

The integrated coal gasification combined cycle furnace (IGSC) operates along the above lines except that gas from coal gasification is used instead of natural gas. A demonstration unit rated at 300 MW is in operation.

4.7 CLASSIFICATION OF GENERATING UNITS

4.7.1 Base-Load Generators

Thermal generating units can be categorized as the base-load generators, peak-load generators, and intermediate-load generators. In a large electricity supply system, a

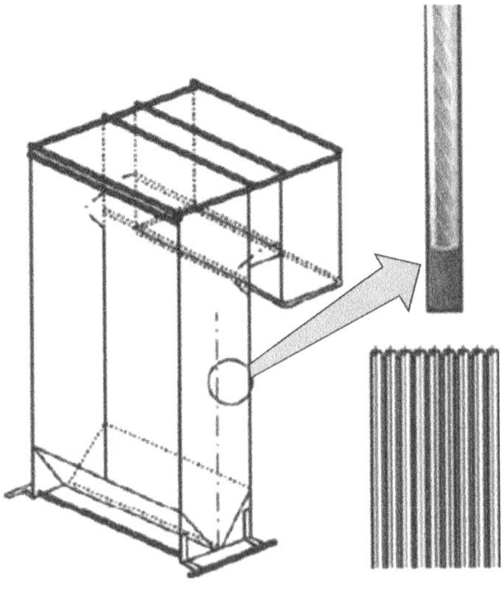

Figure 4-5. Structure of vertical water-wall furnace with rifled tubes. (From [5], © IEEE 2001.)

fairly large portion of the daily load is more or less constant. Time-variable loads ride on this average load. The base-load generators cater to this steady load. These generators have higher rated capacities (1000 MW and above). They take time to start as well as to close down, due to high inertia, mechanical as well as electrical, of the system. Once started, they continue running. They are normally coal based and have the lowest cost of producing electricity energy. Among the three categories, this category has the lowest and steadiest fuel costs.

4.7.2 Peak-Load Generators

On the other hand, peak loads are demanding. They must be serviced as they arise and also over a limited period over which they last. The peak-load generators catering to these loads must be quick starting and quick stopping. Normally, gas turbines serve well as peak-load generators. Their cost of production is the highest partly because they operate only part-time and partly because the fuel costs are comparatively high. The ratings range around 200 MW.

4.7.3 Intermediate-Load Generators

These are brought online between the base-load and peak-load periods. They use both gas and oil. Quite often, old system generators of low and medium output ratings are put in service as intermediate-load generators.

4.8 COMBINED-CYCLE POWER PLANT (CCPP)

Combined-cycle power plants have come to the fore very fast. As of 2003, Japan claims to have about 20,000 MW capacity under this technology and Malaysia claims 5760 MW capacity [6].

The biggest advantage of this technology is an overall thermal efficiency of 60% [7]. It complies with ease with the strict emission requirements on NO_x and other gases. It has a compact layout and requires less space. The construction period is short. It can be easily operated with respect to frequent startups and shutdowns and with respect to quick load following, in comparison with large fossil-fuel-based plants and nuclear power stations.

Figure 4-6 gives a sketch of the workings of a CCPP. An air compressor feeds compressed air into a combustor at a pressure of about 12 atmospheres. A jet of fuel is directed against an igniter there. The resultant flame heats up the compressed air to a fairly high temperature. The hot, high-pressure air expands over the rotating blades of a gas turbine, which drives an electric generator. The high-temperature exhaust from the gas turbine enters the heat recovery steam generators (HRSGs) which are actually steam boilers generating steam at high temperature and pressure. This steam is now fed into a follow-up steam turbine, which, in turn, contributes to driving a generator on a common shaft. The exhaust from the steam turbine is fed back to heat the original air at the starting point.

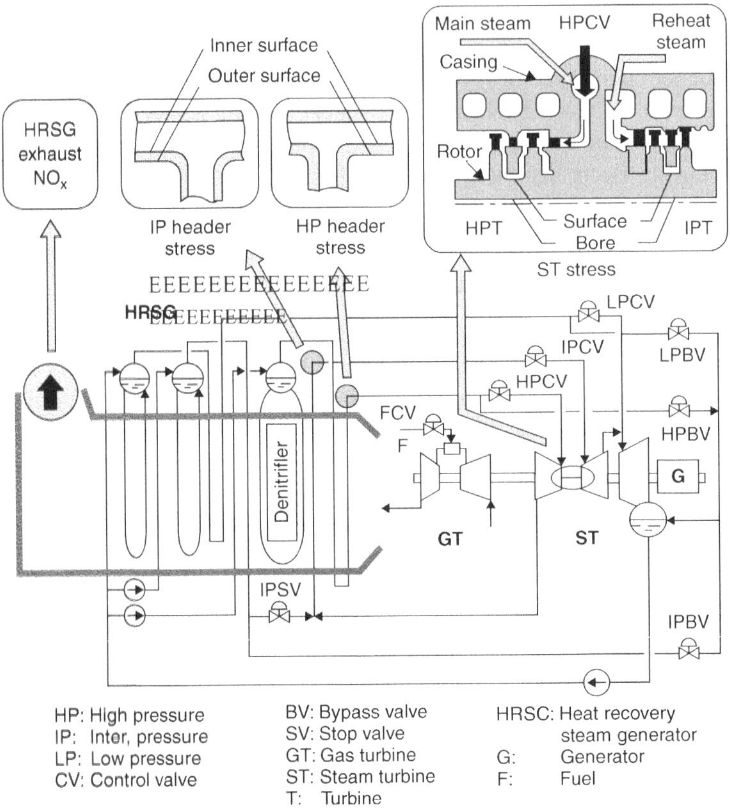

Figure 4-6. Plant configuration and operational constraints of a CCPP. (From [8], © IEEE 2006.)

This system attains high thermal efficiency on two counts. First, the higher the inlet temperature to a gas turbine, the higher is the efficiency. Second, it recovers a major portion of the heat input with the help of the steam turbine.

4.8.1 A Denitrifying Arrangement

A complicating factor is the formation of nitrogen oxides (NO_x), which are environmental pollutants. The higher the air temperature, the larger is the amount of NO_x produced.

In the HRSG, we have a denitrifier. This denitrifier has a catalyst that causes a reaction between NO_x and NHz (ammonia), introduced in the denitrifier, and reduces NO_x to N_2 and O_x to H_2O with the Hz part of ammonia. This is an important and nonviolable feature of a CCPP.

The gas turbine and its connected generator and the steam turbine (or turbines) and its generator (or generators) are all mounted on a single shaft, making it a very significant part of the whole system. A continuously operating lubricating pump for lubricating the bearing of this shaft is highly critical. It cannot be allowed to fail.

The following data typifies the workings of a CCPP (see Table 4-2).

4.8.2 Typical Rating Ratios Between Gas and Steam Portions

A 700 MW CCPP had six units of 117 MW each. Each 117 MW CCPP had 76 MW on the gas generator and 41 MW on the steam generator [9]. Another unit, used as a distributed generator, was rated at 45 MW. In this case, the gas turbine accounted for 2 × 25 MW and the steam turbine accounted for 2 × 3 MW [10]. The combined rating is scaled down to 45 MW.

4.8.3 Advances in Synchronous Generators

We have studied in detail how higher efficiencies have been gained on the fuel side of CCPPs. Here, we will study what developments emerged on the generator side. Output efficiencies, both in air-cooled and hydrogen-cooled generators, have been claimed at as high as 60% and 65% [11].

Coal-Based Thermal Generation Cannot be Left Out. Combined-cycle process-based electricity generation has no doubt a high degree of thermal-to-electricity conversion efficiency. It is projected to occupy a position of importance among prime movers. However, it is dependent on gas, a volatile commodity with respect to price and supply. Comparatively, coal is more dependable. Research is being undertaken to improve thermal efficiencies in coal use.

The latest developments concentrate on improving the steam temperatures, to around 600°C, and also the pressures. Toshiba Corporation of Japan specifies a 1000 MW steam turbine set with steam temperatures at 566/593°C and another 700 MW turbine set at 31 MPs and 566/566/566°C [12].

Table 4-2. Gas turbine parameters

Parameter	Value
Compressor inlet temperature	30 (°C)
Compressor discharge temperature	390 (°C)
Gas turbine inlet temperature	1085 (°C)
Gas turbine exhaust temperature	532 (°C)
Compressor pressure ratio	11.5
Ratio of specific heat	1.40
Compressor efficiency	0.85
Turbine efficiency	0.85
Gas turbine output coefficient	0.003030
Steam turbine output coefficient	0.000428

Source: [6].

These high temperatures are very demanding of the metallurgy of turbine parts, with particular reference to turbine blades rotating at high speeds. Temperature transients during startups or slowdowns are critical. This makes these steam turbines desirable to be used on base-load-serving applications.

Another point to note is the gradual raising of steam temperatures in two stages. This improves heat transfer efficiencies. It also has an effect on formation of nitrogen oxides at higher temperatures. More sophisticated environmental protection measures are called for.

Developments in Sizes. Old models with air cooling fell into a region of 50–300 MVA generally, and up to 600 MVA rating with class F insulation. Replacing air cooling with water cooling of cores increased the sizes further to around 550–650 MVA. Hydrogen-cooled and water-cooled unit sizes ranged between 300–1200 MVA for two-pole generators and up to 1200 MVA for four-pole generators. Nuclear power plant generator unit sizes reached 2500–3500 MVA.

However, there has been a significant turnaround in sizes. Power marketing and spreads in electricity networks created demand for small-to-medium sizes in power-peaking units as well as in distribution networks (250 MVA is common).

Generator Conductor Temperatures and Hot Spots. Let us begin with the problems associated with conductor temperature. There has been no significant breakdown in the magnetic materials so that saturation point in these materials is more or less fixed. When one aims at standardizing cross sections of generators rated at 50/60 Hz, the way to increase the unit sizes is by increasing the length. MVA/m becomes a design factor. With increasing lengths conducting heat away from intermediate sections, a challenge is created. A lot of development in cooling media—air, hydrogen, water, and so on, alone or in combination, has taken place. The next point of consideration is the insulation, around the conductors and within the slot. The traditionally used mica with resin has proved itself. The resin grades have raised the insulation level to class H (180°C), at which the conductor can safely operate at a continuous class F (155°C) temperature.

This temperature control through cooling increases the MVA output for a given rating and size of a generator. By holding down conductor temperatures, it also limits the resistance losses at the increased rating. Superconductors have also been experimented with.

Roebling Method of Bar Insulation. Another method developed for keeping down the losses is the Roebling method, based on extruding the bars through a thermoplastic insulation instead of taping the insulating tapes around them.

Power Formers—HV Generators. A development taking place, not as a part of improving efficiency and holding down temperatures, is winding the stator with PVC-XLPE cables rated at medium to high voltages. Thus, EH voltages up to 155 kV have been obtained as a direct generator output. This enables a generator to be directly connected into a grid, doing away with investment in large-sized step-up transformers

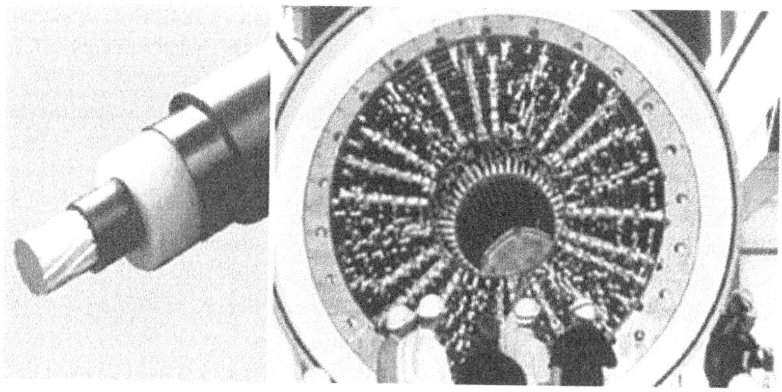

Figure 4-7. Power former HV generator with XLPE cable winding. (From [12], © IEEE 2002.)

and losses involved therein. As per report, two hydro generators rated at 45 kV and 155 kV and one turbo generator rated at 136 kV have already been commissioned.

This development has a special significance for windmill generators located in deep offshore sites. They can send their output via EHV cables directly into transmission. Reduced transmission losses and doing away with a step-up substation are the benefits. Figure 4-7 shows a typical power former with XLPE winding.

Variable Frequency Excitation Control. Another significant development along this line has been use of cyclo-convertors for energizing the rotating magnetic field. The rotating frequency of the field is controlled. With a tie-up to the grid frequency through a feedback loop, synchronizing the output of a distributed-induction-type generator becomes easy and reliable. This also does away with feeding capacitive KVAR to a generator, to cover magnetizing KVARs of the connected induction generators. This is a "booster" in the spread of medium-to-small-sized thermal, hydro and windmill generators. The effect of this rotating field on the mechanical resonance of the rotor has to be watched.

A small but significant development is the mechanical and metallurgical development in the coil retainer ring. It operates under high temperature and significant radial forces.

Developments in generators have kept in pace with overall developments in thermal generation.

REFERENCES

1. Autell, Independent Power Generator—A US Utilities Experience, *Power Engineering Review,* Number 3, 1994, pp. 13–14.
2. S. Azuhata, Advanced Clean Coal Technologies, *IEEE Power Engineering Review,* No. 3, 2001, pp. 6–7.

3. Department of Energy/Energy Information Administration, . DOE/EIA -0562(00), The Changing Structure of the Electric Power Industry 2000: An Update, Energy Information Administration, Washington, D. C., 2000.
4. F. P. de Mello and J. C. Westcott, Steam Plant Startup and Control in System Restoration, *IEEE Transactions on Power Engineering,* February 1994, pp. 93–101.
5. S. Sato, Improvements in Steam Generation Technology for Power Plants, *Power Engineering Review,* March 2001, pp. 8–9.
6. N. Kakimoto and K. Bala, Performance of Gas Turbine-Based Plants During Frequency Drops, *IEEE Transactions on Power Engineering,* August 2003, pp. 1110–1115.
7. K. Wollard and G. Zorpette, Technology '89: Power and Energy, *IEEE Spectrum,* Volume 26, No. 1, January 1989, pp. 56–58.
8. H. Matsumoto, Y. Ohsawa, S. Takahasi, T. Akiyama, H. Hanaoka, and O. Ishiguro, Startup Optimization of a Combined Cycle Power Plant Based on Cooperative Fuzzy Reasoning and a Neural Network, *IEEE Transactions on Energy Conversion,* March 1997, pp. 51–59.
9. I. Shiroumaru, T. Iwamiya, and M. Fukai, Integrated Operation and Management System for a 700 MW Combined Cycle Power Plant, *IEEE Transactions on Energy Conversion,* March 1992, pp. 20–28.
10. A. Bagnasco, B. Delfino, G. B. Denegri, and S. Massucco, Management and Dynamic Performances of Combined Cycle Power Plants During Parallel and Islanding Operation, *IEEE Transactions on Energy Conversion,* Vol. 13, No. 2, June 1998, pp. 194–201.
11. A. W. Layne, Next Generation Turbine Systems, *Power Engineering Review,* April 2001, pp. 18–23.
12. R. E. Joho, Advances in Synchronous Machines: A Turbo Generator View, *Power Engineering Review,* July 2002, pp. 7–11.

BIBLIOGRAPHY

Arroyo, J. M., and A. J. Conejo, Modeling of Start-Up and Shut-Down Power Trajectories of Thermal Units, *IEEE Transactions on Power Engineering,* August 2004, pp. 1562–1568.

Borghi, C. P. and P. L. Ribani, MHD-Steam Thermal Power Plant Electrical Stations with Zero Stack Emission, *IEEE Transactions on Energy Conversion,* March 1996, pp. 194–199.

Cheres, E., Z. J. Palmor, and J. Tuch, Drum-Type Boiler Following Control Configuration, *IEEE Transactions on Energy Conversion,* March 1994, pp. 199–205.

Cohen , E. and J. Good, The Application of a Superconducting Magnet System to the Cleaning and Desulphurization of Coal, *IEEE Transactions on Magnetics,* Volume 12, Number 5, September 1976, pp. 503–506.

Conejo, A. J., F. J. Nogales, J. M. Arroyo, and R. Garcia-Bertrand, Risk-Constrained Self-Scheduling of a Thermal Power Producer, *IEEE Transactions on Power Engineering,* August 2004, pp. 1569–1574.

Delson, J. K., Thermal Stress Computation for Steam-Electric Generator Dispatch, *IEEE Transactions on Power Engineering,* February 1994, pp. 120–127.

Erwin, S. R., J. T. Wood, J. S. Griffith, K. D. Le, J. T. Day, and C. K. Yin, Operational-Planning Software Realizes Production Cost Savings, *IEEE Transactions in Computer Applications Power Systems,* July 1991, pp. 55–60.

Habib, H. and D. Biggs, Development of Cogeneration and Combined Cycle Power Plants. The New Zealand Experience, *IEEE Power Engineering Journal,* Volume 12, Number 6, December 1998, pp. 261–266.

Hammons, S. K. Lee, and K. Y. Low, Analysis of Torques in Large Steam Turbine Driven Induction Generator Shafts Following Disturbances on the System Supply, *IEEE Transactions on Energy Conversion,* December 1996, pp. 693–700.

IEEE Committee Report, Bulk System Reliability Concepts and Applications, *IEEE Transactions on Power Engineering,* 1988, pp. 109–119.

Kelland, D., A Review of HGMS Methods of Coal Cleaning, *IEEE Transactions on Magnetics,* Vol. 18, No, 3, May 1982, pp. 841–846.

Malik, A. S., Determining Expected Startups of Generating Units by Frequency and Duration Approach, *IEEE Power Engineering Review,* May 2002, pp. 57–59.

Matsumoto, H., Y. Ohsawa, S. Takahasi, T. Akiyama, H. Hanaoka, and O. Ishiguro, Startup Optimization of a Combined Cycle Power Plant Based on Cooperative Fuzzy Reasoning and a Neural Network, *IEEE Transactions on Energy Conversion,* March 1997, pp. 51–59.

Patel, A Novel Control Logic for Fast Valving Operations, *IEEE Power Engineering Review,* October 2002, pp. 43–46.

Pereira, L., J. Undrill, D. Kosterev, D. Davies, and S. Patterson, A New Thermal Governor Modeling Approach in the WECC, *IEEE Transactions on Power Engineering,* May 2003, pp. 819–829.

Prince, W. R., E. K. Nielsen, and H. D. McNair, A Survey of Current Operational Problems of Power Systems, *IEEE Transactions on Power Engineering,* 1998, pp. 1492–1498.

Rouse, S., Energy/Environment Performance Improvement Opportunities, *IEEE Power Engineering Review,* March 2001, pp. 13–15.

Shanechi, H. M., N. Pariz, and E. Vaahedi, General Nonlinear Modal Representation of Large Scale Power Systems, *IEEE Transactions on Power Engineering,* Volume 18, Number 3, August 2003, pp. 1103–1109.

Zorpette, G. Energy Management-Loosening the Bonds of Oil, *IEEE Spectrum,* Volume 28, Number 3, March 1991, pp. 34–38.

5

THERMAL STATION POWER ENGINEERING

5.1 START-UP PROCESS OF A CCPP

Start-up periods for a combined-cycle power plant (CCPP) can last as long as 120 hours, as will be shown later. During a start-up period, steady-state temperatures and stabilized temperatures are not established. As a result, stress levels across critical parts are higher than what they would be under stabilized working conditions. The total life period may be assumed to be a sum of maximum plus average stress-level hours experienced by the critical portion of the equipment. Hours of operation at high stress levels eat up a larger percentage of equipment life than those under average stress levels. Against this context, start-up periods assume significance. They have to be optimized and should be kept as short as possible.

A start-up period lasts from the instant the combustor is lighted to the instant when the CCPP has been synchronized and has started load following on the bus. Exit temperatures from the gas turbines are critically controlled by the fuel-flow and air-flow controllers in a CCPP. However, these will vary during the start-up. They will also vary and take time to stabilize under quick load following and even during a shutdown. This variation creates varying stresses in the critical parts.

The critical parts from the viewpoint of stress levels are steam turbine rotor blades, inlet flanges, valves, and so on in the heat recovery steam generators (HRSG). Rotor blades normally work under an established temperature gradient between the tip and

Copyright © 2011 the Institute of Electrical and Electronics Engineers, Inc.

the root. The blades also are subject to centrifugal forces. They are susceptible and prone to frequent failures. Databases have to be established for these failure–stresses patterns under different working conditions.

Another important factor to watch in gas turbine operation is the generation of NO_x gases. As this exhaust gas flows through the HRSG, it passes over a "denitrifier," a catalyst that, with the help of NH_3, deoxidizes NO_x. The controlling factors are catalyst temperature and the contact period with the catalyst.

Thus, we have two major constraints in optimizing the start-up period. With the data on the stress pattern of blades and NO_x patterns for different gas-to-air ratios in a CCPP available, engineers can work out an optimization schedule for the start-up period.

A CCPP is a high-pressure, high-temperature piece of equipment with rapidly rotating blades. It has a limited life, which must be utilized judiciously. Frequent start-ups affect the working life period.

An example is cited wherein algorithms for start-up optimization are worked out. A fuzzy logic solution is used. It is claimed that this workout for a 1650 MW plant with assumed start-up periods varying from 10 to 120 hours yields 26.3% in energy savings and 35.6% in start-up time [1].

It may be noted that energy input during a start-up period is a total economic loss.

5.2 SHORT-TERM DYNAMIC RESPONSE OF A CCPP TO FREQUENCY VARIATION

With an increasing share of electricity generated by CCPPs having fast dynamic characteristics, control systems become important.

The CCPPs have high efficiencies at high stable temperatures. The blade temperatures of the turbine must stay high and constant. As such, the gas exhaust temperature becomes an important parameter. If it is measured at the gas turbine end and held steady, it ensures healthy rotor blade operation. At the steam turbine end, it maintains steady thermal efficiencies for the recovery steam generator. This improves the working of the steam turbines also.

Suppose a generator on the grid cuts out, increasing the load on rest of the system and reducing the net frequency temporarily. Other CCPPs must respond fast by rapidly adding their reserve capacity.

5.3 CASCADE TRIPPING OF A CCPP DUE TO FREQUENCY EXCURSION

Severe load conditions for steam generation result from crash shutdowns, regular shutdowns, and start-ups after a shutdown. This picture is best illustrated by the example of cascade tripping of a CCPP system in Malaysia.

Malaysia, with a CCPP installed capacity of 5760 MW, reported cascade tripping of CCPPs, one after another. The grid frequency dipped by 1.5 Hz when this happened. The event was investigated. The root causes were traced to the control characteristics of fuel/air flows, the HRSG inlet temperature and the extent of loading of the CCPP.

To understand this, let us take a condition in which the plant is loaded to 0.80 PU of its capacity and a sudden load comes on. The controller opens up the fuel valve. The shaft speed dips. The frequency also drops instantly. With the dipping of the frequency, the air compressor speed slows down and air delivery does not momentarily match the extra fuel delivery. Since the CCPP operates with 20% in reserves, it has spare reserves. The controller opens up the air flow valve and the ratio between two flows is maintained. The system recovers speed and frequency.

The paths followed by the fuel flow (RH side) and by the air flow (LH side) are shown in the LH side of Figure 5-1.

However, the system is already operating at 1.0 PU and its air input is at its maximum. It cannot increase in response to the sudden increase in fuel flow. Thus, the fuel flow is forced to return to its original level, as shown in Figure 5-1 (right-hand side). Due to the lack of an increased fuel and air flow mix, the system slows down further and collapses. It had no reserve capacity to pull itself out.

It is recommended that CCPPs not be run at 100% of output capacity.

5.4 OPERATION PLANNING TO MEET LOAD DEMANDS—FLOW DIAGRAM

The operational requirements for a CCPP can be broadly categorized into various systems:

1. Automatic system
2. Optimum-load operational system as per Figure 5-2
3. Fault diagnostic system

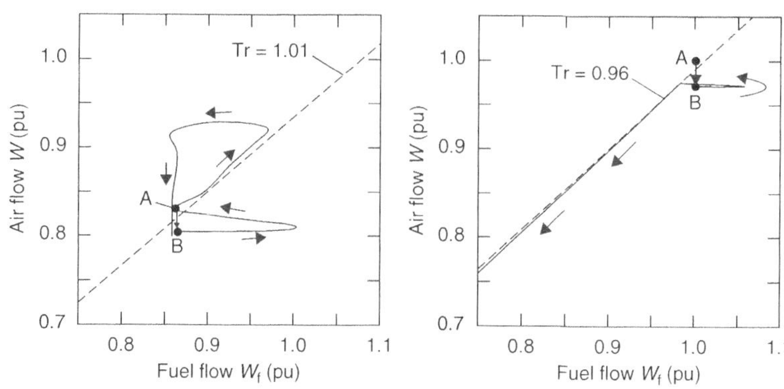

Figure 5-1. Fuel flow and air flow. (From [2], © IEEE 2003.)

4. Routine test work
   - Preventing mistakes in operation
   - Performing scheduled maintenance
   - Using a small number of operators
5. Patrol supporting system

All these systems have to be computerized if the plant is to run efficiently.

5.5 CAPACITY CURVES FOR THERMAL ELECTRICITY GENERATION

A capacity curve is an indispensable guide tool for a plant operator.

The plant operator will be called upon to supply a mix of kWs and KVARs in any unpredictable proportion. He has to stay within the stability limits of the generator. While supplying load, he may be supplying both inductive KVARs as well as kWs, say at a lagging PF of 0.85. The controller might ask him to cut down on kWs and supply compensation-capacitive KVARs to boost up voltages. Figure 5-3 shows a capacity curve that helps the operator to stay within the stable region.

Please note the following. At 50 MW rating, the generators are water cooled. The MW capacity decreases with cooling water temperatures rising from 85°F to 110°F. Under excitation operation, there is a 10% deduction in MVAR to be supplied against the theoretical allowed value. Practical limits of operation for rated voltage and for 90% of the rated voltage differ, falling below the theoretical value.

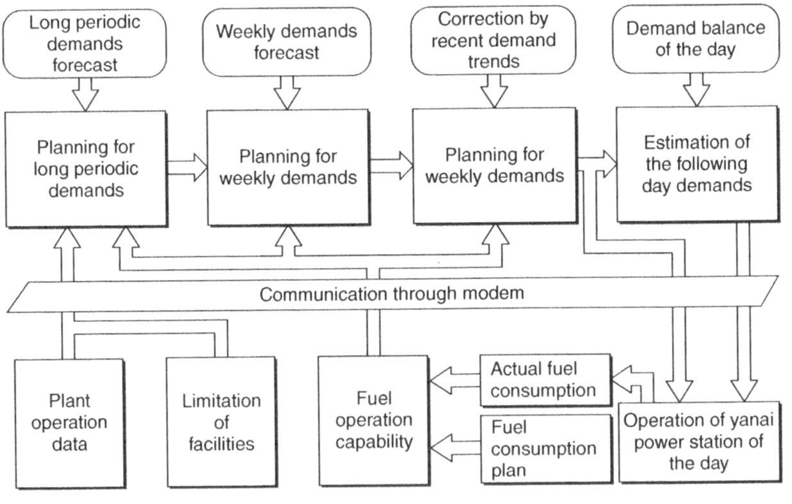

Figure 5-2. Flow diagram of operation for load demands. (From [3], © IEEE 1992.)

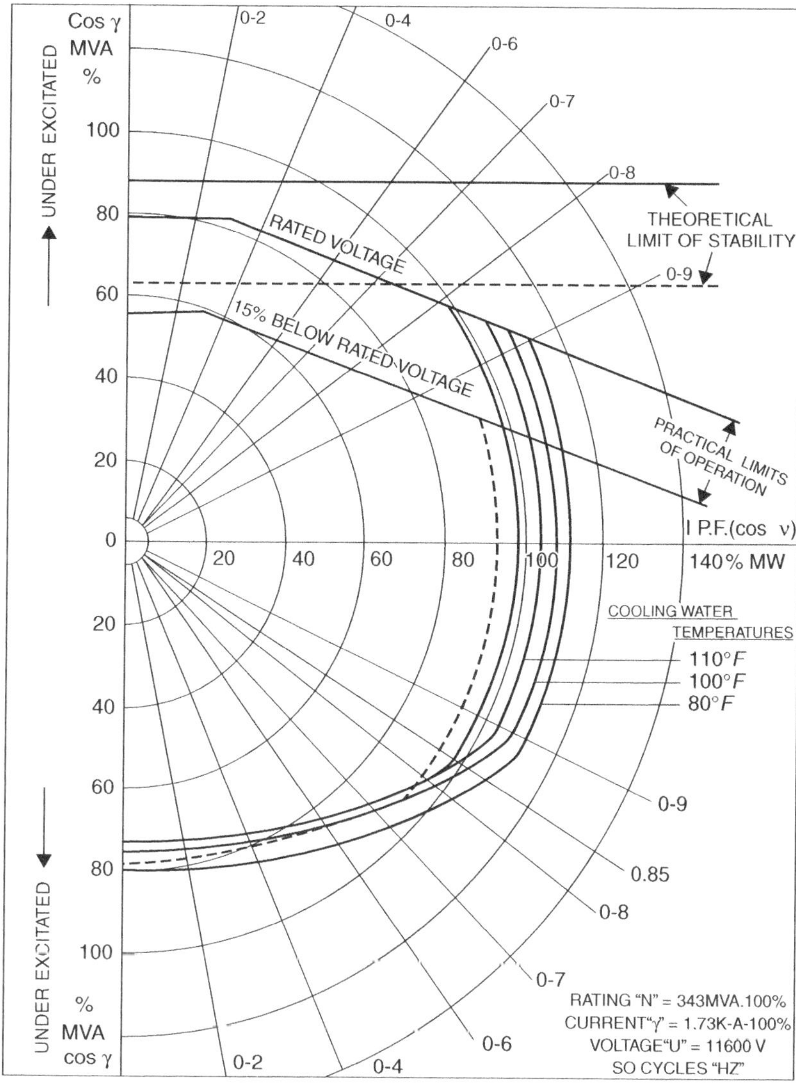

Figure 5-3. Calculated reactive and power capacity curves of Amarkantak generators, 2 × 50 MW each. (Coutesy of A. R. Dwarkanath, former Chief Engineer, MPEB.)

Appendix 5-2 shows capability curves for a 210 MW generator (an increasingly popular size) and for a 500 MW generator (H_2 cooled).

The following points may be noted. With reduction in H_2 pressure from 30 Pst-G by half, the rotor temperature goes up and the delivery capacity comes down. Should the line MVAR, as decided by line impedance and line KV, match the generator KVAR, there will be instability.

5.6 OPERATIONAL ECONOMY INCLUDES FUEL CONSIDERATIONS

5.6.1 Costs

Crude oil and gas prices are most volatile. During the period 2006–2007, crude oil prices went up from US$52–60 to US$140 per barrel and then came back to US$113 per barrel. Crude oil output is controlled by OPEC countries. The prices are also impacted by oil reserves maintained by governments of the United States and Europe.

A generator should have a time-covering contract with its supplier, preferably a long-established refiner.

5.6.2 Reliability of Supply

This is governed both by the supply contract and the minimum fuel stock at site. One of the conditions laid down by NERC for connecting into the grid is that the operator must have a minimum fuel stock (7 days) at all the times at the place of generation.

5.6.3 Emission Caps Considerations

Emission caps influence the choice of mix of input energies. Take the case of a generator with a base plant using oil and supporting generators using gas. It can choose to use these selectively as per load demand.

The first factor in these considerations is the thermal efficiency of a given generator at various percentages (oil:gas) of the loading. At lower percentages, the efficiencies are worst.

Then there are caps on environmental pollution, for which one has to buy emission credits if the pollution is more than permissible. The pollution levels also vary for each type of generator and per load.

Thus, for a given load over a certain time period, the generator has to work out an optimum mix of oil generation and gas generation. Purchase of hydropower also enters into the picture.

5.7 EFFICIENCY IN OPERATING PRACTICES

Considerable savings in daily operations may be realized by adopting measures in thermal energy efficiency and electrical energy efficiency. Scott Rouse of Ontario Power Generation reports energy savings of up to 1.4 billion kW hrs/year in the plants covered under his program [4]. These plants include fossil-fuel-based and hydro-energy-based plants. He provides extensive coverage on how and where energy could be saved. For him this worked out to a saving in O&M costs of Canadian $90 million per year. Over and above this, the group claimed Canadian $10 million in emission credits. Earning these emission credits is a great incentive to improve generation efficiency.

The amount of emissions to be cut has to increase every year. These credits are in great demand and have been rising in value every year.

These savings could be realized in two ways:

1. By reducing the energy input for some of the electricity output or by increasing the electricity output while holding onto the same energy input. This is achieved by increasing conversion efficiency and relates to the boiler, turbine, and so on.
2. By reducing the electricity consumption within the generating station itself. Note that internal electricity consumption is fairly sizeable in these plants. Here we aim at the electrical efficiency.

Conversion efficiency has a higher impact on overall cost reduction than electrical efficiency. On both of these measures, the results can be obtained by improvements in the operating processes and by employing innovative equipment to replace older and absolete equipment.

5.8 ANCILLARY SERVICES COMPULSORILY

FERC (the U.S. Federal Electricity Regulatory Commission) defines ancillary services as those services necessary to support the transmission of energy from resources to load while maintaining reliable operation of transmission in accordance with good utility practices. FERC further identifies these services:

1. Reactive power supply
2. Load following
3. Loss compensation
4. Energy imbalance
5. Scheduling and dispatch services
6. System protection (operating reserves)

Under the regulatory system, these services were treated as free and obligatory, to be provided by the franchised utility.

5.8.1 Reactive Power Supply

From the early days of electrical utilities, generator was supposed to support the load–supply balance and maintain system voltages with variable reactive power support in quantities required, when and where required. Historically, this support was free, but under a new order established in the United States (FERC rule 888) a generator must sell reactive power and a customer must pay for it. The challenge to a generator lies in being ready to meet the requirement as and when asked for and yet to produce the maximum active power and the minimum of reactive power, as the later sells very cheaply.

5.8.2 Load Following

Load following includes

- Matching generation to load within the control area as per allotted load
- Minimizing the differences between actual flows and scheduled flows
- Maintaining the synchronous frequency, within limits

5.8.3 Loss Compensation

There is loss of power from the point at which it is produced by the generator to the point at which a consumer consumes it. It is the responsibility of the generator to replace this lost power. Eventually, the consumer will foot the bill. However, the losses and additional price will pass from one stage to another.

5.8.4 Energy Imbalance

An energy imbalance occurs when a generator cannot supply the required power contracted for and also when a customer fails to consume the power contracted for.

5.8.5 Scheduling and Dispatch Services

Scheduling and dispatch Services have traditionally been "free" ancillary services offered by the generator.

5.9 CHANGING PERFORMANCE REQUIREMENTS FOR THERMAL PLANT OPERATORS

Under the old monopolistic regime, it was sufficient for a thermal plant operator to take care of power dispatches as and when required while keeping an eye on spinning and security reserves, power quality requirements, and so on. He had to fulfill the requirements under ancillary services, as outlined earlier, maintain operating efficiencies, and attend to maintenance. The rest was taken care of by the central office.

In the new and changing setup, the thermal plant is an independent entity, required to make its own decisions and act to meet the unit's commitments. This mean that the operator must:

1. Be cognizant of, cooperate with, and accommodate the green energies, particularly the fastest growing wind energy. He must watch the daily commitment to the system operator and see how wind energy timing is likely to impact his schedule. In particular, he has to cater to the steep ramping rates involved with sudden switch-off and switch-on.
2. Work under lot of constraints originating from environmental concerns. Thus, his fuel mix division hangs on how much extra "pollution" quota he has on hand

and whether it is economical to buy "credits" and maintain production or hold onto reserves.

3. Bid on price, quantity, and time of delivery for the day-ahead commitments on the power exchange. He must know how it works and what are the trends.

Unit commitment is the fastest growing challenge in thermal power production today. See reference [5], which is a list of some 100 articles on this subject. A few of the references pertaining to these articles are given at the end of this chapter.

5.10 EXPANDING GRIDS DEMAND TIGHT FREQUENCY TOLERANCES

The old system of thermal generation imposed a restriction on the acceptable variation on frequency of 50/60 Hz at 1 to 1.5 Hz. This prevented grid-connected local generation from overdrawing from the grid or dumping their idle-time excess power into the grid. Any excursion outside these limits was heavily punished.

Under the old NERC rules, maintaining system frequency under load following was expected to be a free and compulsory ancillary service.

Lately, the regional systems have been interconnected, forming larger power system entries. Take the PSM system in the United States, where, initially, three statewide power systems joined together. Their latest peak hour demand (in 2004) was reported at 10 × 31,000 MWs. The power systems in Western Europe, under the guidance of UCTE (Union for Co-ordination of Transmission of Electricity), have an estimated peak demand at 10 × 20,570 MWs [6]. An error in even a fraction of a hertz in frequency can translate into thousands of MWs of electricity going out of accounting and into objectionable credits. Finely tuned frequency control becomes a paramount need of the day.

The regional transmission systems are connected at a number of points. A slight difference in the frequencies of these systems could overload the intertias to objectionable levels. This has to be watched and prevented.

5.11 RESERVES ARE IMPORTANT IN FREQUENCY CONTROL

In an electrical system, the generation and load must match at all the times. The load varies from a minimum to a peak value, which could increase rapidly. A generator must keep some capacity in reserve for the operational demand of the load. These are "operational reserves." There could also be a contingency also with a generator going out of supply suddenly and overloading the remaining generators in the system. The system as such must have "contingency" or "security" reserves to face this type of situation and prevent the collapse of frequency and the grid that might result.

There is a wide variety of nomenclature for these reserves, such as primary reserves, secondary reserves, regulatory reserves, contingency reserves, reserves beyond 30 minutes, and so on. Wood and Wallenberg [7] define reserves as "the total synchronized capacity minus the losses and the load." Within a contingency period, a generator might be required to work at an overload capacity for a limited period, say 30 minutes, so we must have an overload reserve to cover this period.

In this book, we refer to reserve capacity in a system/generator in two ways. As per the NERC directive, a power system should have sufficient reserve equal to the rating of the largest of its members—generator or transmission—going suddenly out of operation. This still holds true on grounds of security of the system. When the coverage of the system was small, this condition imposed a substantial burden in reserving production capacity for a likely contingency. As the system sizes grew and number of generating units increased, the burden on individual generators decreased. With systems like PJM and UCTE, a question arises as to what extent does a generator have to hold capacity in reserve.

Let us first see how frequencies are rigidly maintained under widespread area networks.

5.12 RESERVES BASED ON DROOP CHARACTERISTIC

Each generator has a characteristic called droop. If a generator rated at an output power P_N is overloaded by additional power ΔP_g, then its synchronous frequency F_n slows down by Δf. Then the droop is given by $-(\Delta f/f_n)/(\Delta P_g/P_n)$ [6].

Under modern frequency control, frequency drop under a load is severely restricted to ±0.200 mHz, that is, 1/5000 of a hertz (compare this to 1.5 Hz earlier). Now ΔP_g becomes the reserve capacity that a generator must hold per its droop characteristic. Droop characteristics limits are specified for generators connected to the line.

The droop characteristic represents the gain of the feedback loop of the frequency controller of a generator. Smaller droop value represent stronger action of the generator. The drift in frequency encountered by a generator can be measured by the change in gain of this feedback loop.

The frequency control in modern networks is exercised at three levels:

1. The primary frequency control
2. The secondary frequency control
3. The tertiary frequency control

5.13 PRIMARY FREQUENCY CONTROL

Primary frequency control is exercised at the load level. When the load goes up, the frequency shifts downward. It is controlled by calling upon the reserves. It can also be controlled by switching off, by preconsent, some loads as well, such as air conditioning loads, for an example.

Primary control basically limits and stops the frequency excursions. Primary frequency control is provided by various systems (see Table 5-1).

Table 5-1. Technical comparison of primary frequency control parameters in various systems

	NERC	UCTE	DE	FR	ES	NL	BE	GB
References*	[24], [25], [26]	[29], [37]	[20], [21]	[18], [19]	[34]	[27], [28], [29]	[17]	[22], [23]
Full availability	No rec.	≤ 30 s	≤ 30 s	≤ 30 s	≤ 30 s	≤ 30 s	≤ 30 s	Pri.: ≤ 10 s Sec.: ≤ 30 s Hi.: ≤ 10 s
Deployment end	No rec.	≥ 15 min	≥ 15 min	≥ 15 min	≥ 15 min	≥ 15 min	≥ 15 min	Pri.: ≥ 30 s Sec.: ≥ 30 min Hi.: as long as required
Frequency characteristic requirement	10% of the balancing authority's estimated yearly peak demand/Hz	20,570 MW/Hz	≈ 4,200 MW/Hz	≈ 4,200 MW/Hz	≈ 1,800 MW/Hz	≈ 740 MW/Hz	≈ 600 MW/Hz	Variable ≈2,000 MW/Hz
Droop of generators	5% in 2004; no rec. anymore	No rec.	No rec.	3–6%	≤ 7.5%	5–60 MW: 10% > 60 MW: 4–20%	No rec.	3–5%
Is an adjustable droop compulsory?	No rec.	No rec.	Yes	Yes	No rec.	5–60 MW: No rec. > 60 MW: Yes	No	Yes
Accuracy of the frequency measurement	No rec.	Within ± 10 mHz	Within ± 10 mHz	No rec.	No rec.	No rec.	Within ± 10 mHz	No rec.
Controller sensitivity	T: ± 36 mHz in 2004; no rec. anymore NI: No rec. I: No rec.	T: ± 10 mHz NI: No rec. I: should be compensated within the zone	T: ± 10 mHz NI: No rec. I: ± 0 mHz	T: ± 10 mHz NI: No rec. I: should be compensated within the zone	T: ± 10 mHz NI: No rec. I: ± 0 mHz	5–60 MW: T: ± 150 mHz; NI: No rec.; I: No rec. > 60 MW: T: ± 10 mHz; NI: ± 10 mHz; I: ± 0 mHz	T: ± 10 mHz NI: ± 10 mHz I: No rec.	T: ± 15 mHz NI: No rec. I: No rec.
Full deployment for or before a deviation of:	No rec.	± 200 mHz	± 200 mHz	± 200 mHz	± 200 mHz	5–60 MW: 30% for ± 150–200 mHz > 60 MW: 70% for ± 50–100 mHz	± 200 mHz	Pri: –800 mHz Sec.: –500 mHz Hi.: +500 mHz

Notes: No rec. = no recommendation; Pri., Sec. or Hi. = primary, secondary or high frequency response; I = intentional; NI = nonintentional; T = total.
*Reference numbers correspond to references in [6], not this chapter.

The droop characteristic represents the reserve power in relation to its rated capacity of a generator for a maximum permissible drift in frequency. In PJM systems, it was 5% before 2004 and abolished later. Some systems make it compulsory to have an adjustable droop reserve. (This is measured with the gain of the feedback loop under isolated conditions.) The reserves must be fully available within a specified period, say $\leq$ 30s. The reserves may be required for as long as 15 min. By then, the controller will have spread the load onto other generators. Note that in modern systems, the controller also has direct control of the generator output.

There is a band of frequencies for a generator over which the controller could be insensitive. A generator with a smaller band will be the first one to respond in the primary control.

Once the maximum permitted drift in the system is declared and the droop characteristic of a generator is known, the reserve to be held by that generator can be calculated. It is compulsory for a generator to hold this reserve. It is at the discretion of a TSO (transmission system operator) to use it. Mostly, it is used under the automatic frequency control employed. When it is not used, the generator gets paid for holding it. When it is used, it is paid for automatically in the increased output. A TSO has to arrange for use of this reserve, under bilateral control, by means of a tendering and bid process under spot prices. This reserve is also termed the "frequency control reserve."

Typically, these are between 3-5% in Great Britain and 5% under NERC (PJM system). All generators collectively in a region like the PJM system do not allow the system to drift beyond say 0.200 mHz of their reserves.

It may be noted that the frequency of a generator tied into a grid is fixed and so is its shaft speed. The governor operation reflects on the possible shift in frequency. This likely shift in frequency due to an overload is measured by the gain in the feedback loop of the governor.

Rebours calls the reserve the "frequency control reserve" [6]. It may be noted that the contingency also includes large generation as well as large load outages.

Primary frequency control basically is a load area control. It stops frequency excursions and brings them to a common level. It also tries to equalize the spread of overload in such a way that it is uniform for all the respondents. The control period is limited; if the generators are overloaded beyond their rated capacities and reserves, they do not continue under this overload beyond 15 min.

5.14 SECONDARY FREQUENCY CONTROL (SFC)

Before the end of these 15 min, the secondary frequency control takes over. It is operated by an interregional TSOs. The region under primary frequency control will have stabilized, usually around the grid frequency. This frequency difference can circulate large amounts of power through the interties. Secondary frequency control (SFC) basically looks after traffic overload and congestion. More than that, it brings the deflected frequency back in tune with the grid frequency by spreading the overload between regions. Table 5-2 gives secondary frequency control parameters for various systems.

Table 5-2. Technical comparison of secondary frequency control parameters in various systems

	NERC	UCTE	DE	FR	ES	NL	BE
References*	[25]	[37]	[20], [21]	[19]	[34]	[27], [28]	[17]
Deployment start	No rec.	$\leq$ 30 s	Immediate or $\leq$ 5 min	$\leq$ 30 s	No rec.	30 s–1 min	$\leq$ 10 s
Full availability	No rec.	$\leq$ 15 min	$\leq$ 5 min	$\leq$ 430 s or $\leq$ 97 s	$\leq$ 300–500 s	$\leq$ 15 min	$\leq$ 10 min
Deployment end	No rec.	As long as required	As long as required	As long as required	$\geq$ 15 min	$\geq$ 15 min and as agreed	As long as required
Control organization	No rec.	No rec.	Pluralistic	Centralised	Hierarchical	Pluralistic	Centalised
Frequency measurement	$\varepsilon \leq 1$ mHz T $\leq$ 6 s	$1.0 \leq \varepsilon \leq 1.5$ mHz T: No rec.	$1.0 \leq \varepsilon \leq 1.5$ mHz T = 1 s	$\varepsilon \leq 1.0$ mHz T = 1 s	ε: Unknown T = 2 s	$\varepsilon \leq 1.0$ mHz T - 4 s	$\varepsilon \leq 1.0$ mHz T: Variable
Exchange measurement	$\varepsilon \leq 1.3\%$ T $\leq$ 6 s	$\varepsilon \leq 1.5\%$ T $\leq$ 5 s	$\varepsilon \leq 1.5\%$ T = 1 s	$\varepsilon \leq 1.5\%$ T = 10 s	ε: Unknown T = 4 s	$\varepsilon \leq 1.5\%$ T - 4 s	$\varepsilon \leq 1.5\%$ T: Variable
Controller cycle time	$\leq$ 6 s	1–5 s	1–2 s	5 s	4 s	4 s	5 s
Controller type	No rec.	I or PI	PI	I	P or PI, depending on the regulation zone	PI, with additional heuristics	PI
Proportional term	No rec.	0–0.5	Unknown	0	Unknown	0.5	0–0.5
Integral term	No rec.	50–200 s	Unknown	115–180 s	100 s	100–160 s	50–200 s
K-factor for measuring the ACE	The frequency characteristic	110% of the frequency characteristic	Unknown	Unknown	Unknown	900 MW/Hz	$\neq$ 660 MW/H

Notes: No rec. = no recommendation; a: accuracy; T = cycle time; P, I or PI = proportional, integral, and proportional–integral controller.
*Reference numbers correspond to references in [6], not this chapter.

Note that the accuracy of measurement for this SFC control is quite high—between 1 to 1.5 mHz. The control is based on a factor called the area control error (ACE) of a zone. This factor includes K, measured in MW/Hz.

5.15 TERTIARY FREQUENCY CONTROL

Tertiary frequency control is a manual control. It comes into play as a last resort if both primary and secondary frequency controls do not operate correctly, for instance, due to a mishap in the control system.

5.16 RIGID FREQUENCY CONTROLS ARE BRINGING IN CHANGES

The following points may be noted regarding frequency controls in wide interconnected systems:

1. Frequencies assume major importance. Stakes involved in frequency control are high. Measurement of frequency must be accurate to the tune of 1 to 1.5 mHz in 50/60 Hz.
2. Old norms are being replaced. There are no more free reserves or too many compulsory and free ancillary services. The services are centrally managed and paid for. This means they are being deployed optimally, both in volume and in cost.
3. With a multitier frequency control, both automatic and manual, and a huge backing of MWs/Hz frequency characteristics, apart from large load balancing, contingencies that led to grid failures in the past are on the way out.
4. Transmission systems with multidirectional and increased traffic are now susceptible to congestion. They can get overloaded beyond their loading capacity when large amounts of energy are forced on them should there be a frequency drift. This looks likely to be the Achilles heel of today's systems rather than the cascading grid breakdowns of the past. A lot of attention is being paid to and many algorithms are being written on this subject.

5.17 VOLTAGE CONTROL SERVICES

Under NERC clarification of ancillary services, voltage control service falls under reactive power supply. Voltage variations are mainly local conditional problems. They are concerned with nuisance tripping of relays and malfunction of computer- and microprocessor-based systems. They are taken care of at the bus terminals with automatic voltage regulators or static voltage controllers, where they are applied. The voltage control is based on reactive power, which inherently costs much less than active power (controlled under frequency control). As such, the standards on voltage-controlling ancillary services are not as rigid. Voltage control measures have to consider bus voltages, load flows, as well as neighboring circuit parameters.

5.18 VOLTAGE MEASUREMENT AT POD INTO THE TRANSMISSION SYSTEM

The voltages that are important from the TSO's (transmission system operators), point of view are not the voltages at the generator terminals but those at the point of delivery (POD), into the transmission system. There is loss of reactive power in the system between the generator terminals and the POD. This is taken at roughly 15% of the generator-rated KVA.

5.19 ATTRACTIVE MARKET PRICES LEAD TO RESERVES OVER AND ABOVE THE COMPULSORY LIMITS

A generator is allowed in the grid subject to his keeping certain reactive power in reserve. This is the basic compulsory reactive power service. Over and above this, the TSO might request enhanced reactive power. Total reactive power that a generator can offer will be determined from the active power-reactive power (P-Q) characteristics of the generator. Note that the generator has to supply reactive power in both modes: absorption capacity and supply capacity. For a power supplier, some systems specify the PF at full load supply at POD as the basic compulsory service. Some others specify a fixed amount in relation to the power rating.

Voltage controls arespecified by the power factors maintained at the point of delivery. Table 5-3 gives practices followed by different systems. Note that unlike active power reserves in a power market, reactive power reserves are localized and do not play much of a role in a power market. Mostly, the transactions are between the parties concerned and it is the responsibility of the players to adhere to the voltage norms.

5.20 IMPORTANCE OF OPERATING FREQUENCY LIMITS FOR A THERMAL GENERATOR

With a large number of generators connected to a system grid, a fixed frequency is a must. Assume that a particular generator shuts down, say due to a mechanical problem or a problem on the energy supply side. Its frequency will fall. The grid pumps energy into this generator and brings up its RPM to the synchronous value. On the other hand, if a generator speeds up and its generated frequency shoots up, the grid will absorb enough energy from this generator so that its RPM stays synchronized. Either way, large transfers of current into or out of an armature will damage a generator. Frequency control relays and overcurrent and unidirectional relays will trip the misbehaving generator.

Table 5-3. Technical comparisons of voltage control parameters in various systems

	NERC	UCTE	DE	FR	ES	NL	BE	GB
References*	[25]	[37]	[20]	[18]	[35]	[27]	[17]	[22]
Absorption capability requirement (lagging)	No rec.	No rec.	pf = 0.95 or 0.975 at the POD for Pn and Un	–0.35-Pn at the POD for Pn and Udim	pf = 0.989 at the POD for Pn and Un	pf = 0.8 at the POD for Pn and Un	–0.10-Pn at the POD for Pn and Un	pf = 0.85 at the terminals for Pn
Production capability requirement (leading)	No rec.	No rec.	pf = 0.925 or 0.9 at the POD for Pn and Un	0.32-Pn at the POD for Pn and Udim	pf = 0.989 at the POD for Pn and Un	pf = 0.8 at the POD for Pn and Un	0.45-Pn at the POD for Pn and Un	pf = 0.95 at the terminals for Pn
Estimated absorption requirement at the POD for Pn and Un	No rec.	No rec.	–0.33-Pn or –0.23-Pn	–0.35-Pn	–0.15-Pn	–0.75-Pn	–0.10-Pn	–0.80-Pn
Estimated production requirement at the POD for Pn and Un	No rec.	No rec.	0.41-Pn or 0.48-Pn	0.32-Pn	0.15-Pn	0.75-Pn	0.45-Pn	0.17-Pn

Notes: No rec. = no recommendation; pf = power factor; Pn = nominal power; POD = Point Of Delivery; Udim = dimensioning voltage; Un = nominal voltage.

*Reference numbers correspond to references in [6], not this chapter.

Generator manufacturers give a range within which the plant operator maintains the frequency. Table 5-4 gives a typical permissible range on generators.

5.21 SYSTEM PROTECTION

System protection through different types of reserves is the responsibility of the generator. All auxiliary services were formerly free services in that they were not charged

Table 5-4. Frequency limits for BHEL thermal/nuclear turbine generator sets

Unit capacity	Frequency	Limit duration
30 MW (GEC design)	48.5 to 51 Hz	For continuous unrestricted operation
60 MW (Skoda design)	49 to 51 Hz	For continuous unrestricted operation
	48 to 49 Hz	2 hours a stretch and 2 hours in a year
100 MW (Russian design)	49 to 50.5 Hz	For continuous unrestricted operation
	50.5 to 51 Hz	3 minutes at a stretch and 500 minutes in a whole life
	47 to 48 Hz	1 minute at a stretch and 180 minutes in a whole life
	46 to 47 Hz	10 second at a stretch and 30 minutes in a whole life
110 MW (Skoda design)	49 to 51 Hz	For continuous unrestricted operation
	48 to 49 Hz	2 hours a stretch and 30 hours in a year
	47 to 48 Hz	30 minutes at a stretch and 2 hours in a year
120 MW (GEC design)	48.5 to 51 Hz	For continuous unrestricted operation
	48 to 48.5 Hz	5 minutes at a stretch and 500 minutes in a whole life
	47 to 48 Hz	5 minutes at a stretch and 180 minutes in a whole life
200/210 MW (Russian design)	49 to 50.5 Hz	For continuous unrestricted operation
	50.5 to 51 Hz	3 minutes at a stretch and 500 minutes in a whole life
	48 to 49 Hz	3 minutes at a stretch and 500 minutes in a whole life
	47 to 48 Hz	1 minute at a stretch and 180 minutes in a whole life
	46 to 47 Hz	10 second at a stretch and 30 minutes in a whole life
210 MW (KWU design)	47.5 to 51.5 Hz	For continuous unrestricted operation
236 MW (Nuclear) (GEC design)	48 to 51 Hz	For continuous unrestricted operation
500 MW (KWU) (BHEL design)	47.5 to 51.5 Hz	10 second at a stretch and 180 minutes in a whole life

Source: IEEMA Journal, Feb. 2000.

for specifically and separately. Now there is debate on fixing the costs of these auxiliary services and also who pays. When these are specifically stated, along with the cost of energy, the central controller has to find out from his software what is the best mix of power and auxiliary services from a generator.

Typical suggestions on achieving these efficiencies as given Rouse [4] are given in Appendix 5-1. It may be noted that Canada has a cold climate and requires considerable energy to keep the stations as well as machinery warm.

5.22 MAINTENANCE PRACTICES

Maintenance practices consist of corrective maintenance, preventive maintenance, and predictive maintenance.

5.22.1 Corrective Maintenance

A defect is attended to after it has occurred. This is a very costly way of maintenance.

5.22.2 Preventive Maintenance

Sudden failure of any equipment in the chain process in a generating station has very costly consequences. In preventive maintenance, equipment is taken down periodically. All the field tests are carried out and if found satisfactory the equipment is put back in service again.

This requires a regular maintenance team. It entails loss of production (for a generator, for example). However, no diagnosis is made on what is happening inside the equipment. For example, hot spots and partial discharges start small and build up to a breakdown stage over time. This period could be a few hours or stretch over a few months into a year. Similarly, the contacts of a circuit breaker making electrical noise with the contact springs create weakness.

5.22.3 Predictive Maintenance

In a predictive maintenance system, developed by EPRI [8], maintenance can have more reliability than before and will do away with lot of manpower. This report explains "just-in-time" servicing.

A variety of sensors perform online monitoring. This data, collected and analyzed against the historical data of the equipment, reveals performance trends and warns of impeding failure. Software developed for power delivery, called the Maintenance Management Workstation (MMW), stores, analyzes, and processes this data and ties in with maintenance termed reliance-centered maintenance.

A SCADA system within the plant can also collect and dispatch regular performance data to a central control station. The central control station in its turn operates distant controls.

All these technological tools and processes cost quite a lot. Their affordability and actual utility for maintaining ones position in competition have to be carefully watched. But the decision on acquiring these should not be unduly postponed.

5.23 CHALLENGES IN MEETING ENVIRONMENTAL OBLIGATIONS

These are covered in detail in Chapters 6, 7, and 8.

5.24 MHD GENERATORS

A CO_2-based plasma under high temperature and pressure produces freed ions. This increases the resistance of the plasma to normal electric current. But under high magnetic and electrostatic field, a flow of ions can be obtained between a cathode and an anode. Externally, this establishes a flow of electric current. The negative electrons gather on the anode plate. In this process, termed magneto hydrodynamic generation (MHD), electric energy is derived directly from the plasma instead of through turbines and generators. The two-stage losses, through turbine and generator, are eliminated. As a result, conversion of heat energy into electric energy has a very high efficiency. The added output of MHD, at sufficiently high temperature and pressure, is used to drive a two-stage system turbine, which drives a generator, which produces electric energy. Overall output energies for different energy-efficiency-improving systems are:

- For a conventional pulverized-coal-fired power plant of 500 MWe with zero stack emission, the calculated overall efficiency is 26.5%.
- For a pressurized fluidized bed power station (PFBC), efficiency is not higher than 30%
- For a integrated coal gasification combustion combined cycle plant, IGCC is 32–34%, for a MHD/steam power plant and zero stack emission it is 40%

The cost of electricity under MHD works out at 20% lower than similar costs under the options mentioned above. However MHD, as of today, is not commercially exploited. The MHD channel sizes are large. At the temperatures and pressures involved, the chamber life is very low. Life realized on side walls and cathodes is around 2000 hrs. Life realized on the anode is 1000 hrs.

One more problem facing MHD for large-sized utilization is the economic and technical viability of a large-sized magnet enveloping the MHD channel. In the illustration worked out, the channel size is 1.05 $m^2 \times 2.4\ m^2 \times 20$ m long. Details of this MHD design are given in Appendix 5-3.

The principle of separating ions from electrons with a membrane and creating an external electric current is used in fuel cells. This is explained in Chapter 12. Fuel cell technology is being exploited commercially on larger and larger scales. Viewed in that context, MHD-based electricity generation feasibility might be a only a few decades away.

APPENDIX 5-1 ENERGY EFFICIENCY PROGRAM [4]

Generation Project Types

In addition to being categorized as either a technology or operations project, a project is also categorized by project type: boiler, turbine, operator improvement, startup, HVAC, lighting, auxiliary, and so on. These project types are combined for both TE (thermal efficiency) and EE (electrical efficiency) saving projects.

Boiler. These projects are related to the combustion process by saving fuel. Examples are temperature control sensor replacement, boiler control replacement, and adding flame stabilizing rings. Electrical energy project examples include optimizing operation of electrostatic precipitators and using synthetic lubricant in the coal pulverizers.

Turbine. These projects involve processes related to the conversion of steam to electricity. Fuel-saving TE projects include turbine maintenance and steam heat optimization. Examples of these electrical energy projects are optimizing the circulating and pump controls and condenser improvements.

Operator Improvement. Significant conversion efficiency savings are achieved by adding computer assistance (feedback-loop software—ECOS upgrade) to improve the efficiency of the thermal conversion process.

Startup. Faster startups for infrequently used turbines result in significant TE savings. This process uses waste from the building heating steam to keep the turbines warm so that they require less fuel to start up. A warm start is faster (3 hours) compared to a cold start (about 6 hours).

Heating, Ventilating, and Air Conditioning (HVAC). TE projects include redirecting warm air from the combustion units and auxiliary boiler efficiency improvements. EE projects include installation of auxiliary heating boilers, free cooling, and insulation.

Lighting, Auxiliary, and Other Projects. These improve electrical efficiency through lighting reductions or reduced lighting levels, replacing fixtures, and adding controls. Also included are projects that change secondary equipment, which do not directly affect the electricity production process. Examples include upgrading service water systems and air compressor systems.

APPENDIX 5-2 CAPABILITY CURVES OF A 210 MW GENERATOR

See Figures 5-4 and 5-5.

APPENDIX 5-3 DESIGN OF AN MHD GENERATOR SYSTEM AND ITS OUTPUT CONVERSION

Professor Borghi of the Institute of Electrotechnics of the University of Bologna has presented a system study for an MHD steam thermal plant to cover an output range between 500 MW to 2500 MWe [9]. It is based on coal as fuel. An additional feature of this study is zero stack emission. While SO_x and NO_x are chemically scrubbed out, CO_2 is compressed and cooled down to a liquid stage. Liquid CO_2 in drums is to be dumped on deep sea beds where it stays liquid, due to high water pressures.

Figure 5-6 shows this system.

A conventional steam generation plant with a steam boiler and two-stage turbines coupled to a generator is taken as the base module. A CO_2 loop-based topper system consisting of a burner, a mixer, and an MHD system has been attached to the base module.

Item 1 in this figure is a burner fed in with coal and a seed. The seed is potassium carbonate. Oxygen mixed with compressed CO_2, supplied by 3, are used by the burner to obtain a high-pressure, pure CO_2 flow. This expands in the MHD chamber.

The MHD channel is 1.05 m^2 at the entrance and 2.4 m2 at the exit. It is 20 m long. The burner temperature is 28000 K and output pressure is 3.5–8.4 bar. Mass flow rates for the fuel and plasma are 43.3 KG/s and 499.3 KG/s. Maximum voltage stresses occur at 5 M, 10 M, and 15 M against electrical power produced at 347.5 MW. Pressure at the exit end of the channel is 1–2.4 bar.

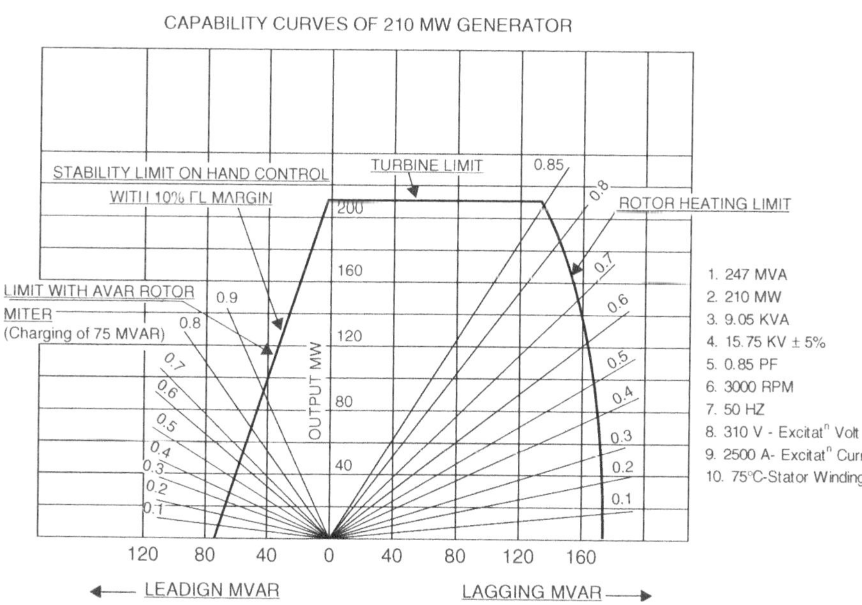

Figure 5-4. Capability curves for a 210 MW generator. (Courtesy of A. R. Dwarkanath, MPEB.)

Generator Type: THDF 115/50

Rated Reactive Power	S_N = 588 MVA		Rated frequency	f_N = 50 Hz	
Rated Active Power	P_N = 500 MVA		Power factor	P.F = 0.85	
Rated Voltage	V_N = 21 kV (1 ± 5 %)		Speed	u_N = 50 s^{-1}	
Rated Active Power	I_N = 16.2 vA		H_2 Pressure	p_S = 4 kg/cm^2	

Stator No: 127368/127369/127370
Rotor No: 127368/127369/127370

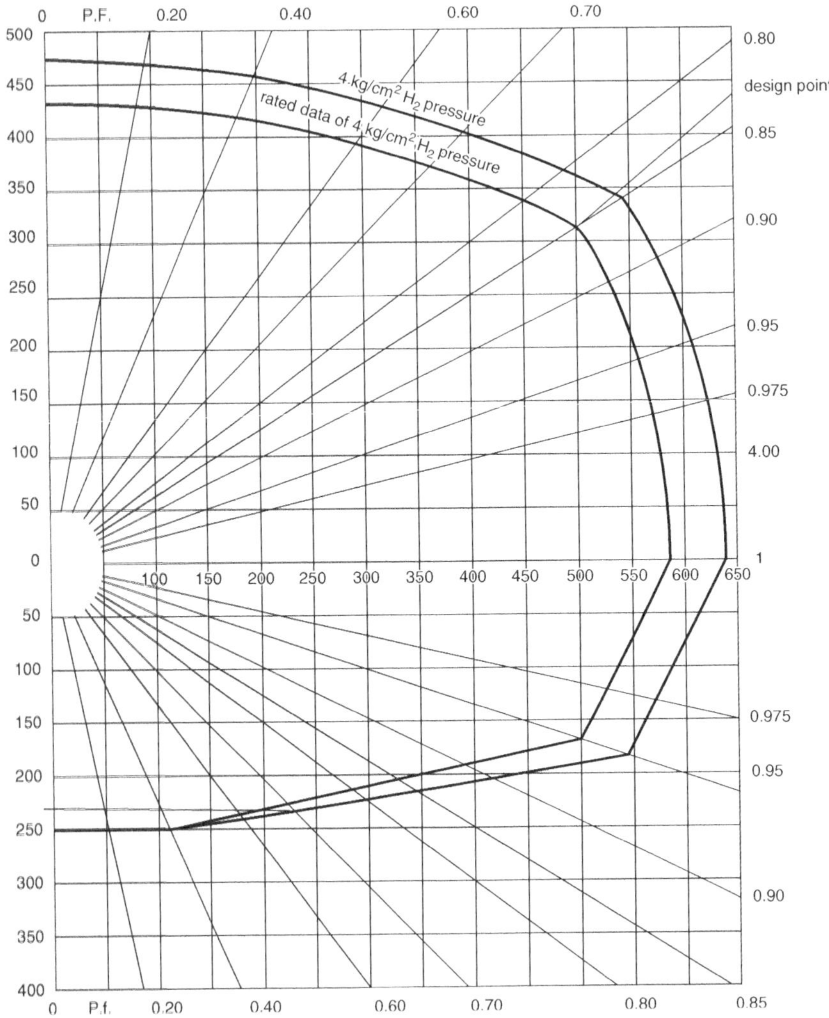

Figure 5-5. Capability curve for a 500 MW thermal generator. (Courtesy of A. R.. Dwarkanath, MPEB.)

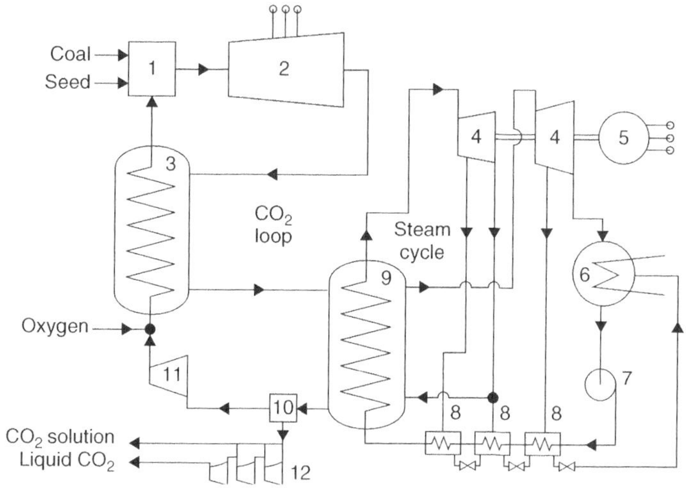

Figure 5-6. Schematic of the plant. (From [9], © IEEE 1996.)

Output from the channel is fed to the steam boiler, 9, via heat exchanger, 3. Part of the CO_2 output from 9 is compressed and fed back into 3 and part is fed to the compressor, 12. Here CO_2 is liquefied.

The steam boiler with absorbed heat from CO_2 raises its own steam. This steam passes through the conventional steam circuit consisting of turbines, 4, condenser, 6, and so on. Steam pressure at entry of the first turbine is 160 bar and for the second turbine it is 42 bar. Maximum steam temperature is 823 K.

Table 5-5 gives calculated efficiencies for thermal power inputs in equivalents ranging from 500 MW to 2500 MW. Please note that energy drawn through the MHD is higher than energy drawn through the steam system by around 20%. In the

Table 5 5. Main results of the system study

P_{Th} (MW)	P_{MHD} (MW)	P_S (MW)	P_{CR} (MW)	P_{CL} (MW)	η (%)
500	121.3	104.5	18.0	22.4	38.22
750	196.0	157.8	27.0	33.6	39.82
1000	271.7	210.5	35.9	44.7	40.70
1250	347.5	264.0	44.9	56.0	41.22
1500	423.5	316.9	53.9	67.1	40.76
2000	575.8	423.6	71.8	89.4	41.40
2250	652.1	477.4	80.9	100.7	41.62
2500	728.4	530.4	89.8	11.8	41.37

Source: [9].

table, P_{Th} is the total rated (input) power for the system, P_{MHD} is the power delivered by the MHD, P_S is the power delivered by the steam system, and ηP_{ThL} is the power delivered by CO_2 to the steam system. P_{CR}, P_{CL}, and P_{OS} are power absorbed by the liquefier for CO_2, compression of the oxidizer, and production of pure O_2, respectively.

Thus,

$$\eta = \frac{P_{MHD} + P_S + \eta_S P_{ThL} - P_{CR} - P_{Os} - P_{CL}}{P_{Th}}$$

Extracting Electricity from the MHD generator

Johnson [10] described the following circuitry. Basically, this is high-current, low-voltage electricity production. The adjoining pairs have intercoupling through Hall effect. In a typical lab experiment, Johnson [10] put switched-gate turn-off inverters across each of these pairs. The chopped DC outputs are fed to pulse transformers whose secondaryies are connected in series. The transformers are inductively joined on the primary sides; otherwise, the secondary's stay isolated. By controlling the switching period and timing, a resultant AC wave can be gathered. This is akin to today's pulse width modulation except that there are a number of DC sources here. Figure 5-7 shows this arrangement.

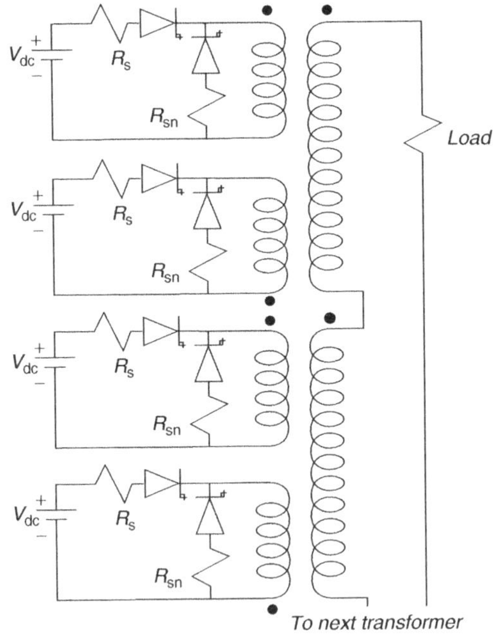

Figure 5-7. Two transformers of a pulse amplitude synthesis and control (PASC) system. (From [10], © IEEE 1993.)

With this arrangement, harmonics, which are inherent in this type of chopped feed, and power factor on the output circuit are taken care of by computer programming to produce a pure sine wave shape. Interelectrode sparking within the tunnel appears as a small, high-varying-frequency addition. This is taken care of with a choke in the output.

A considerable amount of thermal energy can be theoretically recovered from the waste flue gases. In fact Russians have built a prototype. However, this energy recovery process is today not feasible.

REFERENCES

1. H. Matsumoto, Y. Ohsawa, S. Takahasi, T. Akiyama, H. Hanaoka, and O. Ishiguro, Startup Optimization of a Combined Cycle Power Plant Based on Cooperative Fuzzy Reasoning and a Neural Network, *IEEE Transactions on Energy Conversion,* March 1997, pp. 24-31
2. N. Kakimoto and K. Bala, Performance of Gas Turbine-Based Plants During Frequency Drops, *IEEE Transactions on Power Engineering,* August 2003, pp. 1110–1115.
3. I. Shiroumaru, T. Iwamiya, and M. Fukai, Integrated Operation and Management System for a 700 MW Combined Cycle Power Plant, *IEEE Transactions on Energy Conversion,* Volume 7, Issue 1, March 1992, pp. 20–28.
4. S. Rouse, Energy/Environment Performance Improvement Opportunities, *Power Engineering Review,* March 2001, pp. 13–15.
5. G. B. Sheble and G. N. Fahd, Unit Commitment Literature Synopsis, *IEEE Transactions on Power Systems,* Volume 9, Issue 1, February 1994, pp. 128–135.
6. Y. Rebours, A Survey of Frequency and Voltage control Ancillary Services, *IEEE Transactions on Power Systems,* Volumes 1 and 2, February 2007.
7. A. J. Wood and B. Wollenberg, *Power Generation, Operation, and Control,* Second Edition, Wiley, New York, 1990.
8. Esselman, EPRI Development Program for Electrical Utilities, *Transactions in Computer Applications Power Systems,* April 1996, pp. 18 21
9. C. A. Borghi and P. L. Ribani, MHD—Steam Thermal Power Plant Electrical Stations with Zero Stack Emission, *IEEE Transactions on Energy Conversion,* March 1996, pp. 194–199.
10. R. M. Johnson, M. K. Donnelly, and K. E. Marcotte, A Pulse Amplitude Synthesis and Control (PASC) Consolidation/Inversion System for Faraday Connected MHD Generators, *IEEE Transactions on Energy Conversions,* September 1993, pp. 448-454

BIBLIOGRAPHY

Adibi, M. M., D. P. Milanicz, and T. L. Volkmann, Optimizing Generator Reactive Power Resources, *IEEE Transactions on Power Systems,* 1999, pp. 319–326.

Autell, Independent Power Generator—A U.S. Utilities Experience, *IEEE Power Engineering Review,* Volume 3, 1994, pp. 13–14.

Billinton, Operating Reserves Assessment in Interconnected Generating Capacity, *IEEE Transactions on Power Systems,* November 1988, pp. 1479–1488.

Bulk System Reliability Concepts and Applications, *IEEE Committee Report*, IEEE Transactions on Power Systems, 1988, pp. 109–119.

Cohen, A. I. and S. H. Wan, A Method for Solving the Fuel Constrained Unit Commitment Problem, *IEEE Transactions on Power Systems,* Volume 2, Issue 3, August 1987, pp. 608–614.

Conejo, A. J., F. J. Nogales, J. M. Arroyo, and R. Garcia-Bertrand, Risk-constrained Self-Scheduling of a Thermal Power Producer, *IEEE Transactions on Power Systems,* August 2004, pp. 1569–1574.

Daniels, R., Y. Daniels, Krischen, M. Trotigon, and S. Rossiguol, A Survey of Frequency and Voltage Control Auxiliary Services—Part 1—Technical Features, *IEEE Transactions on Power Systems,* February 2007, pp. 350–357. Part 1—Economic Features, pp. 358–366.

Erwin, S. R., J. T. Wood, J. S. Griffith, K. D. Le, J. T. Day, and C. K. Yin, Operational-Planning Software Realizes Production Cost Savings, *IEEE Transactions in Computer Applications,* July 1991, pp. 55–60.

Gargiulo, G., V. Mangoni, and M. Russo, Capability Charts for Combined Cycle Power Plants. Generation, Transmission and Distribution, *IEE Proceedings,* Volume 149, Issue 4, July 2002, pp. 407–415.

Handke, J, E. Handschin, K, Linke, and H.-H. Sanders, Coordination of Long- and Short term Generation Planning in Thermal Power Systems, *IEEE Transactions on Power Systems,* May 1995, pp. 803–809

Handschin, E. and H. Slomski, Unit Commitment in Thermal Power Systems with Long-Term Energy Constraints, *IEEE Transactions on Power Systems,* Volume 5, Issue: 4, November 1990, pp. 1470–1477.

Huse, E. S., I. Wangensteen, and H. H. Faanes, Thermal Power Generation Scheduling by Simulated Competition, *IEEE Transactions on Power Systems,* Volume 14, Issue 2, May 1999, pp. 472–477.

Kirby, B. and E. Hirst, Unbundling Electricity: Ancillary Services, *IEEE Power Engineering Review,* Volume 6, 1996, p. 5.

Lei, W., M. Shahidehpour, and L. Tao, GENCO's Risk-Based Maintenance Outage Scheduling, *IEEE Transactions on Power Systems,* Volume 23, Issue 1, February 2008, pp. 127–136.

Ma, H. and S. M. Shahidehpour, Decomposition Approach to Unit Commitment with Reactive Constraints, Generation, Transmission and Distribution, *IEEE Transactions on Power Systems,* Volume 144, Issue 2, March 1997, pp. 113–1.

Moya, O. E., A Spinning Reserve, Load Shedding, and Economic Dispatch Solution by Bender's Decomposition, *IEEE Transactions on Power Systems,* February 2005, pp. 384–388.

Nabona, N., J. Castro, and J. A. Gonzalez, Optimum Long-Term Hydrothermal Coordination with Fuel Limits, *IEEE Transactions on Power Systems,* May 1995, pp. 1054–1062.

Pappala, V. S., E. Erlich, K. Rohrig, and J. Dobschinski, A Stochastic Model for the Optimal Operation of a Wind-Thermal Power System, *IEEE Transactions on Power Systems,* Volume 24, Issue 2, May 2009, pp. 940–950.

Perez-Arriaga, A. A., Reliability and Generator Adequacy, *IEEE Power Engineering Review,* Volume 12, 1999, pp. 6–9.

Reliability Risks and Probability Applications Subcommittee, Reliability Issues in Today's Electric Power Utility Environment, *IEEE Transactions on Power Systems,* 1997.

Sioshansi, R., R. O'Neill, and S. S. Oren, Economic Consequences of Alternative Solution

Methods for Centralized Unit Commitment in Day-Ahead Electricity Markets, *IEEE Transactions on Power Systems,* Volume 23, Issue 2, May 2008, pp. 344–352.

Tao, L. and M. Shahidehpour, Dynamic Ramping in Unit Commitment, *IEEE Transactions on Power Systems,* Volume 22, Issue 3, August 2007, pp. 1379–1381.

Trudel, G., S. Bernard, and G. Scott, Hydro-Quebec's Defense Plan Against Extreme Contingencies, *IEEE Transactions on Power Systems,* 1999, pp. 958–965.

Willis, L., J. Finney, and G. Ramon, Computing the Cost of Unbundled Services (Power Transmission), *IEEE Transactions in Computer Applications,* October 1996, pp. 16–21.

Yingvivatanapong, C., W.-J. Lee, and E. Liu, Multi-Area Power Generation Dispatch in Competitive Markets, *IEEE Transactions on Power Systems,* Volume 23, Issue 1, February 2008, pp. 196–203.

Zhang, H., Z. H. Liu, and G. Y. Fang, Research on a Super-Small-Scale MHD Power System, *IEEE Transactions on Energy Conversion,* March 1994, pp. 186–191.

Zuyi, L. and M. Shahidehpour, Security-constrained unit commitment for simultaneous clearing of energy and ancillary services markets, *IEEE Transactions on Power Systems,* Volume 20, Issue 2, May 2005, pp. 1079–1088.

6

ENVIRONMENTAL CONSTRAINTS IN THERMAL POWER GENERATION—ACID RAIN

6.1 INTRODUCTION TO ACID RAIN AND CARBON EMISSIONS

Accelerated growth in the use of electricity has pushed up electricity production worldwide. The basic energy supplying sources are

- For thermal electricity production, fossil fuels such as coal, oil, and gas
- For hydroelectricity production, hydro power from lakes and rivers
- For nuclear electricity production, nuclear fission materials
- For green energy production from renewables, solar, wind, biomass, geothermal, and so on

When burned for energy output, fossil fuels produce SO_2, NO_x, and CO_2. The major objectionable gases, SO_2 and NO_x, when absorbed by atmospheric water (clouds) form acids and produce "acid rain," that is, rain with small percentages of these objectionable acids. The effect of the acid rain is visible near oil refineries, thermal power stations, and so on. Tree growths are stunted and negative. Human health is affected. The spread of SO_2 and NO_x is not limited only to around the centers producing these. They spread through the upper atmosphere worldwide. Countries like Canada, whose major national wealth is in forest products, are affected, even though they might not be the major contributors to the acid rain.

Copyright © 2011 the Institute of Electrical and Electronics Engineers, Inc.

CO_2 (carbon dioxide) is not as poisons as SO_2 and NO_x are but its buildup in the atmosphere may bring about dangerous meteorological changes. The atmosphere, including CO_2, acts as a buffer that inhibits radiation of heat from the earth's crust into the atmosphere. Over centuries, a balance has been obtained between the natural and normal production of CO_2 arising from biomass emissions and conversion of CO_2 back into biomass, mainly by plant life. This has maintained the earth's atmospheric temperature in standard ranges within tolerances. With a buildup of CO_2, the balance is getting disturbed and temperatures are likely to rise. It is suspected that if this is not checked, the polar ice caps will melt, flood coastal areas, and drown important harbor cities.

6.2 WORLD CONCERN OVER ENVIRONMENTAL POLLUTION AND AGREEMENTS TO CONTROL IT

Worldwide concern has arisen and international standards have been drawn for limiting and successively bringing down the presence of polluting gases in the atmosphere. The basic thrust is toward control of

- SO_2—sulphur dioxide
- NO_x—a combination of nitrous oxides
- CO_2—carbon dioxide

There is unanimous agreement between most countries on reducing SO_2 and NO_x pollution in stages. This affects financial aspects of countries that have large reserves of fossil-based deposits of higher sulphur content, coal or oil. However, with the use of present technologies, one can reduce the presence of SO_2 and NO_x.

This is not the case with CO_2 pollution. One has to realize that today major electricity production is through burning of carbon contained in coal or oil. Cutting down on this drastically means cutting down the economic growth of a country. This is not acceptable to quite a few countries, including the United States, the largest producer of electricity.

It has been agreed upon by participating countries that a sustained drive should aim at increasing the proportion of noncarbon sources of energy in electricity production by benchmarking this diversion away from carbon. Economic impetus becomes a complimentary but an effective tool in reducing air pollution of each of the above types. We will study these aspects in detail.

6.3 U.S. CLEAN AIR ACT AND AMENDMENTS

The United States took a leading step on air pollution by passing the Clean Air act in 1963. This act established the Environment Protection Agency (EPA), which was initially charged with setting up standards for measurement of air pollution. Several provisions have been added to this act, the last important being the amendment of 1990. This dealt with SO_2 and NO_x. The act dealt separately with the two major polluters using carbon as an energy sources, namely, electricity utilities and transport, including road vehicles. Electricity generating units contributed the largest amount. Table 6-1 gives the yearly emission from this group in the United States.

The 1990 amendment aims at reducing SO_2 and NO_x from electricity generating stations in two phases. Phase 1 started in 1995 and continued up to 2000. It aimed at a total reduction of 10 million tons of SO_2 per year and 2 million tons of NO_x per year. It had identified 261 generating units from 110 power plants all over the country as qualifying for SO_2 reductions. Their SO_2 emissions were taken as of 1985–1987 levels. Their average annual fuel burning rates for this period were also benchmarked.

These units were ordered to cut down their SO_2 emission levels from a benchmark of 2.5 lbs of SO_2/MBtu burnt to 1.2 lbs/MBtu over this period of phase 1. SO_2 production per plant per year was fixed. A plant faced a stiff fine if the emission exceeded this limit. If it could cut down SO_2 below the allotted value, it could get an SO_2 allowance certificate for this achievement. These certificates were tradable. If plant B found it profitable to produce more electricity, increase SO_2 emission, and balance it by buying a certificate from plant A, it was allowed to do so. Plant A was allowed to bank this certificate for next year's use. There was one hitch: Plant A had to burn as much fuel as it did in 1985–1987.

This trading of SO_2 allowance certificates introduced a new factor in optimizing the economic running of power plants. Emission constraints formed an important parameter in unit allocation (which unit to allocate for a given load) and in fuel switching, along with many other factors as explained in the earlier chapters.

Phase 2 began in 2000 and covered all generating units. The reference rate for SO_2 allowance was 1.2 lbs/Mbtu. Now, any unit could improve its performance and be entitled to a tradable certificate.

It may be noted from Table 6-1 that emissions from both SO_2 and NO_x have been reduced progressively, while CO_2 emissions continue to grow up.

6.4 COMPLYING WITH CONSTRAINTS ON THE SO_2 EMISSION RATE

6.4.1 Options Available

The SO_2 allowance market has opened up a variety of options to a thermal electricity generator:

1. Buy low-sulfur fuels.

   Oil:

   low sulfur content—0.5 to 0.25%

   high sulfur content—1 to 3%

Table 6-1. Emission from energy consumption at conventional power plants and combined heat and power plants, 1996 through 2008 (thousand metric tons)

Emission	2008	2007	2006	2005	2004	2003	2002	2001	2000	1999	1998	1997
Carbon dioxide (CO_2)	2,477,213	2,536,583	2,459,500	2,513,609	2,456,934	2,415,680	2,395,048	2,389,745	2,441,722	2,338,660	2,324,139	2,232,709
Sulfur dioxide (SO_2)	7,830	9,042	9,524	10,340	10,309	10,646	10,881	11,174	11,963*	12,843[R]	13,464[R]	13,480[R]
Nitrogen oxides (NO_x)	3,330	5,650	3,799	3,961	4,143	4,532	5,194	5,290	5,638[R]	5,955[R]	6,459[R]	6,500[R]

R = Revised.

Notes: The emissions data presented include total emissions from both electricity generation and the production of useful thermal output. See Appendix A, Technical Notes of the EIA report, for a description of the sources and methodology used to develop the emissions estimates. CO_2 emissions for the historical years 1997–2007 have been revised due to changes in the conventions used to determine fuel combustion. SO_2 and NO_x 2008 values are preliminary.

Source: Calculation made by the Electric Power Division, U.S. Energy Information Administration, *Electric Power Annual 2009*.

Coal:

low sulfur content—<1.3%

high sulfur content—1.3 to 3.5%

Coal with low sulfur content has been reported as having higher ash content, which requires plants to be derated due to ash-handling problems. Low sulfur content fuels are costly, the cost being also controlled by the prices and availability of tradable SO_2 allowances.

2. Adopt clean coal strategies as detailed in Chapter 4. Desulfurize fuel before burning.
3. Switching of fuels. Costlier natural gas is selectively and periodically introduced in coal-fired boilers. The system has to be adoptable. Some old systems are not adoptable. Fuel switching can also be adopted with oil as a partner. This helps to keep within the constraints of SO_2 emission.
4. Flue gas desulphurization. Golden of TVA has reported "limestone forced oxidized" technology removing as much as 95% of SO_2 emission and earning overall as many as 25,000–50,000 emission allowance certificates per year for his utility [1].
5. Emission-constrained dispatching with optimized programs.
6. Just as one has an economy-emission-dispatching program covering all the thermal units under one group's control, one can adopt a similar dispatch program for buying electricity from other (noncarbon) sources, such as pumped-storage systems, nuclear power, or even independent distributed generators. Algorithms exist for working these out.
7. Sale, purchase, or banking of SO_2 allowances.
8. Encourage demand-side management (DSM) with economic incentives to consumers and reduce demand to avoid running into penalties fo SO_2 emissions.

6.4.2 Costs Involved in Reduction of SO_2 Emissions

Volumes of flue gases to be desulfurized are high. The equipment sizes and costs will be considerable. Similarly, operating costs, which involve desulfurizing chemicals, will be high. Table 6-2 gives an idea of these costs. It is an illustrative example based on the following assumptions:

1. Fuel prices

Oil 3% S	Italian Lira 275,000 (US $90/bbl)
Coal 1% S	110,000
Gas 0.23% S	310,000

2. Rate of return 5%. No currency fluctuations.
3. Life

Plant—25 years

FDG—15 years

SCR—15 years

Table 6-2. Percentage increase of cost of electricity (COE) for a 4 × 660 MW plant

	Basic	FGD	FGD + SCR*
Investment + interest (%)	77.7	960	100.8
Personnel (%)	8.3	9.3	9.3
O + M (%)	14.5	18.8	20.1
Total Fixed (%)	100.0	124.1	130.2
Net capacity (MV)	627	613	610
Efficiency (%)	40.2	39.3	39.1
COE fixed costs (%)	100	127	134
COE variable (%)	100	106.2	110.5
COE Total (cents/kW-hr)	5.6	6.4	6.7
(%)	100	115	121

*FGD = flue gas desulfurization; SCR = selective catalytic reduction.
Source: [2].

Note that investment, including interest, shoots up from 77.7% to 100.8% due to emission constraints. Personnel costs dealing with emission aspects at various spots and levels rise to 9.3%. Cost of producing electricity goes up by 21%.

FGD, itself, consumes 14 MW of electric power out of 622 MW produced. SCR hardly consumes anything. Salvaderi gives the following figures on SO_2 production:

Plant heat rate = –2200 kcal/kW-hr as the overall conversion rate

Sulfur at 3% in oil produces 13.8 tons of SO_2/GW-hr

Sulfur at 1% in oil produces 4.6 tons of SO_2/GW-hr

Sulfur at 1% in coal gives 1.7 tons SO_2/GW-hr with a heat rate of 2260 kcal/KW-Hr

So far, we have arrived at costs in adding emission control to the plant, inclusive of both the capital costs and the running costs. Against this, we have to balance some negative costs, such as penalty costs for exceeding the allowed emission rates and cost of buying SO_2 or NO_x emission certificates. The penalty costs are fixed by the constituent states of the United States. The costs of buying certificates are market driven.

Appendix 6-3 gives details on desulphurization and also average flue gas desulphurization costs in the United States over the period 1994 through 2005.

6.5 SURCHARGES ON EMISSIONS

The Massachusetts Department of Public Utility (MDPU) has adopted a specific set of surcharges that the state utilities are required to use in analysis and decisions regarding development of new supply and demand sources. This MDPU valuation has become a reference standard. They have directly valued the costs of pollutants as per Table 6-3.

Table 6-3. MDPU valuation of air pollutants

Pollutant	(\$/lb)
SO_2	\$0.781
NO_x	\$3.383
CO_2	\$0.012
CH_4	\$0.166
CO	\$0.452
N_2O	\$2.082
TSP	\$2.103
VOCs	\$2.787

An externality surcharge is the cost of a given pollutant, say SO_2, per kW-hr of electricity product. It is defined as plant emission rate in lbs/Mbtu for a given pollutant times average plant heat rate in Btu/kW-hr times valuation as per Table 6-4. These surcharges have been individually calculated for various types of power stations. This surcharge, called an "adder," is to be added to the dispatch costs as a shadow price. This exercise helps in arriving at accurate decisions on planning schedules. Table 6-5 shows a set of these calculations for the year 1990.

A number of conclusions have been drawn from the table. If adder-based dispatches are projected for each year, up to 2005, the projected dispatches, including both the

Table 6-4. Components of total externality surcharges (\$/kW-hr)

Resource	SO_2	NO_x	C	Total
Small Coal	2.1	3.0	2.7	7.8
Medium Coal	1.7	2.3	2.3	6.4
Large Coal	1.6	2.2	2.2	6.1
1950s Oil	1.0	2.3	2.1	5.3
1960s Oil	0.9	2.0	1.8	4.7
1970s Oil	0.7	1.1	2.0	3.8
CC	0.1	0.9	1 7	2.8
Peakers	0.1	2.4	2.6	5.2
Thermal Purchases	0.1	2.9	2.5	5.5
Thermal IPPs (1990)	0.3	1.9	0.9	3.2
Thermal IPPs (2005)	0.2	1.2	0.9	2.3
Adv. CCs	0	0.1	1.0	1.1
Adv. CTs	0	0.2	1.7	1.9
RGCC	0.1	0.1	2.2	2.4
Biomass	0	0.8	0	0.8
Refuse	1.0	3.1	1.2	5.3

Notes: Small coal represents old coal-burning plants of small ratings. 1950 oil represents oil-burning plants of 1950 vintage. CC represents combined cycle plants and peakers, both run on distillate oils. IPPs are independent power producers. Adv is advanced. IGCC is integrated gasification combined cycle plant fitted with selective catalytic recovery.

Table 6-5. Adder-based dispatch results in 1990

	Capacity factor (%)	Dispatch cost (¢/kW-hr)	Direct prod. cost (MS)	Emissions (ktons)			Percent change due to MDPU adders				
				SO_2	NO_x	C	Capacity factor	Dispatch cost	SO_2	NO_x	C
Nuclear	72.5	0.6	573	0	0	0	0%	0%	0%	0%	0%
Coal steam	33.7	8.0	164	91	29	2325	–52%	339%	–53%	–52%	–53%
Peakers	0.4	10.4	19	0	0	20	0%	100%	0%	0%	11%
Oil steam	50.1	6.9	1121	250	104	12455	24%	151%	23%	17%	29%
Thermal purchase	6.1	8.2	61	0	1	39	–93%	194%	0%	–89%	–93%
Hydro purchase	90.6	2.6	289	0	0	0	84%	9%	0%	0%	0%
Pumped storage	3.3	0.0	7	0	0	0	–20%	0%	0%	0%	0%
Storage hydro	28.8	0.0	5	0	0	0	0%	0%	0%	0%	0%
Baseload hydro	62.9	0.0	6	0	0	0	0%	0%	0%	0%	0%
IPPs	63.4	8.2	501	17	22	858	0%	50%	0%	0%	0%
CC	82	5.3	74	5	8	1133	5%	111%	0%	14%	5%
Total			2819	363	163	16381			–14%	–14%	–1%

Source: [3].

emission costs and production costs after allowing for price increases, as well as a trend toward gas and new technologies, the dispatches will be as per Table 6-6. Addition of new plants to the existing ones has been considered.

A notable development advantage has been a shift away from coal to oil to new technologies. This reflects improved thermal ratings in kcal/kW-hrH and also lower emission. The adders are tools that yield quick results.

For more details on desulphurization of flue gases in the United States, see Appendix 6-3.

6.6 COMPLYING WITH CONSTRAINTS ON DENITRIFYING

NO_x is formed in a boiler. At high burning temperature, N_2 reacts with oxygen and forms NO_x, which is roughly 95% NO and 5% N_2O gas.

Some fuels also have NO_x inherent compounds. This is fuel NO_x. This fuel-containing NO_x has no relation to the quantity of NO_x produced under high temperature.

Here we come up against contrasting requirements. Turbine efficiency improves as we increase the inlet temperature and pressure. Progress has been made in developing ceramic materials that can work at still higher temperatures. On the other hand, at these tem-

Table 6-6. Adder-based dispatch results (expected) in 2005

	Capacity factor (%)	Dispatch cost (¢/kW-hr)	Direct prod. cost (MS)	Emissions (ktons)			Percent change due to MDPU adders				
				SO_2	NO_x	C	Capacity factor	Dispatch cost	SO_2	NO_x	C
Nuclear	72.5	0.4	530	0	0	0	0%	0%	0%	0%	0%
Coal steam	55.4	8.1	297	148	47	3800	–20%	302%	–24%	–24%	–23%
Peakers	0.2	20.0	18	0	0	10	0%	67%	0%	0%	0%
Oil steam	32	9.0	1271	151	62	7603	–4%	85%	–7%	–11%	0%
Thermal purchase	96.2	8.7	290	3	20	1367	–1%	174%	0%	0%	–1%
Hydro purchase	95.4	2.5	444	0	0	0	0%	0%	0%	0%	0%
Pumped storage	10.3	0.0	6	0	0	0	17%	0%	0%	0%	0%
Storage hydro	28.8	0.0	6	0	0	0	0%	0%	0%	0%	0%
Baseload hydro	62.9	0.0	7	0	0	0	0%	0%	0%	0%	0%
IPPs	63.6	7.6	655	14	20	1226	0%	50%	0%	0%	0%
CC	59.3	7.4	193	7	11	1637	103%	66%	75%	83%	103%
Total existing			3718	3263	160	15,643			–15%	–10%	–2%
IGCC	88.4	4.6	177	7	1	2164	0%	108%	0%	0%	0%
Adv. CC	78.2	5.1	863	0	4	3228	9%	28%	0%	33%	9%
Adv. CT	6.9	8.0	63	0	0	360	590%	17%	0%	0%	567%
Biomast steam	87.5	3.9	59	0	2	0	1%	27%	0%	0%	0%
Refuse	87.6	7.6	31	3	3	77	0%	223%	0%	0%	0%
Total new			1193	10	10	5829			0%	11%	11%
Grand total			4911	333	170	21,472			14%	9%	1%

Source: [3].

peratures we produce more NO_x. This is undesirable and has to be limited. This is done in several ways as reported by Golden of Southern Company Services as per Table 6-7.

6.6.1 Burners Out of Service (BOOS)

Here, the top burner is used as an overflow air port and injects additional secondary air into the furnace, at the cost of fuel. This has the disadvantage that there is capacity reduction of the plant. BOOS is adopted during off-peak seasons mainly to limit the annual NO_x production.

Chapter 3 describes a denitrifier wherein the outlet of the gas turbine reacts with a catalyst like NH3 as it enters the heat recovery generator. Low NO_x burners (along

Table 6-7. Costs of NO_x-removal equipment

NO_x reduction technology	Approximate % reduction: Wall-fired	Approximate % reduction: Tangentially fired	Rough cost, \$/kW-hr
Low NO_x burner (LNB)	50	30	15–25
Overfire air (OFA)	25	N/A	15–25
LNB + OFA	55	35	20–40
Selective catalytic reduction (SCR)	80	80	85
Selective noncatalytic reduction (SNCR)	30	30	11
Burners out of service BOOS)	30	30	0

Source: [4].

with lime injection) also seem to be quite popular. Stage heating is also a useful aid. Here, air temperature is raised in two stages.

NO_x formation and reduction also depend on the type of boiler. The wall-fired boilers have a higher NO_x formation rate (0.5 to 1.5 Lb/Mbtu as reported) than tangentially fired boilers, which have rates reported on average of 1.0 lb/Mbtu. Reduction in NO_x emission with low NO_x burners varies from 50% for wall-fired to 30% for tangentially fired boilers. The low NO_x burners give only a small reduction in NO_x. Catalytic reduction gives reduction up to 80%.

As a result of loss in capacity and also as a result of using costly catalytic reducers, the cost of NO_x reduction is much higher than the cost of SO_2 reduction. MDPU gives the valuation for NO_x at 3.383 \$/lb as against 0.781\$/lb of SO_2.

The 1990 amendment limits the NO_x emission at an average of 0.5 lb/Mbtu varying from plant to plant. Against this, emission for SO_2 is to be brought down from 2.5 lb/Mbtu over a 5 year period to 1.2 lb/Mbtu. The volumes of SO_2 emission are also large compared to volumes of NO_x. Considering all these factors, one can easily see that emission allowances for NO_x available in the market will be costly and will be available in small quantities only.

The NO_x emissions allowed are tied to ozone, O_3, a product of oxygen, generally present in stratosphere at heights above 10 km from the earth. The ozone layers prevent Sun's ultraviolet radiation from reaching the earth. This radiation is unhealthy and is associated with cancer and skin diseases. This O_3 layer varies per season and per location. NO_x emission causes reduction in ozone. Thus, cities are categorized as ozone nonattainment ones where NO_x emissions are highly constrained. Figure 6-1 shows the daily NO_x limits for the Los Angeles Basin as required under Rule 1135 which are more stringent than the national average norms (0.15 against 0.50).

The ozone layer concentration varies between May 1 to September 30 for the northern hemisphere. The NO_x limits in summer have to be kept low there.

6.6.2 NO_x Variation with Load

Partially loaded gas-based generators produce more NO_x than when they are loaded 50% or more. Idling times, start-up times, and so forth are to be watched and aggregat-

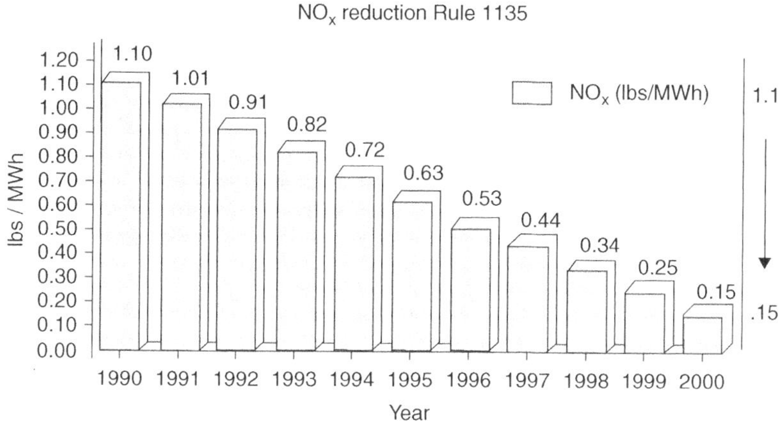

Figure 6-1. Daily NO_x limits for the Los Angeles basin. (From [4], © IEEE 1995.)

ed for checking on the annual allowance for NO_x. Similarly, peaking units mostly small, gas-based generators without the paraphernalia of elaborate denitrifying systems, add substantially to NO_x emission. Southern California Edison Co. has reported a restriction of 200 hours of operations per year on several of their peakers, in order to contain NO_x emissions [1]. Figure 6-2 shows the variation of NO_x emission per load for several differently rated gas-based generators.

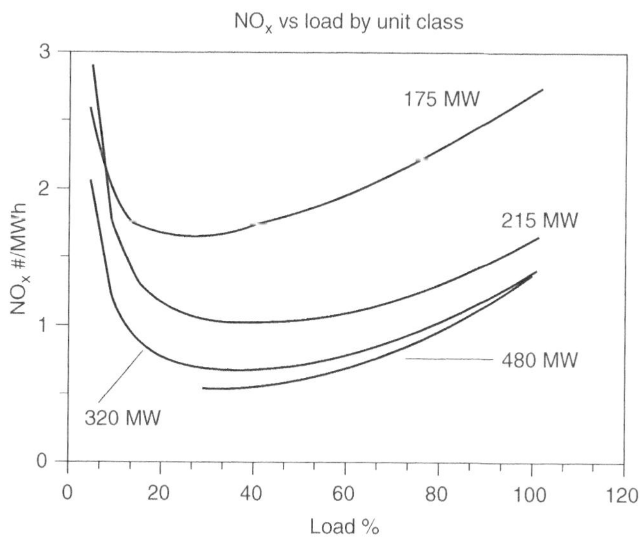

Figure 6-2. Average NO_x curves for Southern California Edison's gas units. (From [4], © IEEE 1995.)

6.7 CONTINUOUS-EMISSION MONITORING SYSTEMS (CEMS)

When we talk of emission constraints in terms of lb/Mbtu, it implies almost continuous monitoring, analyzing, reporting, and storage of data on what is passing through the flue exhausts. Figure 6-3 shows the locations for emission monitoring as well as some readings actually recorded. Please note that as many as six parameters are involved.

A 10% accuracy and 85% availability is expected from CEMS. Typical monitoring intervals are 10 minutes. The data goes to the plant's own control room. This data can help in another way. By working out shadow prices with the help of adders, one can plan the next day's operations, including switching of fuels, allocating units, their loading, and so on.

6.8 THE EUROPEAN SYSTEMS: HELSINKI PROTOCOL ON SO_2 AND SOFIA PROTOCOL ON NO_X

European countries were equally sensitive to the problems of air pollution. Air pollution due to emission from thermal power plants has been addressed. The Helsinki protocol of 1985 first established constraints on SO_2 emission. This was followed by the Sofia protocol of 1988, which established constraints on NO_x emission. The Figure 6-4 gives a comparison between international standards adopted by the European countries.

The 1980 SO_2 level was to be reduced by 30% by the end of 1993. SO_2 is reduced in the coal-fired system by using a wet limestone–gypsum system in flue gas desulphurization (FGD).

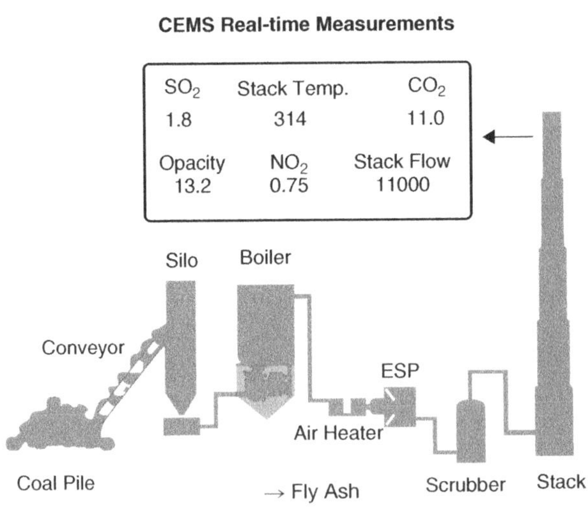

Figure 6-3. Continuous-emission monitoring systems. (From [4], © IEEE 1995.)

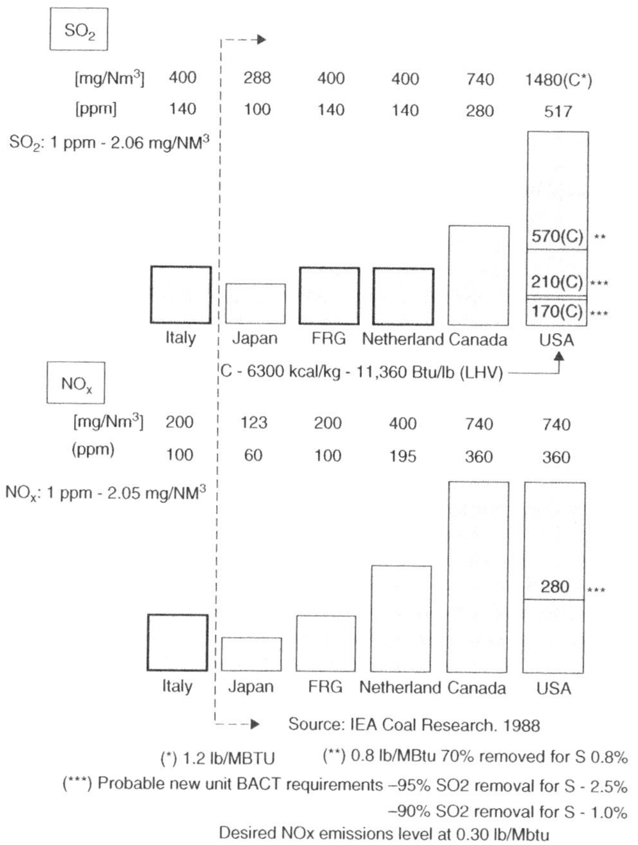

Figure 6-4. Comparison of international standards for new large plants. (From [2], © IEEE 1992.)

A "recovery" process follows desulfurization of flue gas. With limestone and gypsum, it recovers sulfur as sulfur, sulfuric acid, or ammonium sulfate.

NO_x reduction from 1986 levels was to be 30% by the end of 1998. The countries that have joined in this endeavor are known as the 30% Club. For NO_x reduction, selective catalytic reduction is universally used.

6.9 THE JAPANESE EXAMPLE—CITY-WISE AND COMPREHENSIVE

As reported, the Japanese thermal plants were converted from coal to oil fuels when air pollution controls started. This was followed by considerable additions of nuclear plants as well as natural gas plants. With dean coal technologies, the shift is getting redirected to coal again. Emission levels are prescribed by local municipalities.

One notable future of the Japanese system is a combined constraint on emissions (Table 6-8). Thus, if a city has smog on the streets due to air pollution by vehicles, the thermal plants have to further lower their emissions.

6.10 A PLANT RUNNING OUT OF EMISSION ALLOWANCES

A plant may have stopped production because it has exhausted various emission allowances. Although it has capacity to produce, it cannot categorize this as operating reserves. However, it can use it as an emergency reserve. Again, unplanned/undeveloped load shedding is not accepted if one has run out of permits and is forced to cut down production.

6.11 NO_X PERMITS IMPORTANT PLAYERS IN PRICE FIXING OF POWER IN A FREE MARKET

Breipohl et al. [5] project the data for three important northeastern states in the United States to bring out the rising position of NO_x in a free market. In a free market, all types of constraints have a say in the price of electricity power.

The trading licenses in NO_x are small in number. Only 70% of these are available for generation. These available licenses have to cover restraints during summer or low-ozone months. In a worst-case scenario, a market player, who is also a large producer, corners all the tradable NO_x permits and forces smaller units out of production. He can play with the market by fixing his prices, either high or low. A similar situation developed due to transmission congestion in 2000 in the California crisis, when electricity prices shot through the roof and state legislation had to intervene.

It may be noted that in a free market, operating reserves also become tradable commodities.

Air pollution constraints were introduced into the electricity field voluntarily for social and moral reasons. Soon enough, they assumed a larger role, whereby they assumed an important position as optimized economic controllers in the local dispatch programs. As these constraints become stiffer, they can move to a position wherein they can influence pricing of electricity.

Table 6-8. Sample emission limits for Japanese power plants

Plant name	Primary fuel	Service area	Capacity (MW)	Annual SO_x Limit (Tons)	Annual NO_x Limit (Tons)
Osaka	Oil	Urban	624	615	320
Tahagawa 2	Oil	Rural	1200	3020	2100
Nanko	Gas	Urban	1800	None	400

Source: [4].

6.12 AIR POLLUTION BY CARBON DIOXIDE—CO_2

We have already explained how accumulation of CO_2 in the atmosphere affects the environment. Between 1980 and 2020, CO_2 in the air is expected to double from 15.6 ppm to 27.8 ppm. Estimates of the temperature change resulting from this increase vary from 1.5 to 5°C. This can melt polar ice caps and cause havoc all around the world. We will deal with CO_2 emissions in detail in the next two chapters.

APPENDIX 6-1 AMBIENT AIR QUALITY STANDARDS FOR RESIDENTIAL AREAS

	Unit							
	Hour*		8 Hours		Day**		Year	
Pollutant	ppb	$\mu g/m^3$	ppb	$\mu g/m^3$	ppb	$\mu g/m^3$	ppb	$\mu g/m^3$
Sulfur dioxide (SO_2)	170	444	—	—	60	157	30	80
Hydrogen sulfide (H_3S)	140	200	—	—	30	40	6	8
Nitrogen dioxide (NO_2)	100	225	—	—	50	112	30	67
Carbon monoxide (CO)	30000	34000	10000	11500	8000	9000	—	—
Ozone (O_3)	80	157	60	120	—	—	—	—
Ammonia (NH_3)	800##	850	—	—	—	—	140	148
Nonmethane hydrocarbons	1/10 from specified rate in works environment (TLVs) 0.24 ppm for three hours from 6:00–9:00 morning (a.m.)							
Suspended particulate matter (PM-10)	—	—	—	—	—	350	—	90
Dust—fall out matter	—	—	—	—	—	—	7.5 ton/km^3	
Lead	—	—	—	—	—	—	1.5 mg/m^3	
Chlorine##	30 (30 min)	100	—	—	10	30	—	

Notes:

*Average hour should not occur more than twice during the period of 30 days on the same site.

**Daily average (24 hours) should not occur more than once during the year.

##Should not occur more than once per year.

—These standards should apply in residential dominated areas that lie on the border of industrial areas.

APPENDIX 6-2 AMBIENT AIR QUALITY STANDARDS FOR INDUSTRIAL AREAS

	Unit							
	Hour*		8 Hours		Day**		Year	
Pollutant	ppb	μg/m^3	ppb	μg/m^3	ppb	μg/m^3	ppb	μg/m^3
Sulfur dioxide (SO_2)	300	782.5	—	—	200	523.3	65	157
Hydrogen sulfide (H_3S)	—	—	—	—	130	173.3	—	—
Nitrogen dioxide (NO_2)	100	225	—	—	50	112	30	67
Carbon monoxide (CO)	30000	34000	10000	11500	8000	9000	—	—
Ozone (O_3)	80	157	60	120	—	—	—	—
Ammonia (NH_3)	800##	850	—	—	—	—	140	148
Nonmethane hydrocarbons	1/10 from specified rate in works environment (TLVs) 0.24 ppm for three hours from 6:00–9:00 morning (a.m.)							
Suspended particulate matter (PM-10)	—	—	—	—	—	350	—	90
Dust—fallout matter	—	—	—	—	—	—	7.5 ton/km^3	
Lead	—	—	—	—	—	—	1.5 mg/m^3	
Chlorine##	30 (30 min)	100	—	—	10	30	—	

Notes:
*Average hour should not occur more than twice during the period of 30 days on the same site.
**Daily average (24 hours) should not occur more than once during the year.
##Should not occur more than once per year.
—These standards should occur in industrial dominated areas to protect residents permanently.

APPENDIX 6-3 DETAILS ON DESULPHURIZATION PLANTS IN THE UNITED STATES

Number and capacity of fossil-fueled steam-electric generators with environmental equipment, 1996 through 2007

	Flue gas desulfurization (scrubbers)		Particulate collectors		Cooling towers		Total[1]	
Year	Number of generators	Capacity[2] (megawatts)	Number of generators	Capacity[2] (megawatts)	Number of generators	Capacity[2] (megawatts)	Number of generators	Capacity[2] (megawatts)
1996	182	85,842	1.134	352,154	477	166,749	1,299	377,144
1997	183	86,603	1.133	352,068	480	166,836	1,301	377,195
1998	196	87,783	1,130	351,790	474	166,896	1,294	377,112
1999	192	89,666	1,148	353,480	505	175,520	1,343	387,192
2000	192	89,675	1,141	352,727	505	175,520	1,336	386,438
2001	236	97,988	1,273	360,762	616	189,396	1,485	390,821
2002	243	98,673	1,256	359,338	670	200,570	1,522	401,341
2003	246	99,567	1,244	358,009	695	210,928	1,546	409,954
2004	248	101,492	1,217	355,782	732	214,989	1,536	409,769
2005	248	101,648	1,216	355,599	730	217,646	1,535	411,940
2006	NA	NA	NA	NA	NA	NA	NA	NA
2007	NA	NA	NA	NA	NA	NA	NA	NA

[1]Components are not additive since some generators are included in more than one category.
[2]Nameplate capacity.
NA = Not available. Form EIA-767 data collection was suspended in the data year 2006.
Notes: These data are for plants with a fossil-fueled steam-electric capacity of 100 megawatts or more. Data for independent power producer and combined heat and power plants are included beginning with 2001 data. Beginning in 2001, data for plants with combustible renewable steam-electric capacity of 10 megawatts or more were also included. Totals may not equal sum of components because of independent rounding.
Source: Energy Information Administration, Form EIA-767, "Steam-Electric Plane Operation and Design Report."

Average flue gas desulfurization costs, 1996 through 2007

Year	Average overhead and maintenance costs (mills per kilowatthour)[1]	Average installed capital costs (dollars per kilowatt)
1996	1.07	128.00
1997	1.09	129.00
1998	1.12	126.00
1999	1.13	125.00
2000	.96	124.00
2001	1.27	130.80
2002	1.11	124.18
2003	1.23	123.75
2004	1.38	144.64
2005	1.23	141.34
2006	NA	NA
2007	NA	NA

[1]A mill is one tenth of one cent.
NA = Not available. Form EIA-767 data collection was suspended in the data year 2006.
Notes: These data are for plants with a fossil-fueled steam-electric capacity of 100 megawatts or more. Beginning in 2001, data for plants with combustible renewable steam-electric capacity of 10 megawatts or more were also included. Data for independent power producer and combined heat and power plants are included beginning with 2001 data. Totals may not equal sum of components because of independent rounding.
Source: Energy Information Administration, Form EIA-767, "Steam-Electric Plane Operation and Design Report."

REFERENCES

1. T. V. Golden, Potential Impacts of Clear Air Regulation on System Applications. *IEEE Transactions on Power Systems,* May 1995, pp. 647-656.
2. L. Salvaderi, An International Perspective on the Future of Power Generation and Transmission World Wide: The Italian Case, *IEEE Transactions on Energy Conversion,* March 1992, pp. 83–92.
3. J. F. Busch, Jr. and F. L. Krause, Environmental Externality Surcharges in Power System Planning: A Case Study of New England, *IEEE Transactions on Power Systems,* August 1993, pp. 789–795,
4. K. D. Le, J. L. Golden, C. J. Stansberry, R. L. Vice, J. T. Wood, J. Ballance, G. Brown, J. Y. Kamya, E. K. Nielsen, H. Nakajima, M. Ookubo, I. Iyoda, and G. W. Cauley, Potential Impacts of Clean Air Regulations on System Operations, *IEEE Transactions on Power Systems,* May 1995, pp. 647–656.
5. Breipohl, A.M., F. N. Lee, M.-Y. Lin, H. D. Wang, and J.-Y. Chiang, An Adaptive Production Simulation Structure for Operational Planning with Annual Emission/Fuel Constraints, *IEEE Transactions on Power Systems,* August 1994, pp. 1634–1642.

BIBLIOGRAPHY

Abido, M. A., Environmental/Economic Power Dispatch Using Multiobjective Evolutionary Algorithms, *IEEE Transactions on Power Systems,* November 2003, pp. 1529–1537.

Aki, H., T. Oyama, and K. Tsuji, Analysis of Energy Pricing in Urban Energy Service Systems Considering a Multiobjective Problem of Environmental and Economic Impact, *IEEE Transactions on Power Systems,* November 2003, pp. 1275–1282.

Aki, H., Problem of Energy Pricing in Urban Energy Service Systems Considering Multiobjective Problem of Environmental and Economic Impact, *IEEE Transactions on Power Systems,* 2004, pp. 2775–2787.

Borghi, C. A., and P. L. Ribani, MHD-Steam Thermal Power Plant Electrical Stations with Zero Stack Emission, *IEEE Transactions on Energy Conversion,* March 1996, pp. 194–199.

Chao-Ming, H. and H. Yann-Chang, A Novel Approach to Real-Time Economic Emission Power Dispatch, *IEEE Transactions on Power Systems,* February 2003, pp. 288–294.

Conroy, G. H., Industrial Applications and Results of Low NO_x Precalciner Systems, *IEEE Cement Industry Technical Conference 1997, 1998,* pp. 297–318.

El-Keib, A. A., H. Ma, and J. L. Hart, Economic Dispatch in View of the Clean Air Act of 1990, *IEEE Transactions on Power Systems,* May 1994, pp. 972–978.

Gjengedal, T., Emission Constrained Unit-Commitment (ECUC), *IEEE Transactions on Energy Conversion,* March 1996, pp. 132–138.

Gjengedal, T., S. Johansen, and O. Hansen, A Qualitative Approach to Economic-Environment Dispatch-Treatment of Multiple Pollutants, *IEEE Transactions on Energy Conversion,* September 1992, pp. 367–373.

Greer, W. L. and T. L. Matz, Cement Kiln Dust Enforceable Agreement—Yes or No?, *IEEE Cement Industry Technical Conference 1996,* pp. 309–314.

Harju, J. A., M. D. Jensen, E. N. Steadman, J. A. Sorensen, and E. M. O'Leary, The Plains CO_2

Reduction (PCOR) Partnership—Identifying CO_2 Sequestration Opportunities for the Cement Industry in the Central Interior of North America, *IEEE Cement Industry Technical Conference 2006,* 7 pp.

Hobbs, B. F., J. C. Honious, and J. Bluestein, Estimating the Flexibility of Utility Resource Plans: An Application to Natural Gas Cofiring for SO_2 Control, *IEEE Transactions on Power Systems,* February 1994, pp. 167–173.

Horton, J., A. Linero, F. M. Miller, Use of SNCR to Control Emissions of Oxides of Nitrogen from Cement Plants, *IEEE Cement Industry Technical Conference 2006,* 29 pp.

Jakl, F., K. Bakic, and L. Valencic, Combined Use of the Air Monitoring System in Production and Transmission of Electricity, *IEEE Transactions on Power Systems,* August 1997, pp. 1068–1075.

Jepsen, O. L. and S. W. Miller, New Generation of Low NO_x Calciners, *IEEE Cement Industry Technical Conference 1997,* pp. 281–295.

Kamas, J. and J. Keeler, Predictive Emissions Monitoring Systems: A Low-Cost Alternative for Emissions Monitoring. In *IEEE Cement Industry Technical Conference 1995,* pp. 497–509.

Kamiya, M., H. Ikeda, S. Shinohara, and H. Yoshida, Dala-Collection and Transmission System for Automatic Environmental Tests, *IEEE Transactions on Industry Application Magazine,* pp. 39–48.

Lang, T. A., A Perception of the Greenhouse Problem and the Possibilities in the Cement Industry for Action, *IEEE Cement Industry Technical Conference 1993,* pp. 445–465.

Lee, F. N., J. Liao, and A. M. Breipohl, Coordination of SO_2 Emission Allowance Trading, Energy and Spinning Reserve Transactions, and Consumption o Take-or-Pay Fuels, *IEEE Transactions on Power Systems,* August 1994, pp. 1243–1252.

Leppitsch, M. J. and B. F. Hobbs, The Effect of NO_x Regulations on Emissions Dispatch: A Probabilistic Production Costing Analysis, *IEEE Transactions on Power Systems,* November 1996, pp. 1711–1717.

Matsumoto, H., Y. Ohsawa, S. Takahasi, T. Akiyama, and O. Ishiguro, An Expert System for Startup Optimization of Combined Cycle Power Plants Under NO_x Emission Regulation and Machine Life Management, *IEEE Transactions on Energy Conversion,* June 1996, pp. 414–422.

McQueen and D. M. Long, Greenhouse Gas Reduction/Recovery from Portland Cement Operations—A Case Study on the Solicitation Process, System Evaluation, Funding and Status of Demonstration, *IEEE Cement Industry Technical Conference 2006,* 10 pp.

McQueen, A. T., S. J. Bortz, M. S. Hatch, H. J. Buening, D. E. Shore, R. L. Leonard, and E.F. Bouse, Cement Kiln NO_x Control, *IEEE Cement Industry Technical Conference 1993,* pp. 241–261.

Mirolli, M. D., Cementing Kalina Cycle Effectiveness, *IEEE Transactions on Industry Application Magazine,* July 2006, pp. 60–64.

Nisbet, M., The Clean Air Act Amendments: Regulator's Dream, Cement Industry's Nightmare, *IEEE Cement Industry Technical Conference 1995,* pp. 307–326.

Palanichamy, C., N. S. Babu, and C. Nadarajan, Municipal Solid Waste Fueled Power Generation for India, *IEEE Transactions on Energy Conversion,* December 2002, pp. 556–563.

Rahman, S. and A. de Castro, Environmental Impacts of Electricity Generation: A Global Perspective, *IEEE Transactions on Energy Conversion,* June 1995, pp. 307–314.

Rau, N. S. and S. T. Adelman, Operating Strategies Under Emission Constraints, *IEEE Transactions on Power Systems,* August 1995, pp. 1585–1591.

Ravishankar, N. R., Relevance to Electrical Industry, Carbon Credits, IEEMA Journal, February 2007, pp. 54–58.

Roy, S., Cost-effectiveness of Emission Control at Fossil-Fuel Units for Different Cumulative Load Patterns, *IEEE Transactions on Power Systems,* February 1997, pp. 321–328.

Shearman, E., Energy Responses—How Energies are Coping with New Environments, Power Engineering Review, November 1994, pp. 13–17.

Shimoda, M., A. Sugano, T. Kimura, Y. Watanabe, and K. Ishiyama, Prediction Method of Unburnt Carbon for Coal Fired Utility Boiler Using Image Processing Technique of Combustion Flame, *IEEE Transactions on Energy Conversion,* December 1990, pp. 640–645.

Srinivasan D. and A. G. B. Tettamanzi, An Evolutionary Algorithm for Evaluation of Emission Compliance Options in View of the Clean Air Act Amendments, *IEEE Transactions on Power Systems,* February 1997, pp. 336–341.

Steuch, H. E., J. Hille, W. H. Sun, M. J. Bisnett, and D. K. Kirk, Reduction of NO_x Emissions from a Dry Process Preheater Kiln with Calciner Through the Use of the Urea Based SNCR Process, *IEEE Cement Industry Technical Conference 1995,* pp. 465–483.

Tsuchiya, Y. and T. Sakakibara, A Model for Evaluating the Reverse-Power-Flow Energy of Refuse-Fired Power Generation in Japan, *IEEE Transactions on Energy Conversion,* December 1993, pp. 681–687.

Ula, A. H. M. S., Global Warming and Electric Power Generation: What is the connection? *IEEE Transactions on Energy Conversion,* December 1991, pp. 599–604.

Venkatesh, P., R. Gnanadass, and N. P. Padhy, Comparison and Application of Evolutionary Programming Techniques to Combined Economic Emission Dispatch with Line Flow Constraints, *IEEE Transactions on Power Systems,* May 2003, pp. 688–697.

Vickers, V.L., W. J. Hobbs, S. Vemuri, and D. L. Todd, Fuel Resource Scheduling with Emission Constraints, *IEEE Transactions on Power Systems,* August 1994, pp. 1531–1538.

Walsh, M., V. C. Ramesh, and K. Ghosh, Impediments to Markets for SO_2 Emission Allowances, *IEEE Transactions on Power Systems,* May 1996, pp. 996–1001.

7

ENVIRONMENTAL CONSTRAINTS IN THERMAL POWER GENERATION—CARBON AND THE KYOTO PROPOSALS

7.1 CONTINUING GROWTH OF CO_2 IN THE AIR

In the preindustrial era, CO_2 contamination in the atmosphere was about 250 ppm. Presently, it is estimated at about 350 ppm. As per the Energy Information Administration (EIA) 2008 release, it will reach 450 ppm by 2030. If this goes on unchecked, it is likely to reach 560 ppm by the middle of the 21st century. The spurt in the growth of CO_2 is attributed to growing use of fossil fuels—coal, oils, and gas. The major consumers of these fossil fuels are electricity generators and automobiles. Electricity generation tops the list.

Amongst the fossil fuels, coal, since it is abundant, has been the top favorite of electricity generators. For reasons explained earlier, liquid fuels have been under watch. Gas has become a favorite because of higher thermal efficiencies obtained with it. Gas has relatively less harmful emissions. However, unlike coal, oils and gases have volatile prices.

Let us first see in what proportion various fuels—commonly used as well as synthetic ones—produce CO_2.

7.2 CO_2 FROM DIFFERENT FUELS

CO_2 emissions from various fuels is shown in Table 7-1. Appendix 7-4 gives energy consumption by sector and sources, projected over the period 2005–2020 in the United States. It also projects GDP figures. The major polluters are clearly the electricity utilities.

Electricity Power Generation. By Digambar M. Tagare

Copyright © 2011 the Institute of Electrical and Electronics Engineers, Inc.

Table 7-1. Carbon dioxide emissions from direct combustion of various fuels

Fuel	CO_2 emission rate (kg C/10^9J)	Ratio relative to methane
Methane	13.5	1
Ethane	15.5	1.15
Propane	16.3	1.21
Butane	16.8	1.24
Gasoline	18.9	1.4
Diesel oil	19.7	1.46
No. 6 fuel oil	20.0	1.48
Bituminous coal	23.8	1.73
Subbituminous coal	25.3	1.87

Source: [1].

7.3 CO_2 EMISSION BY FUEL TYPE

Figure 7-1 and Table 7-2 show world energy-related carbon dioxide emissions by fuel type, actual and projected over the period 1990–2030.

7.4 COAL HAS THE HIGHEST RATE OF GROWTH AMONG ENERGY SUPPLIERS

The rate of growth in use of fuels is:

- Liquid fuels (as judged by CO_2 emission) is (14.9 – 9.0%)/9 or 65.55%
- Natural gas is (8.7 – 4)/4 = 117.5%
- Coal is (18.8 – 8.3)/83 = 126.5%

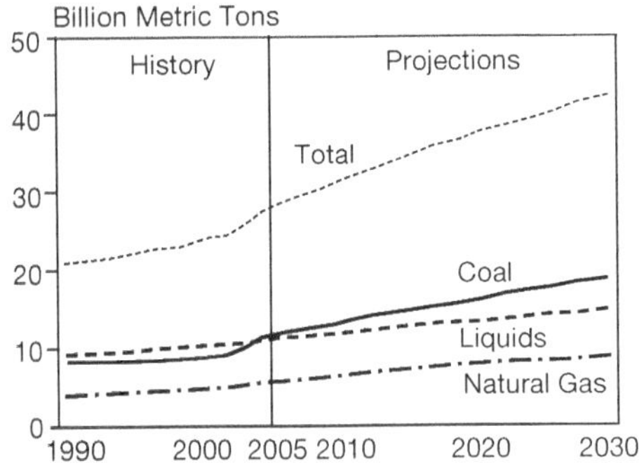

Figure 7-1. World energy-related carbon dioxide emissions by fuel type, 1990–2000. *Source:* Energy Information Administration (EIA), *International Energy Annual 2005* (June– October 2007), www.sia.doc.gov/. Projections: EIA, World Energy Projections Plus (2008).

Table 7-2. Billion metric tons of CO_2 emissions from various fuels

	Liquids	Natural Gas	Coal	Total
1990	9.0	4.0	8.3	21.2
1995	9.3	4.3	8.2	21.8
2000	10.1	4.8	8.7	23.6
2005	11.0	5.7	11.4	28.1
2006	11.2	5.8	11.7	28.6
2007	11.3	6.0	12.0	29.3
2008	11.4	6.1	12.3	29.9
2009	11.6	6.2	12.7	30.5
2010	11.7	6.3	13.0	31.1
2015	12.6	7.1	14.7	34.3
2020	13.3	7.8	15.9	37.0
2025	14.0	8.2	17.3	39.6
2026	14.2	8.3	17.6	40.1
2027	14.4	8.4	17.9	40.7
2028	14.5	8.5	18.2	41.2
2029	14.7	8.6	18.5	41.8
2030	14.9	8.7	18.8	42.3

Even with a strong bias and a campaign against liquid fuels, these have a positive growth rate.

Coal is the universal favorite, as explained in Chapter 3. In spite of a campaign for reducing CO_2, coal, the largest contributor, continues to grow in use and coal produces the highest amount of CO_2 per unit of electrical energy. A percentage reduction in use of coal represents a much larger percentage reduction in CO_2 emission as compared to other energy sources, like natural gas, for example. Capture and storage of CO_2 is being experimented on in various countries. However, capture and storage is the only aim for an industrial unit as and when it is developed. There is no other byproduct that can add to GDP. At present levels of financing reduction in CO_2 emission rates, this activity will find it difficult to take root.

There is a sidebar to this scenario that may help us from becoming over-alarmed, as explained below.

7.5 EARTH'S OCEANS AND SEAS ABSORB CO_2

As per the EIA report referred to earlier, the earth's oceans and the soils absorb CO_2. While energy-related CO_2 emissions have increased at the rate of 2.1% per year, CO_2 concentrations have increased at the rate of 0.6% per year only. The oceans and the soils have absorbed 42% of the emissions, at the current rates of emission, leaving 58% back. However, as the earth temperatures are expected to rise, it is questionable whether they will absorb the same proportionate amount. We should be concerned but not overalarmed.

7.6 DEVELOPMENTS ON THE FRONT OF REDUCTION IN GREENHOUSE GAS EMISSIONS

Leaders and experts from different nations have taken the lead on checking atmospheric pollution. The International Panel on Climate Change (IPCC) was formed jointly by the World Meteorological Organization (WMO) and United Nations Environment Program (UNEP) in 1988. After IPCC released its first report on climate change in 1990, this issue got a lot of attention. It was discussed at the United Nations conference on environment and development in 1992 in Rio de Janeiro. The United Nations framework convention on climate change (UNFCCC) came into existence in 1994. The cornerstone of this convention was common but differential responsibilities (CBDR) between the developed countries and developing countries. The UNFCCC then came out with a protocol in 1997 at a convention in Kyoto. It has become known as the Kyoto Protocol. A meeting in Montreal, Canada dealt mainly with emission caused by the automobiles. Automobiles are the biggest consumers of liquid fossil fuels.

Many countries, particularly the developing countries, were not happy with the initial protocol. Their views were discussed at subsequent annual meetings of UNFCCC. A largely acceptable accord was presented at a conference in Merrakech in 2005. This accord, the Merrakech Accord, was finally accepted in 2007. All new proposals are now formulated under this accord.

7.7 KYOTO PROPOSALS

The Kyoto proposals are a significant milestone in the evolution of electricity power generation because they:

1. Turn the focus on noncarbon sources of energy, particularly the renewable energies.
2. Aim at controlling and reducing the use of thermal power production.
3. Try to equalize electricity dispensed per head between developed and the developing countries.
4. Emphasize on using electricity more efficiently.

The main points in the protocols are:

1. The protocols identified the polluting or greenhouse gases and the sector/source categories. See Appendix 7-1 on UNFCCC-1998 at the end of this chapter.
2. They obtained quantified emission limitations or reduction commitments on greenhouse gases by some 39 countries.
3. Various policies and measures to limit these emissions were recommended.
4. Various means were recommended for containment of CO_2.

The key means of reducing CO_2 emissions are:

1. Reduction in growth of energy demand via energy efficient architecture, lighting, home appliances, industrial machinery, and automobiles
2. Increase in nuclear generation
3. Emphasis on renewable energy resources, including hydro power
4. Emphasis on biofuels for transport
5. Carbon capture and storage
6. Anthropogenic sequestration—enlarging the plant world

7.8 CLAUSE 1 OF KYOTO PROTOCOL OF 1998

This clause states [2] that countries should:

> Implement and/or further elaborate policies and measures in accordance with its national circumstances, such as:
>
> - Enhancement of energy efficiency in relevant sectors of the national economy
> - Protection and enhancement of sinks and reservoirs of greenhouse gases not controlled by the Montreal Protocol, taking into account its commitments under relevant international environmental agreements; promotion of sustainable forest management practices, forestation and reforestation
> - Promotion of sustainable forms of agriculture in light of climate change considerations
> - Research on, and promotion, development and increased use of, new and renewable forms of energy, of carbon dioxide sequestration technologies and of advanced and innovative environmentally sound technologies
> - Progressive reduction or phasing out of market imperfections, fiscal incentives, tax and duty exemptions and subsidies in all greenhouse gas emitting sectors that run counter to the objective of the Convention and application of market instruments
> - Encouragement of appropriate reforms in relevant sectors aimed at promoting policies and measures which limit or reduce emissions of greenhouse gases not controlled by the Montreal Protocol
> - Measures to limit and/or reduce emissions of greenhouse gases not controlled by the Montreal Protocol in the transport sector
> - Limitation and/or reduction of methane emissions through recovery and use in waste management, as well as in the production, transport and distribution of energy

7.9 ORIGINAL KYOTO PROPOSALS

A 1990 level of emission of greenhouse fases (GHGs), was chosen as a reference level and a proposal was drafted that required the countries that accepted responsibilities under it to reduce their emission of GHGs by 5% below this level over a period from 2008–2012.

This arrangement was favorable to developed countries which, were already experiencing high growth in energy consumption. As a result, their 1990 levels were comfortably high. The percentage growth required to be made up between 1990 and 2012 was relatively low. They could make it up with renewable energy sources and nuclear power. They had the investment potential and technical infrastructure to achieve this.

The case with developing countries was exactly the opposite. Many countries, particularly India and China, with large populations, did not accept this. The United States also abstained. Many modalities remained to be detailed and finalized. With a view to this and to accommodating more countries, UNFCCC met several times and finalized the Kyoto Protocol by 2005. After a meeting at Marakesh, it finalized a broadly accepted accord by 2007.

Table 7-3 lists the parties to this protocol and their subsequent reduction levels.

7.10 PROPOSALS FOR PARTIES TO THE 2007 PROTOCOL

Parties were to undertake domestic policies and measures to reduce GHG emissions and to enhance removal by sinks. These policies and measures should not affect adversely the developing countries. On the other hand, parties were to provide financial resources and technical guidance towards developing the emission reduction policies of the developing countries, including emission trading, joint implementation, and a clean development mechanism (CDM).

Table 7-3. Parties to the Kyoto Protocol of 2007

	Emissions target (expressed in relation to emissions in the base year or period*)
Austria, Belgium, Bulgaria, Czech Republic, Denmark, Estonia, European Community, Finland, France, Germany, Greece, Ireland, Italy, Latvia, Liechtenstein, Lithuania, Luxembourg, Monaco, Netherlands, Portugal, Romania, Slovakia, Slovenia, Spain, Sweden, Switzerland, United Kingdom of Great Britain and Northern Ireland	–8%
United States of America**	–7%
Canada, Hungary, Japan, Poland	–6%
Croatia	–5%
New Zealand, Russian Federation, Ukraine	0
Norway	+1%
Australia**	+8%
Iceland	+10%

*This base year is flexible in the case of countries with economies in transition.

**Countries that have declared their intention not to ratify the Protocol.

7.10.1 Emission Trading with ERUs and LULUCF

Parties were allocated emission reduction units (ERUs) for achieving emission reductions below the target levels and also with regard to forestry. These could be traded or banked among themselves. Land use, land use change and forestry (LULUCF) was to help in absorbing CO_2 and converting it into vegetation. On the other hand, solid wastes and stagnant waters emit methane. Thus, parties could gain credit allocations under LULUCF. This sort of trading created a market force promoting emission control measures. It need not be done under pressure from rules and regulation only.

7.10.2 Joint Implementation

Parties could implicate their emissions on their own or jointly with others. Party no. 1, which helps party no. 2 to implement its program, could receive the emission reduction units for itself. The process was divided into Track 1 and Track 2, with or without joint participation. There are differences in requirements for acceptance, inspections, and approval between the two tracks.

7.10.3 Clean Development Mechanism (CDM)

This mechanism covers mainly countries not party to the Protocol. Later on, it was extended to small-scale industries and private parties. These entities got more liberal conditions for getting certified emission reduction units (CERs). CDM credits could be generated from emission reduction in projects as well as from forestation and reforestation projects. Separate rules cover registration, validation and verification by designated authorities. This will be covered in details in the next chapter. Parties could buy out CERs generated by nonparties; this helped both.

7.10.4 Certified Emission Reductions (CERs)

Participants under a CDM can earn CER units. Each CER unit stands for reductions in emission of CO_2 by one ton. These units can be traded or sold. This creates additional resources for participants in the developing countries. At the same time, it creates a relief system for participants in countries party to the Protocol, in that they can buy these CERs to cover their transgressions.

7.10.5 CER to the Rescue of Protocol Parties

Only the developing countries and the private sector (noncommitted) were allowed to produce CERs. The CERs were tradable and bankable.

On the other hand, the committed countries had to honor their commitments, spread out by the respective countries among their energy producers. Should a producer fail to meet these commitments, he had a viable alternative to reducing or closing down his production. He could buy CERs at the prevailing market rate to cover his deficit. Thus, CERs provided an escape route for the committed producers, while at the same time encouraging CO_2 reduction by underdeveloped countries and private entities therein.

One CER unit is priced at one ton of CO_2 removed. NO_x and SO have dedicated metering equipment which continuously measure the percentage of these gases in the exhausts flues of the energy related sources. This is not the case with CO_2 and methane; their emissions have a universal base. Instead, the CO_2 measure is indirectly worked out from the electricity production as well as electricity consumption. One may ask why we include electric consumption. This is because savings in electric consumption by improving efficiencies also helps reduce CO_2 production.

The workout of CO_2 from the data on electrical energy is an important aspect of electricity power generation. We will deal with it in the next chapter.

7.10.6 Passage of the CDM Proposal

The interested party has to prepare a project report on prescribed forms. These forms for different projects have been standardized and are put on the Web. This project report has to be validated by a designated carbon management services group. This form goes to the designated operational entity (DOE) in that region. The DOE scrutinizes it and, if satisfied with answers and corrections asked for, asks the party to pay registration fees. This then goes to the secretariat. The secretariat scrutinizes it and, if found OK, puts it on a website along with an announcement for public comments as well as for comments by the stakeholders to the project. The period allocated for this (say, 8 weeks) is also put up on the Web. If the project sails through, it goes to a working group in the secretariat. The working group refers this to experts from an experts committee formed for this subject. After approval from this committee, the project goes to the executive board. After following a set process the executive board will announce its approval. The applicant party then becomes eligible for issuing CERs.

This takes quite some time. Once approved and registered, the permission issued to the party issuing CERs remains valid for a period of some eight years, termed the first track period.

7.11 PROJECT REPORT NEEDS

7.11.1 Eligibility Criteria

Clause no. 3.1 of the Kyoto Protocol states that:

- The party must be a party to the Kyoto Protocol.
- The party's initial assigned amount must be recorded in the CAD
- The Party's national registry must be in compliance with the requirements under Kyoto Protocol.

There are further requirements on compliance to the national GHG system and registry with Kyoto Protocol articles. The party has to declare initial stocks and yearly consumption of carbon.

7.11.2 Additionality Factor

Aditionallity is an important aspect for applying and getting registered for CERs. An application for approval under the CDM will be rejected if it does not score on this point. Basically, the project proposed under the CDM should be an additional project. It must not be just an alternative on a different scale. In other words, alternatives to the proposed project should be verified and the additional nature of the proposed project should be established.

Additionality under a CDM project has to be proved correctly. Over the last two years, the CDM-EB rejected 8–9% of the projects submitted during 2007. It gave automatic approval to only 57% of the projects it received.

The CDM–CER process is expected to give a big push to electricity production in the developing countries.

Certified emission reduction units, until the recent economic slowdown all over the world, were a high-priced commodity with a keen demand from developed countries. A CER unit was traded at about 10 Euros per unit. This has encouraged emission reductions through reducing losses and adding carbonless electricity generation.

7.12 AN ILLUSTRATIVE VALIDATION REPORT

We give below an extract of a validation report for a project seeking authority to issue CER certificates in Brazil, BT Geradora de Energia Electrica SA, Ferradura Small Hydro Power Plant. The project aims at the substitution of electricity from fossil power plants in the grid. The selected baseline methodology is eligible for the relevant project category and is applicable to the project being considered. The application of the baseline methodology and the discussion and determination of the chosen baseline is transparent and realistic.

The generated electricity is fed into the regional grid, in this case the Rio Grande do Sul South–Southeast–Midwest interconnected subsystem of the Brazilian grid. The contracts and licenses concerning the operation of the plant and the supply of electricity to the grid therefore could be presented during the on-site visit and have been submitted via different e-mails to the validator.

An important step when assessing the baseline approach is to prove that the project itself does not represent the baseline scenario. To demonstrate this, the Executive Board established on its sixteenth meeting the "Tool for the demonstration and assessment of additionality." The project uses that tool for demonstrating its additionality, although it would not be completely necessary for small-scale projects.

In order to demonstrate the need for the CDM, the project owner and developer explained the difficulties in the Brazilian finance sector for project financing and the existing institutional barriers for such projects. The difficulties are due to insufficient or missing financing options from banks. In order to get loans, the evidence of high guarantees is necessary.

Before getting informed about the Kyoto Protocol, BT Geradora de Energia Elétrica SA tried to find other possibilities to finance the project, but financing of the project proved to be impossible. This could be demonstrated by the project developer via a document showing the refusal of a bank credit in September 2001. So without the possibilities of CDM projects under the Kyoto Protocol and the possibility to use the revenues from selling electricity for financing the project, a decision to go ahead with the project was not possible.

Moreover, the official statement of the government that the CDM should also be accepted for PROFINA projects clearly proves that also the Brazilian authorities are also recognizing the existence of barriers to such projects, even with the existence of subsidies. So it could be demonstrated transparently and retraceably that the project was not a business-as-usual one.

Below, we provide an illustrative workout for the project.

7.13 WORKOUT FOR EMISSION FACTORS AND EMISSIONS FOR HYDRO AND WIND ENERGY INSTALLATION

Project information	Project A	Project B
Types	Hydro station	Wind farm
Size	5 MW (small-sclae according to CDM criteria)	100 MW (large-scale accoding to CDM criteria)
Projected generation (net)	17,500 MW-hr/yr	912,600 MW-hr/yr
Commissioning year	2009	2009
Year of CDM registration	2008	2008
Grid	NEWNE	Southern
CDM methodology	AMS-I.D/Version 13	ACM0002/Version 07
Baseline Emission Factor Calculation		
Calculation method	Weighted average	Combined margin
Data vintage for projection of emission reductions	2007–2008 (most recent available at time of PDD validation)	For OM: 2005–2006, 2006–2007, 2007–2008 (most recent 3 years available at time of PDD validation) For BM: 2007–2008
Data vintage for verification of emission reductions	Actual year of generation, i.e., 2008–2009, 2009–2010 etc. (emission factor fixed ex post)	Same as for projection (emission factor fixed ex ante)
Accounting of imports	Not mandatory, but done	Mandatory

Project information	Project A	Project B
Weights for combined margin	Operating margin, 50% Build margin, 50%	Operating margin, 75% Build marging, 25% (default for intermittent sources)
Emission Reduction Calculations		
Values in tCO_2/MW-hr	0.81, Weighted average	1.00, Operating margin 0.7, Build marging 0.93, Combined margin
Projected emission reductions	14,125 tons CO_2 per year	290,625 ton CO_2 per year
Actual emission reductions	Monitored net generation × monitored weighted average	Monitored net generation × fixed combined margin

7.14 OPEN SKIES DIVIDED IN TONS OF CO_2 PER NATION

Regional and areawise baselines are established on the basis of immediate past thermal power production. These establish a comparator in tons of CO_2 per MW-hr of electricity produced.

As per directions coming out of the MEF declaration of July 2009, the open skies are drawn in terms of CO_2, and tonnage of CO_2 allowed to be produced is to be allocated by country. There are no specific agreements on this so far. However, this will allow the regional authorities to fix emission norms and maximum CO_2 emission per plant. The plant/region gets penalized when the norms are exceeded. They will also get tradable credit points if they improve on this norm. This affects thermal power generation in unit commitments in the evolving electricity power market. It will affect their planning on future expansion and raising capital for this.

The baseline parameters are briefly discussed in the following sections.

7.15 AN EXAMPLE OF BASELINE AND EMISSION REDUCTIONS

See Table 7-4. Please note that these factors are specific with respect to location, period, and type of facility registered under a CDM. Determining these factors is a complex requirement. A number of approved baseline and monitoring methodologies have been published. As many as 56 such large-scale methodologies have been approved. There are a further 14 approved consolidated methodologies. We will study the tool to calculate the emission factor for an electricity system in the following section.

Table 7-4. Baseline emission factors, and baseline reductions, Kemphole Mini Hydel Scheme

Year	October 2003–March 2004	April 2004–March 2005	April 2005–March 2008	April 2006–March 2007	April 2007–March 2008	April 2008–March 2009	April 2009–March 2010		
On-site project emission reductions		—	—	—	—	—	—		
No. of units replaced in the grid, millions	6.7	41.0	52.0	52.0	52.0	52.0	52.0		
Emission factor considered, tons CO_2/MW-hr	0.814	0.814	0.814	0.814	0.814	0.814	0.814		
Carbon emission reductions in a year (tons of CO_2)	5,413.4	33,373.5	42,328.9	42,328.9	42,328.9	42,328.9	42,328.9		
Baseline emissions (tons of CO_2)	5,413.4	33,373.5	42,328.9	42,328.9	42,328.9	42,328.9	42,328.9		
Project emissions (tons of CO_2)	—	—	—	—	—	—	—	Total	Average
Emission reductions (tons of CO_2)	5,413.4	33,373.5	42,328.9	42,328.9	42,328.9	42,328.9	42,328.9	250,426.9	35,775.3

7.16 METHODOLOGICAL TOOLS TO CALCULATE THE BASELINE AND EMISSION FACTOR

We have to convert electricity saved, either through more efficient use or through its creation from sources other than thermal, into equivalent tons of CO_2 reduced from emission by a thermal energy source based on carbon. We require a baseline against which to calculate and an emission factor to convert MW-hrs into tons of CO_2. The following section based on the Kemphole Mini Hydel Scheme illustrates this.

7.17 TOOL TO CALCULATE THE EMISSION FACTOR FOR AN ELECTRICITY SYSTEM

There are three margins determined:

1. OM is operating margin, consisting of a cohort of power plants that reflect the existing power plants whose electricity would be affected by the proposed CDM project activity.
2. BM is build margin, consisting of a cohort of power plants whose construction would be affected by the proposed CDM activity.
3. CM or combined margin combines the results of the OM and BM.

A project electricity system is defined by the spatial extent of power plants that are physically connected through transmission and distribution lines to the project activity, that is, the renewable-power-plant location or the consumers whose electricity is being saved, and that can be dispensed with without causing significant transmission constraints.

Significant transmission constraints exists if

1. In a system with spot markets for electricity there are differences in electricity prices (without transmission and distribution costs) of more than 5% between system during 60% or more of the hours of the year.
2. Transmission lines are operated at 90% or more at rated capacity during 90% or more of the hours of the year.

Electricity transfers from connected electricity systems to the project electricity system, defined as electricity imports and electricity transfers to connected electricity systems, are defined as electricity exports.

There are four operating methods—the simple OM, simple adjusted OM, dispatch data analysis OM, and average OM.

7.18 SIMPLE OPERATING MARGIN (OM)

Data required is a three-year generation weighted average, based on the most recent data available at the time of submission of the CDM-PDE for validation.

Emission factor is calculated as generation weighted average CO_2 emission per unit net electricity generation (tons CO_2/MW-hr) of all generating plants serving the system, but not including low-cost, must-run power plants. It may be calculated based on (see Table 7-5):

1. Data on fuel consumption and net electricity generation of each power plant (option A)
2. Data on net electricity generation, the average efficiency of each generator unit, and the fuel types used in each power unit (option B)
3. Data on totals of net electricity generation, fuel types, and total fuel consumption (option C)

7.19 INCENTIVES FOR EMISSION REDUCTION

For the underdeveloped countries, the goals set up were reduction or stabilization of CO_2 emissions by a given year. The main economic incentive was carbon credits that could be traded and were in demand all over the world.

The developed countries embarked on a different route. Rather than limiting CO_2 emissions, they stipulated the percentage participation of renewable energy in the total generation and sale of electricity to be achieved by a targeted year. Hydroelectricity, a major contributor to renewable energy, was already established. It was removed from the list of renewable energies. So was natural gas, even with its relatively low CO_2 emission. Initially, only wind generation and dedicated closed-loop biomass generators formed renewable energy sources [3]. Later on, it was extended to include geothermal, solar energy and landfill gas. Nuclear energy, though not an atmospheric pollutant, is not included in the incentives for a nonpolluting renewable energy source.

In the United States, there were three types of incentives:

1. Renewable Energy Production Credit (REPC). This extended tax exemptions to the renewable energy generator. The initial production tax credit was 1.5

Table 7-5. Weighted average emission factor, simple operating margin (OM), build margin (BM), and combined margin (CM) of all grids for FY 2007–2008 (intergrid and cross-border electricity transfers included), in tons CO_2/MW-hr

Grid	Average	OM	BM	CM
NEWNE	0.81	1.00	0.60	0.80
Southern	0.72	0.99	0.71	0.85
India	0.79	1.00	0.63	0.81

Notes: Average is the average emission of all stations in the grid, weighted by net generation. OM is the average emission from all stations excluding the low cost/must run sources. BM is the average emission of the 20% (by net generation) most recent capacity addition in the grid. CM is a weighted average of the OM and BM (here weighted 50:50).

cents/kW-Hr of "green" energy produced. This was limited to an initial period of 10 years from start of their operations. This incentive increased in absolute value to accommodate effects of inflation. There was a further tax credit of up to 10% on capital investment of the project. The REPC policy was floated by and adopted by the federal government. It has been periodically renewed.

2. Renewable Energy Portfolio Standards (REPS). Various state governments as well as some European states adopted what is known as renewable energy portfolio standards or REPS. Under the REPS, a state requires that a certain portion of electricity generated or sold must come from renewable energy sources by a stipulated year. In the United States, these standards varied from 5% to 20% by deadlines ranging from the years 2010 to 2020. The European Union fixed targets by various nations at 12% of total energy and 22% of all electricity to come from renewables by the year 2010. Any shortfall from the target could be made by buying tradable energy credits (TREC), equivalent to carbon credits from the developing countries.
3. Tradable Energy Credit (TREC). On the cash incentive side, the renewable energy generator was to sell his output to a utility at a contracted price or at a ruling market price. In addition, he could get a tradable renewable credit of one unit for 1 kW-Hr generated. He could sell his TREC.

On the tax incentive side, the renewable energy generator has a myriad of incentives.

On the operating side, the renewable energy generator has exemptions on sales tax of his electricity, as well as on the purchase tax of his requirements, at the state level.

On the yearly income side, he is allowed an accelerated depreciation rate. Thus, if a power generator is allowed a life of 25 years, in this case depreciation allowance is based on a life of 5 years. Capital investment is financed through participating equity capital, but to a larger extent through borrowed capital as loans. Renewables are uneconomical, in that they cannot stand the rigors of the open market and compete. As such, these are risky projects, for which commercial bank will charge a higher rate of interest than normal.

A generator is helped in two ways:

1. The Government guarantees the return of this loan, that is, it provides the security cover. With this, the bank loan rates come down.
2. The Government itself offers a limited starting loan at low interest.

The interest portions on these loans are tax exempt for renewables.

Property sales tax concessions play an important role. Property or property use rights over the contracted period form a sizeable chunk of the investment portfolio. Exemption of sales tax for this, a large amount at the beginning year, when the production at start is low, goes a long way to help. There is curious angle to acquiring "property" for an offshore wind power installation. Only a royalty payment might be made.

Table 7-6. Overview of possible incentive programs

Incentive category	Incentive	Brief description
Customer choice incentives	Renewables portfolio standard (RPS)	TRECs are allocated to renewable energy appliers for kW-hr generated and may be used to fulfill the RPS.
		Tradable renewable energy credits (TREC) Utility green pricing program
Direct cash incentive	Fixed tariffs	Utility providers are required to purchase renewable energy as fixed prices, set by some regulating agency.
	Direct production incentive	Renewable generators are given cash payments based upon the amount of electricity generated by the facility.
	Direct investment incentive	A generator receives cash for investing in renewable energy much like a grant.
Indirect cash incentive	Production tax credit	The developer receives an annual tax credit linked to the amount of electricity produced.
	Investment tax credit	The developer receives a one-time tax credit for renewable energy investments.
	State and local sales tax reduction	Reduced sales tax on the components of a renewable energy facility reduces the installation and overall levelized cost of the project.
	Property tax reduction	This reduces the overall cost of land for a renewable energy installation.
	Accelerated depreciation schedule	This allows companies to claim the loss of asset value as a noncash expense which may be deducted from taxable income and thus decrease annual income tax.
Low-cost debt financing	Government subsidized loans	The lower interest rates of these loans help project developer finance their projects.
	Project loan guarantees	The government guarantees that a loan will be repaid to the lender, consequently making it easier to obtain project financing.

Source: [4].

With all these tax incentives, the investor may not be able to fully absorb them all and might have to pay an alternative minimum tax (AMT).

Table 7-6 gives an overview of possible incentive programs. The goal in thermal power production is to return to 1990 levels of CO_2 emission for whatever maximum MW-hr they can generate and sell in a competitive market by 2050.

APPENDIX 7-1 DEFAULT EFFICIENCY FACTORS FOR POWER PLANTS

	Old (before 2000)	New (after 2000)
Coal		
Subcritical	37%	39%
Supercritical		45%
Ultrasupercritical		50%
IGCC		50%
FBS	35.50%	
CFBS	36.50%	40%
PFBS		41.50%
Oil		
Steam turbine	37.5	39%
Open cycle	30%	39.50%
Combined cycle	46%	46%
Natural Gas		
Steam turbine	37.50%	37.50%
Open cycle	30%	39.50%
Combined cycle	46%	60%

REFERENCES

1. A. H. M. S. Ula, Global Warming and Electric Power Generation: What is the Connection? *IEEE Transactions on Energy Conversion,* December 1991, pp. 599–604.
2. Kyoto Protocol to the United Nations Framework Convention on Climate Change (UNFCCC), United Nations Publications, 1998. (http://cdm.unfccc.int/about/index.html)
3. K. Palmer and D. Burtraw, Cost-Effectiveness of Renewable Electricity Policies, *Energy Economics*, Volume 27, Number 6, November 2005, pp. 873–894.
4. J. G. Vining and A. Muetze, Economic Factors and Incentives for Ocean Wave Energy Conversion, *IEEE Transactions on Industrial Applications,* Volume 45, March/April 2009, pp. 547–554.

BIBLIOGRAPHY

Aki, H., T. Oyama, and K. Tsuji, Analysis of Energy Pricing in Urban Energy Service Systems Considering a Multiobjective Problem of Environmental and Economic Impact, November 2003, pp. 1275–1282.

Bakia, J. F., Combined Use of the Air Monitoring EIA International Energy Outlook 2008—What Will it Take to Stabilize CO_2 p. 4.

Government of India, Clear Development Mechanism—Central Electricity Ministry of Power, Database CnCO_2 Emission for All Grid Connected Power Stations in India, 22 pp.

Le, K. D., J. L. Golden, C. J. Stansberry, R. L. Vice, J. T. Wood, J. Ballance, G. Brown, J. Y. Kamya, E. K. Nielsen, H. Nakajima, M. Ookubo, I. Iyoda, and G. W. Cauley, Kyoto Manual 3.1. The Eligibility Criteria, May 1995, pp. 64–656.

Methodological Tool for the Demonstration and Assessment of Additionality, E339, Annex 13, p. 10.

Mozumder, P. and A. Marathe, Gains from an Integrated Market for Tradable Renewable Energy Credits, *Ecology Economics,* Volume 49, Number 3, July 2004, pp. 259–272.

Participation in Green Power Purchases Exceeds 13%, Refocus Weekly [Online], http://www.sparksdata.co.uk/refocus/showdoc.asp?docid=70524233andaccnum=1and-topics=,

U.S. Dept. of Energy, Energy Efficiency and Renewable Energy, Renewable Energy Production Incentive [Online], http://apps1.eere.energy.gov/repi/, October 2006.

U.S. Dept. of Energy, Energy Efficiency and Renewable Energy, Top Ten Utility Green Power Programs [Online], http://www.eere.energy.gov/greenpower/resources/tables/topten.shtml, May 2006.

U. S. Environmental Protection Agency, Carbon Dioxide Emission from the Generation of Electric Power in the United States, July 2000, pp. 18.

U.S. Small Business Administration, SBA Basic 7(a) Loan Program [Online], http://www.sba.gov/financing/sbaloan/7a.html, October 2006.

Whole Foods Market IP, L.P., Whole Foods Market Makes Largest Ever Purchase of Wind Energy Credits in United States [Online], http://www.wholefoodsmarket.com/company/pr_01-10-06.html, January 2006.

Wood, A. J. and B. F. Wollenberg, *Power Generation, Operation, and Control,* Second Edition, New York: Wiley, 1990.

8

NUCLEAR POWER GENERATION

8.1 NUCLEAR POWER GENERATION PROCESS IN BRIEF

In nuclear power generation, energy is derived from controlled fission of heavy elements, principally uranium. This energy is converted through thermal generation into electricity. This thermal generation consists essentially of removing heat optimally and at desired temperature and pressure from a reactor core. The steam generated drives steam turbines coupled to electricity generators. Except for optimally removing steam from the reactor vessel, the rest of the process is similar in scope to thermal generation (see Chapter 3), except that steam turbines are used in place of gas turbines.

8.1.1 Risks Involved

The nuclear fission portion is accompanied by serious risks to society. The fission materials have a long-duration radioactivity—hundreds of years. The radioactivity has the capacity to attack and change basic biological structures in human, animal, and plant species. The spread of these materials in air, along the ground, and through water sources that might result from an accident at a nuclear reactor has to be guarded against, at whatever cost. The fission produces isotopes, including plutonium. Plutonium forms the basis of atomic bombs, which have a mass destructive capacity. Knowledge about and access to plutonium in the wrong hands would be suicidal to a society.

Copyright © 2011 the Institute of Electrical and Electronics Engineers, Inc.

8.1.2 Scattered Designs and Systems

This danger has resulted in the siting of nuclear power stations in remote sites with low population densities. Facilities are tightly guarded and only skeletal information of a general nature reaches the public. This secrecy has also resulted in these stations being the only ones in their class. This increases their cost for want of a modular approach and makes it difficult to confine and define the scope.

The whole system—nuclear and thermal together—has a large inertia. It is best operated as a supplier of steady base load for a utility.

8.2 RISE, FALL, AND RENAISSANCE OF NUCLEAR POWER PLANTS

That a steady electricity output could be produced by controlled fission of uranium was proved at Argonne National Laboratories in the United States in 1951. This was followed by the use of this development for a U.S. Submarine in 1954. Russia built up its commercially exploitable nuclear power plant rated for an output of 5 MW at Obninsk in 1954.

In the period 1950 to 1970, there was rapid development in the design, construction and operation of nuclear power plants (NPPs) in different countries. The initial fears regarding the risk of the unknown were overcome one by one. The plant sizes gradually went up to 1000 MW. By 1980, the capacity factor, that is, the ratio of actual production to the rated capacity, improved from 56.3% to 90.5% in the United States. World-wide, it improved from 72.2% to 88.1% in this period. The plant costs were quite high—between US$4000 to 5000 per kW compared to US$50 per kW for coal plants. Present Indian costs for coal-based thermal power plants average Rs 4000 cr per MW (1 crore = 10^7).

The factors in favor of NPPs were very low fuel costs (practically nil since the fuel has a steady, long life) and a steady, high output to cover the base load of the utilities. Plant orders kept on rising until 1980.

The oil embargo by the oil producing countries in 1973 changed the scenario in the electricity world significantly. Efficiency in electrical apparatus gained prominence. Demand-side load management leveled the peak loads for which the system capacities were designed. As a result, there was an overcapacity in electricity production. Orders for new NPPs were cancelled.

This period of recession was further extended by deregulation and reconstruction of the electricity industries. Generation, transmission, and distribution were segregated and open competition took place. Utilities that found transmission more profitable sold their NPPs and moved out of power production. However, the established plants continued to work satisfactorily and proved to be commercially viable.

There were two serious accidents in this period: one at Three Mile Island in the United States and the other at Chernobyl in the Soviet Union. These raised safety concerns.

Within the period 1960–2000, there were more than 300 NPPs in 34 countries, with generating capacities of more than 250,000 MW. These proved to be safe and reliable in operation. There was also an improvement in capacity factor. Today, there are 440 NPPs producing 10% of the world supply of electricity [3].

By 2000, the interest in NPPs revived. There were many factors in their favor. The NPPs were environmentally clean in operation. Environmental concerns, including CO2 accumulation in the atmosphere, raised the construction and operating costs of coal-based thermal units on one hand. On the other hand, technical improvements, particularly with accident safety, increased yields, life of usable uranium, and modular design and construction, favored the consideration of new NPPs. The capital costs are now expected to go down to US$1450 per KW. The oil prices controlled by OPEC kept on increasing, accelerating the race toward independence from oil as a fuel for electricity generation.

Gradually, the excess plant capacity of the 1970s to 1980s was absorbed by the increase in demand. The reorganization of the electricity business helped to consolidate nuclear plants into the hands of few owners.

There was significant improvement in the power output from the existing NPPs, up to 20% over the previously licensed rating. The licenses had to be, and were, uprated by the regulating bodies.

8.3 POWER-UP RATES

The established NPPs continued to show a steady, safe, and financially sound performance. These establishments had time-bound operating licenses, specifying their production capacities. Over the years, the control systems and operations changed for better. As a result, there was significant improvement in the old licensed capacities. There was a movement toward getting old licenses uprated and extended for another 20 years.

The power-up rates were classified into two categories:

1. Stretch power-up rates, based on changes in instrumentation set points and minor plant modifications.
2. Extended power-up rates, based on significant modifications to the balance of plant equipment.

All this has resulted in the existing NPPs asking for extensions to their present licensed period by 20 years. They got it.

Another notable event that encouraged the revival of NPPs was their acceptance as green power projects. The European Union had agreed to produce green power to the extent of 20% of total electrical power by the year 2010.

The number of licenses for executing NPPs went up substantially in almost all the countries in the world. However, getting a license and establishing an NPP is a long, drawn-out process, as will be shown later.

While this was progressing, thermal generation based on natural gas took the lead. New gas fields were discovered, increasing the supply of gas. Gas-based thermal plants for electricity have now become more attractive (see Figure 8-1).

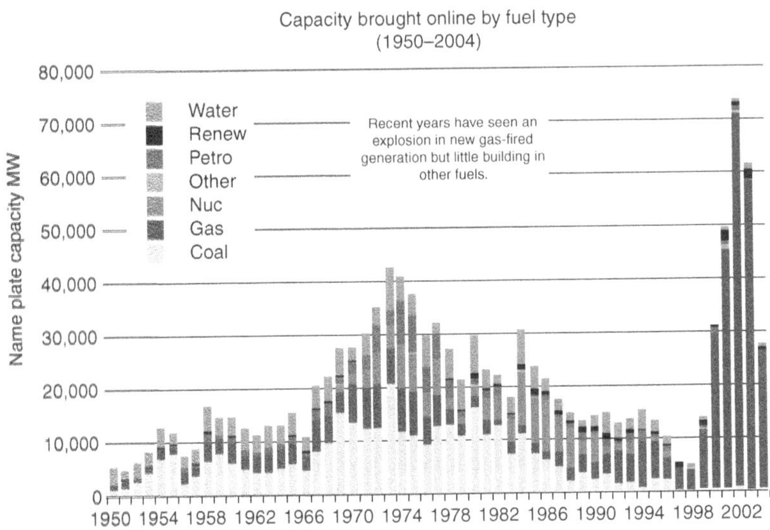

Figure 8-1. New generation placed in service from 1950–2004. (From [2], © IEEE 2006.)

8.4 ADVANTAGES OF NUCLEAR PLANTS

The advantages of nuclear plants are:

1. Steady, reliable supply of electricity
2. Environmentally acceptable
3. Moderate and limited space requirements—when compared to other types of production such as hydroelectric
4. Fuel costs are low and fuel life is long
5. Frees a society from overdependence on fossil fuels (oil and coal)

8.5 SOME TYPES OF NUCLEAR POWER REACTORS

American NPPs are light water reactors of three types:

1. PWR–Pressurized water
2. BWR–Boiling water
3. ABWR–Advanced boiling water, which can withstand operational transients better. They have stream-driven, high-pressure injection systems in place of electrical-driven motor–pump systems, to provide reactor cooling in case of accidents. In the latest version, reactor cooling is provided by water flowing under

gravity from a high level in place of water circulated by electrically driven pumps. The safety factor became independent of position in electric supply. This is being adopted universally.

8.6 OTHER TYPES FROM DIFFERENT COUNTRIES

Other types of NPPs are:

1. Canadian NPPs. CANDU reactors are heavy water moderated reactors. These can use a variety of mixtures with uranium. A light water stream cycle operating at higher temperature and pressure in the process improves efficiency.
2. Russian NPPs. VVER type, pressurized water with enhanced passive safety features. Rating up to 1500 MW.
3. British NPPs. Graphite moderated, advanced gas cooled (AGR), MAGNOX, and AGR varieties.
4. French NPPs. Gas cooled initially but now standardized on pressurized water. France produces 80% of its total requirements of electricity form NPPs.
5. Japanese in collaboration with Westinghouse Electric Corporation. Advanced pressurized water reactors (APWR) up to 1500 MW.
6. Indian NPPs. Advanced heavy water reactors. Primary fuel is a mix of uranium and plutonium. Research is ongoing to incorporate thorium, which is abundantly available in India.

8.7 PLANNING OF NP PLANTS

Considerations of public and worker safety and concerns on environmental issues impact the planning and execution of NPPs. The periods involved are fairly long. In the past, there was no standardization of the processing of reactor cores, moderents for reactors, and operational safety, mainly on account of security. Each project was individually designed and constructed.

8.7.1 U.S. Planning Process for an NP Plant—Stages 1 to 3

Stage One—Reactor Design Approval. This involves approval of the reactor design by the reactor suppliers. This is independent of site considerations. The Nuclear Regulatory Commission (NRC) prepares a preliminary safety analysis report.

The design is approved at this stage with the acceptance of the above report from the NRC. The completion period is prolonged. New developments will keep on coming, raising the temptation to add or to correct the original designs and layouts. If this also involves reworking a project under construction, the cost will shoot up and completion will be further delayed.

EPRI has produced a document called a Utility Requirement Document toward this end. The document was first offered in 1990 and is continually amended.

Stage Two—Site Approval. This involves the following considerations:

1. Safety of site: seismic considerations (check for earthquake zones), frequency of lightning strikes, meteorological considerations (frequency and intensity of storms)
2. Safety during construction and operation.
3. Safety in case of an accident—decommissioning aspects
4. Environmental issues—waste handling, storage, and disposal; wastewater disposal
5. Plant staffing
6. Availability and costs of a distribution system of an electricity utility, also of water supply
7. Defense and aviation safety clearances

Stage Three—Conditional Construction and Operating License (COL). Additional details on plant operations and site-specific designs are now approved. During construction, inspection, tests, and analysis and acceptance criteria (ITAAC) are checked and carried out for clearance to the next stage of construction activity. Upon completion of the ITAAC, plant commissioning operations may commence.

8.7.2 Periods Involved at Each Stage

Table 8-1 gives an idea of periods involved at each stage.

8.8 FINANCIAL RISKS IN PLANNING

Each stage stretches over a long period. Public scrutiny and approval are involved in each stage. These can add to the construction costs.

During the transitional period from 1980–1990, when the electricity business was restructured, the interest in NPPs dropped. Costs of projects cancelled at different stages had to be borne by the promoters. The U.S. government has offered to share these costs with the promoters.

Although the total costs of electricity production from NPPs are going down, actual costs are fairly high (see Figure 8-2). Support from government regulators on this front is necessary to make the projects commercially viable.

8.9 OPERATION OF NP PLANTS

After the Three Mile Island accident in the United States, operation and management of NPPs was looked into seriously. An Institute of Nuclear Power Operators was es-

Table 8-1. Sample nuclear plant construction schedule

Activity	Time frame	Best estimated completion date
Develop COL application	24 months	31 December 2007
COL application review by NRC and completion of plant engineering	42 months	30 March 2011
Site development and preparation for construction	Approximately 12 months after issuance of LWA-1*	30 March 2011 (assumes LSW-1 issued 12 months prior to COL)
Plant construction	36–48 months	30 March 2014
Fuel load and startup testing	Up to 12 months for first of a kind plants, 6 months for all others	30 March 2015

*An LSW-1 allows site development but does not allow the beginning of any safety-related work. An LWA-1 is issued by the NRC under several conditions, including the approval of an adequate site restoration plan to assure restoration of the site in the event of rejection of the COL application.
Source: [2].

tablished. Steady improvement in operation on NPPs was helped by the following activities.

8.9.1 Personnel

Training programs for coordinated action and teamwork operations were laid out for operating personnel. These were given hands-on training in the plant as well as simultaneously in training classes. The training was repetitive and reviewed at three differ-

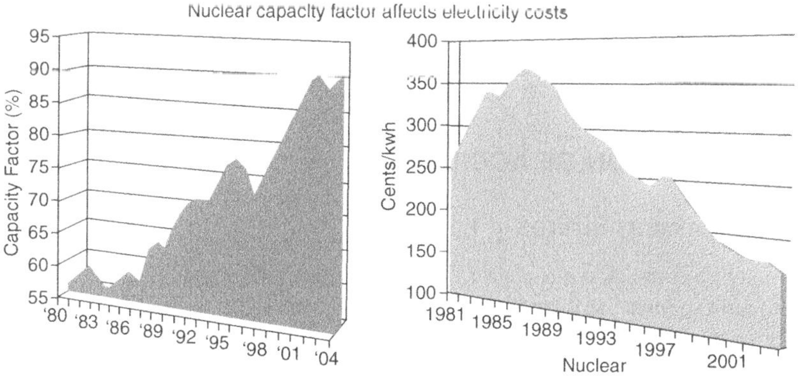

Figure 8-2. As reliability increases, operating costs decline. (Source: EIA, updated April 2005.)

ent levels. The managerial personnel also had a training program that was repeated and assessed. Simulators were used for training and were a great help in predicting the impact of new equipment on the existing ones.

Personal safety measures helped to reduce the radiation exposure rate, particularly in the boiling water reactors, from 859 persons per unit in 1980 to 42 for AWR and 61 for pressurized water reactors.

The Nuclear Regulatory Commission issued a Federal Regulation called "Requirements for Monitoring the Effectiveness of Structures, Systems and Components for Impact on Availability and Reliability."

Probability risk assessment techniques identified conditions that would make a plant drift into a risk condition. The unplanned shutdowns of NPPs came down to 0% by 2004 from 2.3% in 1980.

8.9.2 Technical

New core designs were compact. These addressed leaking fuel elements and gave higher power density. This raised the steam pressures and temperatures and resulted in improvements in steam turbines and steam generators. Adding moisture separation also helped.

Measurement of feed water flow is a critical parameter. More accurate flow meters were developed.

8.10 SAFETY MEASURES TO PREVENT EXPLOSION IN A REACTOR VESSEL

Uranium and other heavy elements release atom-bound energies in large quantities. This energy, under controlled fission, can be converted into a steady flow of electricity. If the control of the fission is lost, large amounts of energy are released over short time periods, creating pressure waves as well as heat waves. This is how an atomic bomb works.

Layers of safety are built around a rector to prevent runaway fission and powerful explosions in a nuclear reactor (Figure 8-3).

8.11 PREVENTION OF ACCIDENTS

8.11.1 Lightning Strikes

A NPP may be connected into a grid through one or more transmission lines. A lightning hit on these lines can cut off the line and the power supply; short the ground, with possible rise in ground voltage level; cause high-frequency (HF) injection into the power system.

In case of a complete cutoff, the NPP must have enough power for a safe auxiliary shutdown. Under the license terms, an NPP must have captive diesel generators of suf-

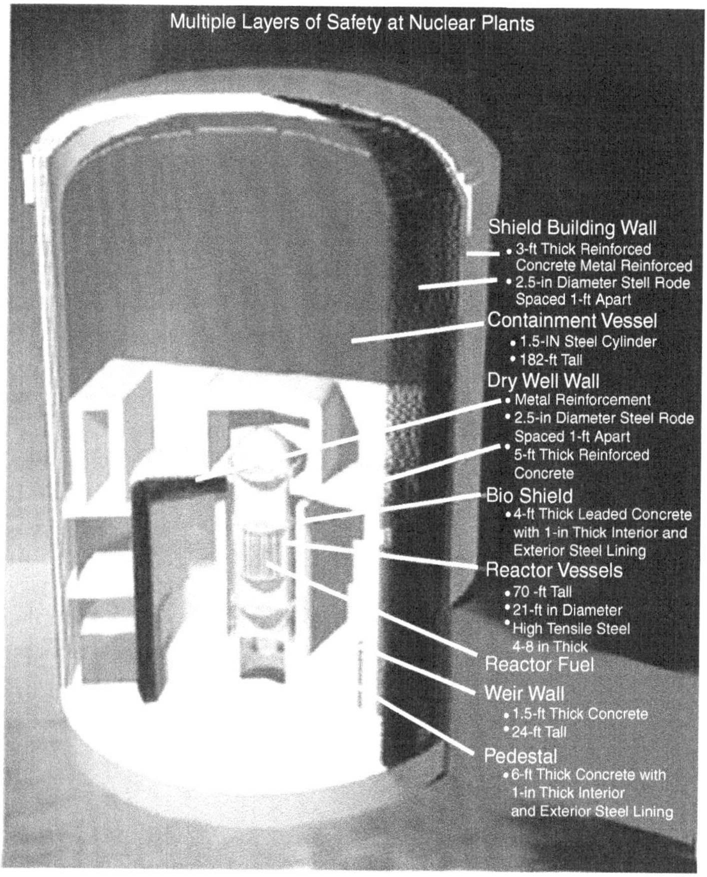

Figure 8-3. Layers of safety. (From [3], © IEEE 2006.)

ficient power to undertake this and also must have sufficient battery power in an uninterruptible power supply to start the generators. A station "black-out" occurs if the generators fail to start.

A local lightning strike can cause sufficient ground voltage rise on healthy phases to neutral so as to cause spurious maloperation of various relays and contactor coils. Station grounding systems have to be especially reliable and regularly renovated.

8.11.2 Utility Bus Voltage Dips

A voltage-graded relay sensing the dip in bus voltage on the interconnection may cut off the site supply if the connecting transmission line voltage dips below a safe level for prolonged periods due to a short in the off-site system or due to an overloaded line. This will result in unhealthy overloading of the NPP power system.

A voltage-graded relay has two set points. The first set point activates a caution signal in the control room. The second set point trips the interconnecting CB.

It may be noted that the voltage-graded relay is insensitive to transient voltage dips and that it works over and above the overcurrent and earth fault protection incorporated in the CB protection circuitry.

8.11.3 The Generator Output Trips

Reactor core rod assemblies drop down into the moderator pool under no-voltage conditions. Reactor core cooling has to be augmented with increased water flow. Water-circulating pumps and their driving motors form part of the very essential and critical class IE machinery.

Reactor vessel cooling is also a critical operation. In case an offsite supply trips and an essential supply is to be transferred, the latest trend is to put the critical water circulation under a gravity circulation system having raised level water storage tanks and larger pipe diameters. This does away with the dependence on availability of electric supply.

8.11.4 Off-Site Supply Trips

When a full supply trips, various induction motors keep on rotating under their own inertia. With their residual magnetic field, they act as induction generators producing a three-phase electric voltage at diminishing voltages and lowering frequencies (Figure 8-4). The large motors put in enough energy into this diminishing voltage and falling frequency voltage.

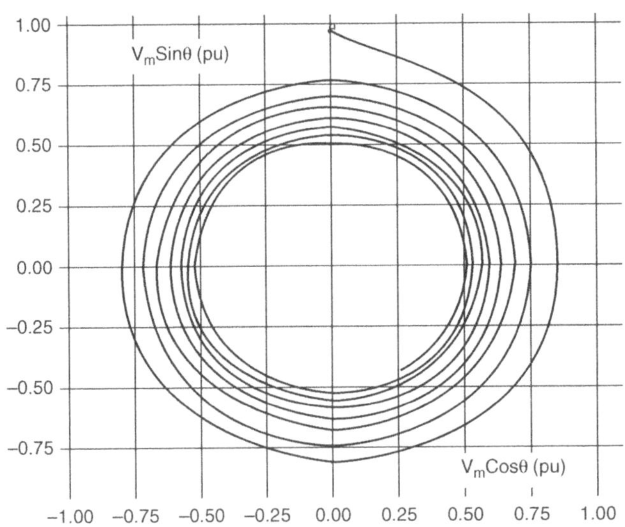

Figure 8-4. Polar plot of typical motor bus voltage decay characteristics. (From [3], © IEEE 2006.)

The diesel generators (DGs) now start up. Their voltages are brought up to within 95% of the rated system voltage and the class IE (highly critical) power system is transferred to the DG sets. The difference in voltages and frequencies will cause a transient, creating an unhealthy transient torque and shock on the rotor shafts. Differences in the phase angles of the two systems also play an important part. Thus, the bus transfer has to be fast enough to stay within recommended phase angle differences.

Figure 8-4 below shows the graph of decaying voltages on an induction generator cut off from its driving voltage. Figure 8-5 shows phase angle characteristics of a motor bus.

8.12 CLASS IE EQUIPMENT AND DISTRIBUTION SYSTEMS—UNGROUNDED EARTHING SYSTEMS

Class IE equipment must be in operation in case of any emergency such as when the external supply from a utility to run the essential equipments gets cut off. This can happen when the utility system trips due to a fault within the utility.

A distribution system for class IE must be distinct and must be in working condition at all times. When the external supply gets cut off, the emergency diesel generators are started by an independent battery power supply and are connects to class IE distribution systems and help to keep class IE equipment working. In this scenario, a single line-to-earth fault in the distribution system is acceptable as the system will continue to run. If a second line-to-earth fault develops, the distribution system must trip or else the generators and interlinking equipment will be damaged.

There are two operations to achieve this. The first leaves the system ungrounded so that the first fault does not draw a very damaging current. It also prevents transient down the lines. Transients are essentially line-to-ground phenomena. The second option is to have a limited ground resistance incorporated into the earthing system. This reduces the first fault current. However, it does not prevent transients.

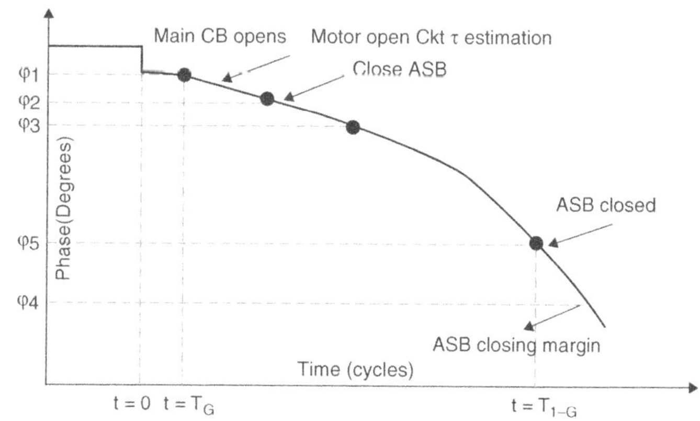

Figure 8-5. Phase angle characteristics of a motor bus. (From [3], © IEEE 2006.)

In a fully ungrounded system, when one line is grounded due to a fault, the voltage across the insulation of equipment in that system will jump to 1.73 times the rated voltage. A full line-to-line voltage is impressed across the insulation. Therefore, it is very essential to incorporate equipment insulation voltage rating at 1.73 times the rated system voltage for all the equipment in this system. Without this, class IE equipment will operate for a certain period and then start failing. Take power cables, for example. The first fault on an improperly specified cable cannot be passed off under the category of single, tolerable, line-to-earth fault. The supply of the cable could be faulty.

The same caution applies in the case of equipment in resistance-grounded distribution systems.

In case of a direct lightning strike when the ground voltages are likely to experience a raised transient voltage, class IE systems and equipment are relatively unaffected.

8.13 ENVIRONMENTAL CONSIDERATIONS—RADIATION HAZARD

Fission materials radiate harmful rays. This radiation affects almost all forms of life and lasts for years. Safety measures have to be undertaken from the uranium mining stage to the disposal of used uranium and its isotopes to prevent harmful exposure to this radiation. The intensity of radiation and the duration (in hours) for which a worker comes in contact with it are monitored and controlled. When a worker has accumulated his allowed dose, he is relieved of that duty and placed elsewhere. Personal safety records on workers who are thus exposed to radiation have been satisfactory.

Uranium ore is mined and transported to a crushing or grinding shop. Mine workers and mill workers are monitored. The uranium is separated from stone materials called "ore burden" and is formed into a "yellow cake." The yellow cake is enriched, formed into pellets, and then into rods. This is a highly secret process. Not only are the workers monitored, but they also are under constant surveillance for their habits, acquaintances, and so on.

Cores thus formed become the core of the reactor vessel. Care is taken to see that radiating materials do not spill into the plant system and outside, say through circulating water.

8.14 WASTE MANAGEMENT

Plant operators monitor temperature, pressure, water chemistry, and so on of the reactor core to optimize its output. The fuel fission capability of the core rods gradually reduces until it is ready to be taken out and added to the "spent fuel" list. At this stage, hardly 5% of the uranium has been "used" [3].

8.14.1 Reprocessing

In reprocessing, usable uranium and plutonium are removed and reconstituted in a mix with enriched uranium, then put into service again. This method consumes 6% of the fuel. The lot then goes to "spent fuel" stage. In the United States, it is transported to a single depository at Yucca Mountain. Special containers are designed and tested to prevent release of their radioactive contents in the event of any accidents. These containers are also resistant to acts of sabotage and have to withstand a 30 ft drop, puncture by a 6 in rod, 30 minutes under fire, and immersion under 650 ft of water [3].

8.14.2 Underground Storage Tanks

Special waste-storage tunnels receive these containers. Figure 8-6 shows construction of one such tunnel.

8.15 ENVIRONMENTAL BENEFITS

Macdonald [3] lists the following benefits of nuclear power electricity generation for year 2004 in the United States:

1. Prevented release of 3.43 million tons of SO_2, 1.11 million tons of nitrogen oxide, and 696.6 million tons of CO_2, which would have come from coal- and oil-based thermal plants.
2. Created 2000 tons of high-level waste per year as against a total of 40 million tons of other hazardous wastes.

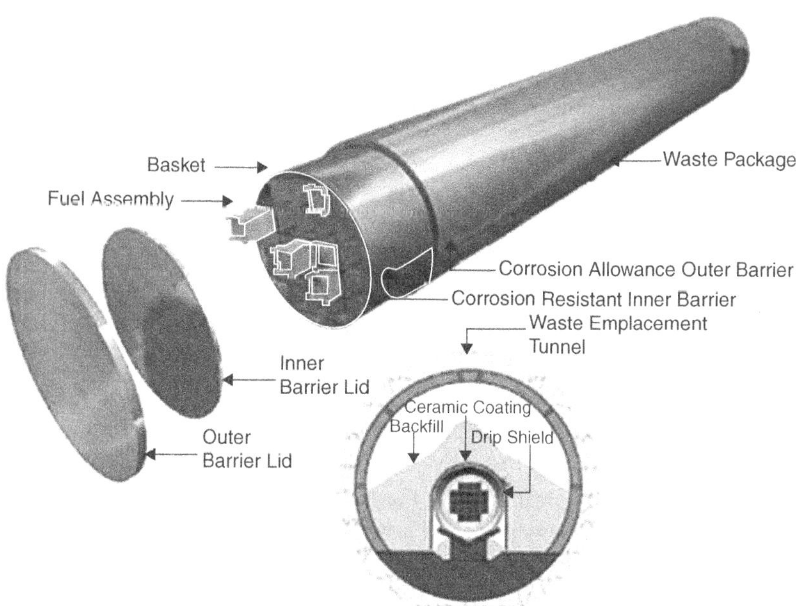

Figure 8-6. Waste container storage tunnel. (From [3], © IEEE 2006.)

3. 696.6 million tons of CO_2 released is equivalent to the exhaust of 134 million cars. Use of uranium instead of oil releases oil for use in cars, easing shortages as the number of cars increase and the oil supply decreases.
4. Equivalent of CO_2 emission quantity in CO_2 emission quality (CeQ) per kW-hr of electricity produced is 2.6 for nuclear fuel, 2.13 for wind energy, 8.77 for thermal energy [7]. These figures are worked out over the total lifetime cycle of each type of plant, from the construction stage to the removal/replacement stage.

8.16 CHALLENGES FOR RESEARCH

The Generation IV International Forum (GIF) of NPP operating countries has the following members: Argentina, Brazil, Canada, France, Japan, Republic of Korea, Republic of South Africa, Switzerland, United Kingdom, and the United States. The subjects for research by this GIF are [1]:

1. Electricity Power Generation. To be more safe, secure, and economical. To provide NPPs for small grids and also for remote applications.
2. Produce Hydrogen Era. Go to high enough temperatures to produce hydrogen directly after splitting water molecules and passing the product through a filter. Use the balance of the heat for thermal production of electricity.
3. Management of Actinides. Actinides are the heaviest elements found in nuclear fuel and include isotopes of uranium, plutonium, neptunium, americium, curium. Develop methods to utilize actinides as a mix with normal NPP fuel. It is expected that 94% of the uranium energy can thus be consumed and only 1% will become waste material. This lowers the fuel cost and takes care of the disposal problem. It reduces the risk of mischief by terrorist organizations.

8.17 RAPID INCREASE IN POPULATION EXPECTED

By 2050, the population of the world is expected to double. Average living standards also are expected to rise. Electricity requirements will increase in all the countries of the world. Nuclear-power-based electricity is an alluring solution. It has now become fairly safe and risk free. NPP will be catching up fast (see Appendix 8-3).

8.18 FAST BREEDER REACTORS

In a normally working nuclear reactor, the uranium rods emitting fast neutrons get slowly converted into and covered with a layer of plutonium. This plutonium slows down the neutron output. A stage is reached at which the utilization efficiencies fall to a level where the fuel rods have to be reprocessed or discarded.

In a fast breeder reactor, uranium rods clad with this plutonium are subjected to fast neutrons at high temperatures. This reconverts plutonium into useful uranium. In other words, the life of original rods is almost doubled. At one stage, fast breeder reactors were thought to be the future for nuclear power.

In a fast breeder reactor, there is no place for moderating media. Instead, cooling of the reactor is essential. The most commonly used coolant has been liquid sodium. This sodium becomes radioactive and has to be cooled a second time with sodium. This second sodium liquid heats up water and produces steam that drives the cascading set of steam-turbine-coupled electric power generators.

In place of plutonium, thorium is being experimented with. Thorium also gets converted into uranium. However, other products, actinides, besides uranium are produced. These are highly radioactive, so thorium is still under investigation.

With increased supplies of uranium, the activity in the fast breeder reactor field seems to have slowed down.

APPENDIX 8-1 NUCLEAR REACTOR ACCIDENT AT THREE MILE ISLAND

The Three Mile Island nuclear generating station in the United States had two high-pressure water reactors. Both the reactors had unblemished records, continuous power output for over a year, plant load factors over 90%, and nil accident records. However, plant no. 2 had a serious accident on March 28, 1979. The damage could have turned out to be catastrophic but was contained.

The reactors core rods were contained and controlled under circulating high-pressure water. This high pressure raised the boiling point temperature of water, so no steam was formed. This high temperature water was cooled down by a secondary water circulating system that conveyed the thermal energy thus liberated.

There was a malfunction in the secondary system. As a result, the primary water system went to a high temperature. The automatic safety system shut down the reactor. This involved dropping down of core rods into the moderator and releasing the pressure in the primary circuit to normal design levels. Thus, the pressure relief valve had to open instantly when the reactor shut-down process started. It also had to shut after a set time, of 12 seconds in this instance, so that pressure would be maintained in the primary water circuit and reactor vessel at the designed value.

This pressure relief valve failed to close but defective instrumentation showed it as closed on the panel. The pressure continued to drop, exposing the core rods, which went into high heat and began melting. This produced high temperatures and steam. The water circulating pumps started vibrating dangerously when they had to handle steam along with water. They were shut down and cooling under gravity was restored. Panic broke out but the plant came under control, with trials, within 48 hours.

The core meltdown contaminated the circulating water. It generated hydrogen that could have caused a disastrous explosion. It did not, since the closed vessel did not have enough oxygen to create a hydrogen-based explosion. The core vessel and other structures were not affected.

The contamination was restricted to the drainage system and drainage pit. It did not spill into the surroundings or into the atmosphere.

The clean up took nearly 12 years and cost US$973 million. Fishing out the half-burnt core rods, encasing them, and transporting them to the storage site was a big job.

The causes for the accident were mainly attributed to faulty control and indication systems and inadequate staff training in handling emergencies.

No human tragedies have been attributed to this accident. However, it caused ripples around the world for a long time. Safety systems and repetitive staff training have become important and turned into internationally adopted standards. It also increased the costs of nuclear power installations. Further information on this event can be found in Nuclear Issue Briefing Paper no. 48, March 2001 (www.uic.com.au/nip14.htm).

APPENDIX 8-2 CHERNOBYL ACCIDENT

Chernobyl, located in Ukraine, USSR, had four nuclear generator plants. One of these plants blew its top on April 25, 1986. The plant was designed around graphite-moderated, enriched uranium. The reactor cores were water cooled.

On April 25, a routine test to determine the power output when the generator was disconnected from the grid was being carried out. At low outputs, the reactor becomes unstable and liable to emit spurts of energy. The cooling water was probably inadequate and emergency accident-prevention equipment did not operate, which probably increased water cooling of the reactor and automatic dropping of the reactor core into the moderator. The steam built up. After the first accident, small fires broke out, which the operators tried to put out. This was followed by a second large accident that blew off the top and spewed out steam, gases, reactor core material, and graphite from the moderator.

The reactor fire took nine days to be put out. Some 5000 tons of sand and mineral clays were dropped on the burning reactor from helicopters. The radioactive gases had limited half lives. Most of the released material was deposited nearby as dust and debris. The accident was blamed on flawed reactor design and on inadequately trained operators.

The explosion killed 30 people, including 28 from radiation exposure. Another 209 people on-site doing cleanup work were exposed to radiation and 134 of these were found to be poisoned by radiation; 19 of these died subsequently.

Some 45,000 people were evacuated within 8–10 days of the accident from a contaminated area of 10 km radius. Another 116,000 people were evacuated and resettled elsewhere from within a 30 km radius (2800 km^2). This exclusion zone was further extended to 4300 km^2.

The fission materials had half-life periods ranging from 8 days to 30 years. Some of these that were airborne are suspected to have increased radiocontamination in far off areas in Russia and Europe.

A UN agencies report (UN SCEAR 2000) claims that "apart from this [thyroid cancer increase] there is no evidence of a major public health impact attributable to radiation exposure 14 years after the accident. There is no specific evidence of increase in overall cancer incidence or morality or in nonmalignant disorders that could be related to radiation exposure."

APPENDIX 8-3 WORLDWIDE CAPACITY AND GENERATION OF NUCLEAR ENERGY

	In operation in 1999		Under construction at end 1999		In operation in 2010		Net generation in 1999, TW-hr	Nuclear share of electricity generation in 1999, %
	Units, number	Capacity, MW_e	Units, number	Capacity, MW_e	Units, number	Capacity, MW_e		
South Africa	2	1800			2	1800	11.6	7.1
Total Africa	2	1800				1800	11.6	
Canada	14	9998			22	14,902	69.3	12.4
Mexico	2	1364			2	1364	9.6	5.5
United States of America	104	97,557			98	93,730	727.7	19.8
Total North America	120	108,919				109,996	806.6	
Argentina	2	935	1	694	3	1629	6.6	9.0
Brazil	1	617	2	2506	3	3123	4.0	1.4
Total South America	3	1552	3	3200		4752	10.6	
Armenia	1	376					2.1	36.4
China	3	2167	7	5420		9587	14.1	1.2
India	11	1897	3	606		4013	11.5	2.7
Japan	52	43,445	5	4761	64	58,000	303.3	34.7
Korea (Democratic People's Rep.)			2	1900	1	950		
Korea (Republic)	16	12,990	4	3820	26	21,400	97.8	42.8
Pakistan	1	125	1	300	2	425	0.1	0.1
Taiwan, China	6	4884	2	2630	8	7514	36.9	25.3
Total Asia	90	65,884	24	19,437		101,889	465.8	
Belgium	7	5713			7	5713	46.7	57.7
Bulgaria	6	3538				2314	14.5	47.1
Czech Republic	4	1648	2	1824	6	3472	12.5	21.1
Finland	4	2656			4	2676	22.1	33.1
France	59	63,183			59	64,420	375.0	75.0
Germany	19	21,142			13	17,412	160.4	31.2
Hungary	4	1729			4	1729	14.0	39.0
Lithuania	2	2370			1	1185	8.7	73.0
Netherlands	1	449					3.3	3.9
Romania	1	648	4	2592	2	1296	4.8	10.4
Russian Federation	29	19,843	3	2825		21,336	110.9	14.4
Slovakia	6	2408	2	776		1592	13.1	47.0
Slovenia	1	632			1	632	4.5	36.0

(continued)

	In operation in 1999		Under construction at end 1999		In operation in 2010		Net generation in 1999, TW-hr	Nuclear share of electricity generation in 1999, %
	Units, number	Capacity, MW_e	Units, number	Capacity, MW_e	Units, number	Capacity, MW_e		
Spain	9	7459			9	7798	56.4	28.3
Sweden	11	9452			10	8852	70.2	46.5
Switzerland	5	3127			5	3127	23.5	35.3
Ukraine	14	12,115	2	1900	15	13,090	67.4	43.8
United Kingdom	33	12,742			13	7750	88.0	26.0
Total Europe	215	170,854	13	9917		164,394	1096.0	
Iran			1	1000	2	2000		
Total Middle East			1	1000		2000		
Total world	430	349,009	41	33,554		384,831	2390.6	

Notes:
1. The estimates of nuclear generating capacity in 2010 for Armenia, China, India, Korea (Democratic People's Republic), Pakistan, Russian Federation and Slovakia reflect the Reference Case projections in *International Energy Outlook 2001*.
2. As the number of reactor units in operation in 2010 is not available for all countries, no regional and global totals have been computed for this item
3. The capacity and output of the Krsko nuclear power plant, shown against Slovenia in the table, is shared 50/50 between Slovenia and Croatia

Source: Nuclear Issue Briefing Paper no. 22 (www.uic.com.au/nip22.htm).

REFERENCES

1. J. Disosway, Generations: The Nuclear Family Comes of Age, *IEEE Power and Energy Magazine,* November 2006, pp. 18–24.
2. S. W. Semmes, A Nuclear Renaissance, *IEEE Power Engineering Review,* December 2006, pp. 34–42.
3. J. D. Macdonald, Safe and Secure, *IEEE Power Engineering Review,* December 2006, pp. 49–55.

BIBLIOGRAPHY

ASME, 2001 ASME Boiler and Pressure Vessel Code—An International Code. Section, 3, Division 3, Containment Systems for Storage and Transport Packagings of Spent Nuclear Fuel and High Level Radioactive Material and Waste. ASME, 2001.

Balamourougan, V., T. S. Sidhu, B. Kasztenny, and M. M. Thakur, Robust Technique for Fast and Safe Transfer of Power Plant Auxiliaries, *IEEE Transactions on Energy Conversion,* Volume 21, Issue 2, June 2006, pp. 541–551.

Baxter, F. D., Questionable System Grounding Practices at Nuclear Power Plants, *IEEE Transactions on Energy Conversion,* Volume 17, Issue 2, June 2002, pp. 295–298.

Carter, J. P., The Transformation of the Nuclear Power Industry, *IEEE Power Engineering Review,* December 2006, pp. 25–33.

Dent, R. A. and Z. Dicic, Adjustable Speed Drives Improve Circulating Water System, *IEEE Transactions on Energy Conversion,* Volume 9, Issue 3, September 1994, pp. 496–502.

Disosway, J. Nuclear's Comeback Greater Demand, More Plants, New Hope, *IEEE Power Engineering Review,* Volume 4, Issue 6, November 2006, pp. 4–6.

Disosway, J., The Nuclear Comeback—An Industry in Hot Standby, *IEEE Power and Energy Magazine,* Volume 4, Issue 6 November 2006, pp. 14–16.

Jancauskas, J. R., A Recommended Approach for Calculating Degraded Voltage Relay Setpoints for Nuclear Generating Stations, *IEEE Transactions on Energy Conversion,* Volume 9, Issue 1, March 1994, pp. 173–178.

Kaffashan, A., P. Kelly, and R. Sullivan, and Circuit by Circuit Power Cable Separation in Nuclear Power Plants, *IEEE Transactions on Energy Conversion,* Volume 8, Issue 4, December 1993, pp. 596–601.

Kueck, J. D., G. E. Attarian, H. C. Leake, and T. R. Sims, Risk Factors Regarding the Application of Degraded Voltage Relaying at Nuclear. Generating Stations, *IEEE Transactions on Energy Conversion,* Volume 16, Issue 4, December 2001, pp. 374–379.

Malcolm, J. S. and D. L. Harmon, Advanced Control Room Design, *IEEE Power Engineering Review,* December 2006, pp. 43–48.

Roman, H. T., Robotic Applications in PSEG's Nuclear and Fossil Power Plants, *IEEE Transactions on Energy Conversion,* Volume 8, Issue 3, September 1993, pp. 584–592.

Rourk, C., A Review of Lightning-Related Operating Events at Nuclear Power Plants, *IEEE Transactions on Energy Conversion,* Volume 9, Issue 3, September 1994, pp. 636–641.

Yangping, Z, R. M. Edwards, and K. Y. Lee, Hybrid Feedforward and Feedback Controller Design for Nuclear Steam Generators Over Wide Range Operation Using Genetic Algorithm, *IEEE Transactions on Energy Conversion,* Volume 12, Issue 1, March 1997, pp.100–105.

Yeager, K. E. and J. R. Willis, Modeling of Emergency Diesel Generators in an 800 Megawatt Nuclear Power Plant, *IEEE Transactions on Energy Conversion,* Volume 8, Issue 3, September 1993, pp. 433–441.

Zhengyu, H. and R. M. Edwards, Power Generation Efficiency Improvement Through Auxiliary System Modifications, *IEEE Transactions on Energy Conversion,* Volume 18, Issue 4, December 2003, pp. 525–529.

9

WIND POWER GENERATION

9.1 INTRODUCTION TO WIND

Wind power has been utilized by mankind since historical times. It has brought distant lands together, from the early days when the Greeks set sail on their ventures to later when Colombus "discovered" America. The famous character "Sancho Panza" made windmills immortal in literature.

Of late, windmills have been getting increasing attention on account of wind energy being available "free" of cost and also on account of being the most nonpolluting source of electricity. On the electricity front, it started with a wind turbine driving an electricity generator, mainly an induction motor, which is universally available. The output ratings were minuscule: a few hundred watts. Today, it is entering into the club of megawatt-scale electricity producers. Kriegers Flak, an island located in the Baltic Sea between Sweden, Denmark, and Germany, is expected to fully commission its wind farm rated at 630 MW by 2010.

The latest rates of growth are striking. In Europe, wind farm installations grew at a rate of 38% during 2007, compared to 19% during 2006. By 2007, total installed wind power capacity there stood at 67 GW. Germany, Denmark, and Portugal were promi-

Copyright © 2011 the Institute of Electrical and Electronics Engineers, Inc.

nent. In the United States, total wind farm capacity stood at 11 GW in 2007. It is growing fast in China, India, and many other countries [1].

Yet, total contribution by wind energy to production of world electricity energy at the beginning of 2006 has been very minuscule—0.7% of the total of 17,500 TW-hr [10]. It has its drawbacks by its very nature.

9.1.1 Technology Growth in Wind Turbine Generators

Wind turbine generators (WTGs) started as fixed-speed wind turbines with conventional induction generators and capacitor banks as static reactive compensators. Capacitors supplied reactive power for the air gap magnetic flux, which the induction generators could not produce. Denmark initially standardized on this model, terming it the Danish Concept. These turbines contributed 71.6% of the total WTGs there by 2006 [39].

Later, squirrel-cage rotors in induction generators were replaced with wound rotors. Variable rotor resistance, variable speed compatibility with gears, and capacitor banks became standard features.

Doubly fed induction generators (DFIGs) followed with partially rated power electronic convertors. The converter helped to provide independent control of active and reactive power outputs of the WTGs. The PE converter rating was generally at 30% of the WTG rating.

Finally, TWGs with added functions in the PE convertors arrived. The PE portion increased the costs but gave better control and helped in the fault-ride-through facility. This category constitutes just 0.2% of the total WTG population [2].

9.1.2 Nature of Wind

Wind may blow steadily during certain periods, varying by day, season, location, and so on. Let us say the velocities fall within some zones. The wind may die down, falling almost to nil. Then it may rise from a very low speed. There may be a wind lull, when the wind dies out and then rises in short bursts. A wind gust is the opposite phenomenon to a wind lull. A very strong wind is a storm. This nature of wind makes it an unreliable source of power due to its variability and uncertainty.

9.1.3 Components of A Wind Turbine Generator

Figure 9-1 shows the components of a wind turbine generator. These are mainly:

- The rotor blades, whose pitch is adjustable as per wind velocity so as to catch maximum wind energy.
- The gear box which adjusts the rpm of the rotor of the generator as closely as possible to the grid synchronous frequency.
- The generator, which converts mechanical input into an electrical output.

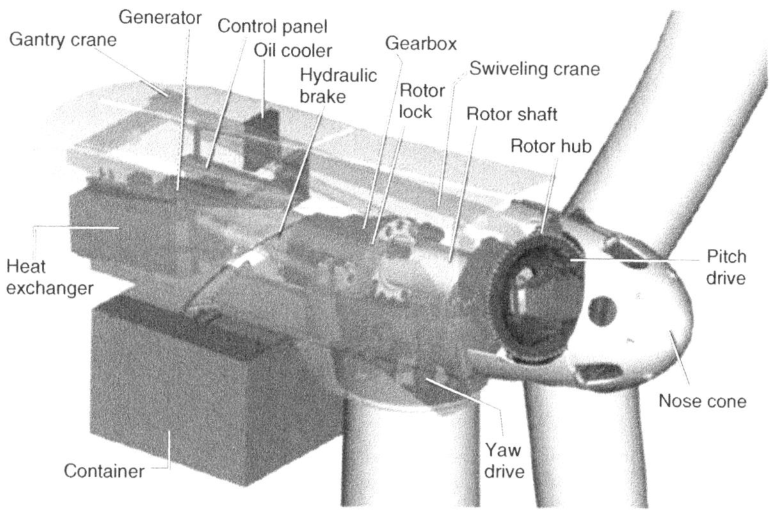

Figure 9-1. A wind turbine generator. (From [3], © IEEE 2005.)

9.2 OPERATION OF WIND TURBINE GENERATORS

9.2.1 Output of a WTG

Power captured by a WTG is given by

$$Pm = \frac{1}{2}\,\rho\pi R^2 C_p(\lambda)V^3$$

where

ρ = air density
R = wind turbine radius = approximately rotor blade length
C_p = power coefficient, which is a function of
χ = tip speed and wind velocity ratio
β = pitch angle
V = wind velocity
λ = $R\Omega/V$ that is, tip speed/wind velocity ratio (Figure 9-2 (a))
Ω = angular speed of the blade × 2π

There are two operating conditions for a WTG:

1. For a given wind condition it should produce maximum possible power. Please refer to Figure 9-2(a). This is possible when λ stands at λ_{opt}.
2. There is a minimum wind condition below which the WTG becomes unstable. λ_{is} represents the crossover point for the stable condition limit. At λ_{stall}, the WTG will stall.

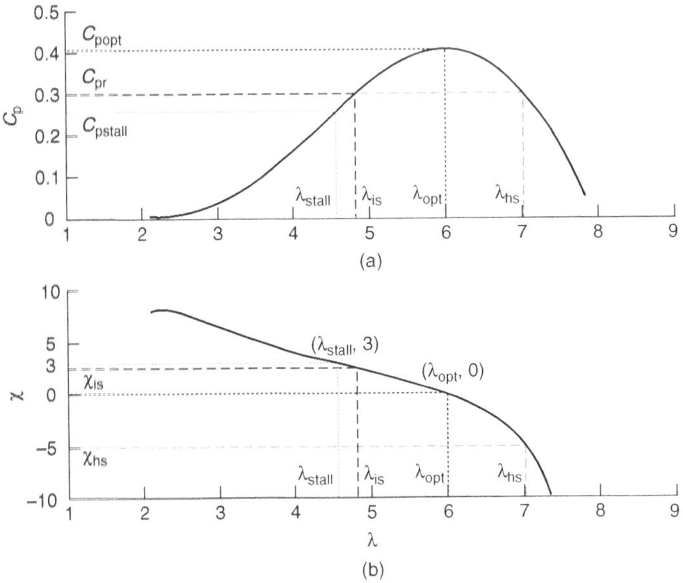

Figure 9-2. Turbine coefficients C (a) and χ (b). (From [4], © IEEE 2004.)

Note that the ratio of WTG blade tip speed to wind speed, λ, plays an important part. The WTG control should perform in such a way that it is at λ_{opt} under different conditions of wind load.

The rate of change in λ is given by another quotient χ:

$$\chi = (dC_p/d\lambda)[\lambda/C_p(\lambda)]$$

Figure 9-2(b) shows the variations of χ with λ.

A WTG is in an unstable condition when χ becomes positive. The figure shows power versus speed curves of a wind turbine with wind velocity as a parameter. The dashed line is a boundary between high- and low-speed regions.

9.2.2 Performance Improvement through Blade Pitch Control

At low speeds, the pitch angle is almost zero. Maximum possible energy is scooped up (maximum power strategy). At high speeds, the pitch angle increases (Figure 9-3). Beyond a certain wind speed, automatic mechanical brakes apply and electrical dumping resistances are used as loads.

9.2.3 Efficiency of a WTG

Average efficiency of a WTG is defined as the ratio of energy delivered to grid to the energy at the turbine rotor shaft. As the energy is transmitted from one member to the

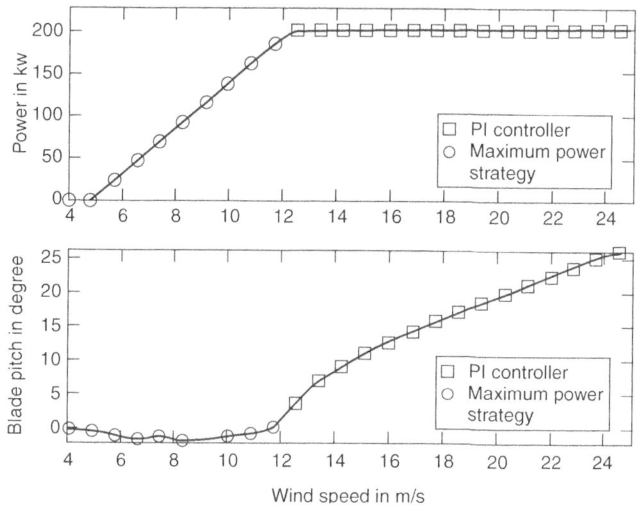

Figure 9-3. Strategy for the variation of blade pitch. (From [5], © IEEE 2004.)

next in the transmission system of a WTG, losses are incurred. Tables 9-1 to 9-3 give a relative idea of these losses.

9.2.3 Losses in a WTG

Average and rated efficiencies for the three different types of WTGs are 82–86% at low wind speeds and 89.7–89.9% at high wind speeds. Thus, weather forecast and past statistical data form important requirements for efficiency and reliability when integrating wind farm energies into today's mega grids.

9.2.4 Flickers in the Output of a WTG

There are two main causes of flicker in the supply from a WTG:

1. Mechanically related causes
2. Wind velocity related causes

Table 9-1. Losses in the gear at rated load

Friction, windage, and oil churning losses	1.0%, including turbine bearing
Gear mesh losses	1.7%
Total losses at rated load	2.7%

Source: [6].

Table 9-2. Losses in the generators at rated load

	Induction generator (grid connected)	Synchronous generator (diode loaded)	Directly driven PM generator (diode loaded)
Core losses	1.5%	1.5%	1.2%
Copper losses, and additional losses	1.5% stator and rotor	1.15% stator	3.5% stator
Friction, windage, and cooling losses	0.5%	0.5%	1.0%, including turbine bearing
Excitation losses	—	0.75%	—
Total losses	3.5%	3.9%	5.7%

Source: [6].

Table 9-3. Losses in the frequency converter at rated load

Voltage drop losses of the diode rectifier	0.4%
Resistive losses of rectifier and inductor	0.2%
Step-up converter transistor losses (at rated current through the transistor)	0.75%
Step-up converter diode losses (at rated current through the diode)	0.25%
No load converter losses	0.1%
Inverter load losses at $\cos(\varphi) = 1$	1.5%
Inverter and inductor resistive losses	0.3%
Total losses at rated generator voltage	2.75%

Source: [6].

Mechanically Related Causes

- Motor turbine imbalance
- Rotor blades passing in front of the wind structure
- Structural modes due to mechanical eigenfrequencies (frequencies at which there is mechanical resonance)
- Rotational sampling

The flickers caused by these mechanical causes have a regular pattern, low amplitude, and a low-frequency range of 0.65 to 0.71 Hz.

Wind Velocity Related Causes. Wind flow has regular bursts that can cause flicker. This flicker has a high amplitude and a range of 0.01 Hz–10 Hz. This flicker is objectionable and has been investigated deeply.

Flickers Irritate. If P_{st}, the short-time flicker intensity, is greater than 6.35% in a range of about 9 Hz it becomes an irritation to the human eye. IEEE Specification 561 lays down rules and limitations on these types of flickers. The flicker rules have their origin in the operation of industrial arc furnaces, which caused voltage dips at these low frequencies.

If we consider *SC*, the capacity of a grid, as its inertia, then the larger the inertia, the lower the flicker and vice versa. If we consider χ/R as grid impedance factor, then larger this factor, lower the flicker intensity and vice versa. Flicker emission also rises proportionately with the wind turbulence intensity.

Flicker intensity does not add up linearly as the number of WTGs in a cluster or wind farm, *N*, grows; it rises proportionately to $\sqrt{\Sigma N^2}$.

Flickers produced by one turbine are not influenced by flickers produced by other turbines. However, they are rather accentuated by flickers produced by the grid itself.

9.3 CONNECTION OF WIND ENERGY PLANTS TO THE GRID—THE GRID CODE

In the early days of wind electricity generation, the plant sizes were small. With an induction generator, there was no problem of synchronizing with grid frequency. External capacitors took care of voltages; when there was a disturbance in the grid leading to low voltages at the point of connection, the wind plants were disconnected and stayed disconnected until the grid disturbance was cleared. Today, wind plant sizes have increased. Should a wind plant get disconnected due to a grid disturbance, it could aggravate the situation. A grid code for interconnection has evolved.

The mean features of the grid codes are:

- A low-voltage ride-through (LVRT(is essential for getting into a grid.
- Accurate power control at a PF of ±0.95 has to be maintained at the point of connection.
- Accurate plant models must be submitted.
- SCADA data must be supplied as agreed with the system operator.

9.3.1 Low-Voltage Ride-through

Wind energy farms are now a sizeable constituent of the power supply. If a fault develops in the grid, with voltages at the point of common coupling between the grid and the wind farm, falling to low levels during the fault clearance time, the wind farm should not disconnect. In other words, it must have a low-voltage ride-through capacity.

If a fault occurs in the grid system to which a wind energy farm is connected, the voltage at PCC dips to a low percentage, depending on the severity and location of the fault. The control system clears the fault within the time frame specified under grid discipline. Figure 9-4 shows an envelope of the voltage profile under the GB Grid Code (CC A 4.3) The right-hand side shows two illustrative faults: fault N, in which PCC voltage dips to 25% for a brief time; and fault M, in which it dips to 62% for a longer time.

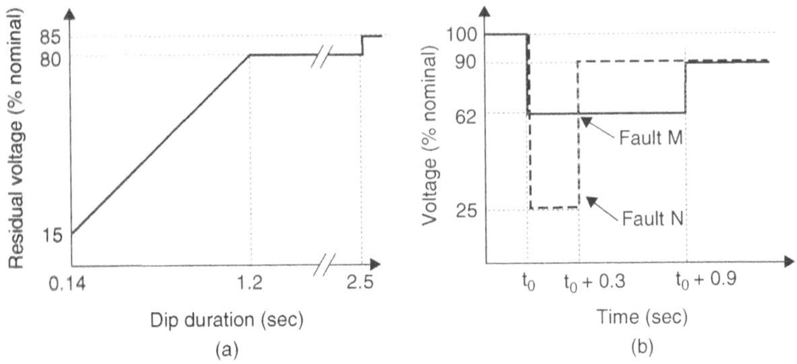

Figure 9-4. Voltage-duration profiles for grid voltage dips. (a) Grid voltage-duration profile. (b) Selected grid fault scenarios. (From [7], © IEEE 2007.)

Within the period during which the fault is being cleared and beyond to a safe limit, the wind energy system should hold on and ride through. At the instant of the start of and during the duration of fault, the voltages on WTGs stay low. These WTGs depend on external capacitors for the air gap magnetic flux as well as reactive power compensation. The reactive power from the capacitors is proportional to the square of voltage and goes down significantly at low voltages. Besides, the grid fault, depending on its location, could also drain out some reactive power from the capacitors. All this tends to speed up and destabilize the WTGs against a weakened flux in the air gap.

A sizeable correction to brake-up is anticipated through blade pitch angle control, which will try to slow the WTG down. However, the activating arms of the blade pitch control have to be really strong to be effective in a very short time. This will also put a strong mechanical stress on the WTG system.

This is best supported with the help of dynamic reactive power control from an SVR on the grid or even synchronous generators at the interconnections of the wind farm system to the grid. Note that the Danish systems have two large synchronous generators at their interconnection to the Nordal system in Sweden. These are rated at 160 MVA, and 100 MVA respectively.

With PE converters on the WTGs, the converters can extend dynamic RP control. Figure 9-5 shows the effect on WTG power output of dynamic RPC.

9.4 AMERICAN GRID CODE

The grid code developed by the American Wind Energy Association (AWEA) requires a low-voltage ride-through with normal fault clearing time of nine cycles for a three-phase fault. A single-phase fault to ground is allowed longer time to clear.

The WTG must stay online during these fault periods, with voltages dipping at PCC to 0.15 PU and later on cleared to 0.0 PU. The grid code also requires reactive power control at the PCC at +0.95. It also requires that accurate plant models be supplied [1].

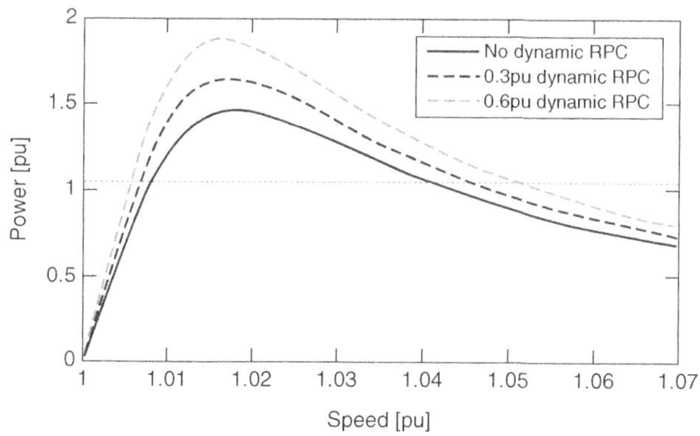

Figure 9-5. Effect of dynamic RPC on steady-state characteristics. (From [7], © IEEE 2007.)

Putting a parallel link with a series bypass resistor in the WTG output helps to hold down the speedup and fallout of synchronization. After voltage restoration, the resistance link is shorted. This is shown in Figure 9-7.

9.5 A RESISTIVE BRAKING OF A WTG

Figure 9-8 below shows power-versus-speed characteristics of a WTG. The top curve represents conditions with a series damping break resistor (SDBR). The bottom curve shows the same without SDBR. A grid fault occurs at point 1 and the SBDR is switched in. The characteristics change over to the bottom one at point 2. It travels along 2–3 within the duration of the fault. At point 3, the SDBR is shorted and the characteristics jump to point 4. However, actual WTG at this changeover is at a low speed, corresponding to point 3, and also has a low power output. So the characteristics travel pack to point 5, which coincides with the original point 1. The SDBR has prevented a runaway increase in WTG speed.

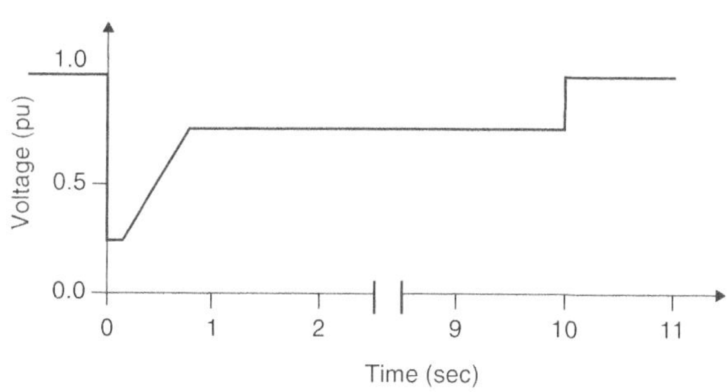

Figure 9-6. Specified voltage profile for the fault-ride-through test according to the Danish Grid Codes. (From [2], © IEEE 2007.)

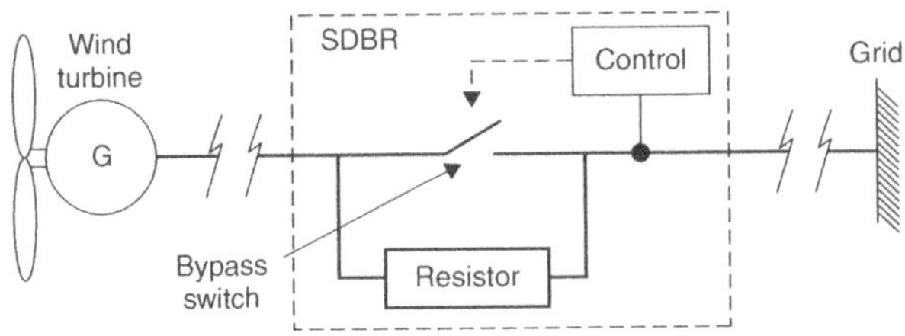

Figure 9-7. SBDR schematic arrangement. (From [2], © IEEE 2007.)

The ride-through periods for a WTG must be naturally much longer than those allowed under fault-clearance codes. Danish grid codes specify testing each WTG as per the specified voltage profile shown in Figure 9-6.

9.6 POWER AND PF CONTROL

Accurate power control on both active and reactive power is now obtained on the DFIG machines with power electronic control, as explained earlier. This enables the WTGs to maintain the required PF automatically at the PCC.

9.7 MODELING OF A WIND TURBINE GENERATOR

It is desirable first to understand how a vastly spread electricity power system operates physically. The following gives a brief sketch.

A transmission system operator (TSO) looks after load following and power quality on a very small scale time scale, say on a minute or 10 minute basis. For this, he has a

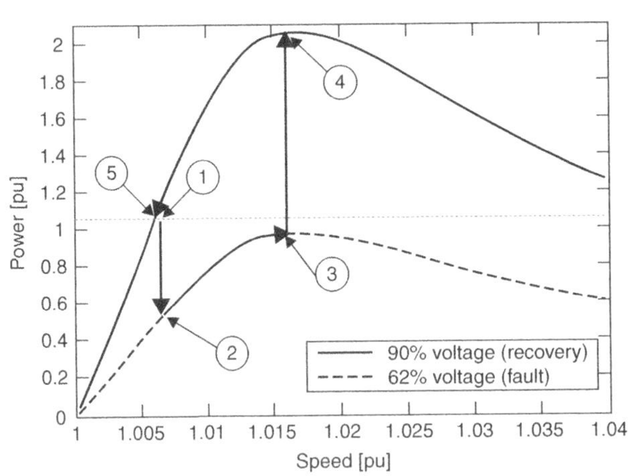

Figure 9-8. Enhanced FRT response with SDBR. (From [7], © IEEE 2007.)

schedule of power offers from various generators. The TSO has also a schedule of time-bound requirements from the customers. He matches these and balances the load. The TSO also has an updated chart of transmission facilities with all their characteristics in his computer. He selects an optimum route for a load dispatch. This route has minimum operating losses and costs. Modern fast-operating computers and accurate data are essential for his work. Supplying of accurate models of WTGs is compulsory for this reason.

9.7.1 Objectives

The TSO must simulate and forecast system stability problems with respect to frequency deviations, voltage variations, transients caused by voltage dip faults, and power system oscillations, which generally occur between 0.1 Hz and 10 Hz.

9.7.2 Method

Electrical and mechanical parameters of a WTG are converted into algebraic quantities. These algebraic notations are used to develop algorithms to arrive at characteristic functions. See a typical illustration in Figure 9-9.

Most of the grids in the world require a WTG dynamic model to be submitted to the transmission system operator for permission to join the grid.

Typical Irish grid requirements are listed below [8]. Any WTG greater than 5 kW must submit a model incorporating the following features:

1. Generator general characteristics
2. Turbine generator and drive train mechanical characteristics
3. Variation of power coefficients and pitch angle to tip speed ratio
4. Blade pitch control
5. Converter controls
6. Reactive components
7. Protection relays

Time per step for simulation should not to exceed 5 microseconds.

Although models for simulations for thermal and hydro generators have long been used and standardized, those for WTGs are still evolving and there are no standards.

9.7.3 Present Problem Areas in Modeling

1. Secrecy by WTG manufacturers who are developing the newest machines.
2. Nonstandardization, beginning with numbering and naming components.
3. Numerical instability arising out of rounding off and truncating practices, among others.
4. A large number of differing models.

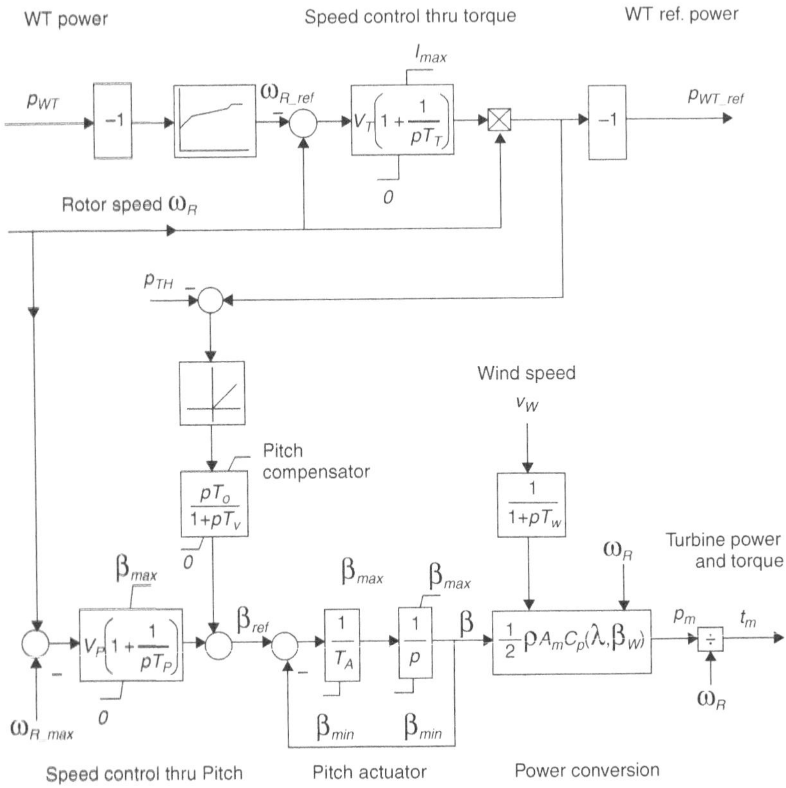

Figure 9-9. Rotor speed and pitch angle control as well as power conversion model. (From [9], © IEEE 2007.)

9.7.4 Model Validations

Models have to be validated for incorporation at the transmission system operator's operating desk. This validation can be partly done at the manufacturers place or at a designated laboratory test setup and, finally, in the field. Simulating a full 3 ph fault of sufficient magnitude will be difficult do. The importance and employment of models has already been commented upon.

9.8 ECONOMICS OF WIND ENERGY

9.8.1 How Does a Modern Power System Operate on the Marketing Side?

Equivalent to a TSO on the physical operation side, there is a system operator on the marketing side. Basically, he matches the offers and bids closely and allots them at the concluding time of day-ahead activities. These allotments are binding. A buyer cannot refuse them or else he risks a penalty plus differentials. A supplier cannot refuse or else the TSO makes the difference from other suppliers and debits the defaulter's account. The TSO also has a stock of his own previously contracted reserves that he sells at market price.

It may be noted that actual transactions take place in the real-time market, where prices and time slots can be negotiated and finalized over fixed time slots, say one hour. The prices at the end of this settlement period are the final prices at which accounts are settled.

Assume for a while that with all the available means in the hands of a TSO, the supply falls short and some load goes without service. This leads to a loss-of-load possibility or LOLP. It also leads to hand an imbalance and a Cost of Imbalance.

How do the evolved marketing practices affect the working and future of wind energy? Daily wind flow characteristics and daily load flow characteristics will vary. Universally, the late night and early periods show low loads. These have to be served by the base-load-supplying thermal generators, which have to run at near uniform output on account of thermal considerations, cost considerations, and so on. The TSO does not take into account wind energy contribution in his daily scheduling over this period. During peak load periods, when electricity gets the highest prices of the day, the cost of LOLP or the imbalance cost are the highest.

Smith [37] has worked out relations between capacity factor (CP) of wind energy, load factor (LF) of a system, and energy penetration (EP) for wind. Typically, for EP of 20%, an LF of 60% and wind capacity factor of 40% are required. Typical values for penetration factors are given in Table 9-4.

Assume that the settlement period is 2 hours. The unit commitment by a WTG operator based on his judgment can go far wide due to uncertainty of wind flows. His cost of imbalance will go up. Should this period be reduced to 1 hour, he can commit his units in smaller quanta and reduce his cost of imbalance. Meteorological data, past and current, over the next 24 hours can greatly help all the market players. This data is made available commercially in the western countries.

9.8.2 Unit Commitment and Scheduling

With power marketing growing in scale, load balancing becomes important. For this, the system operator must know accurately how much unit commitment he can have from the wind farm and how much capacity he should hold in reserve. This is a demanding task, considering the erratic nature of wind and the magnitude of loads to be handled. But the system operator has many tools in his hand, including dynamic scheduling and accurate hourly weather forecasts.

Dynamic Scheduling. With accurate modeling of the system and computerized software he can find out what can happen to the system when an electric component is added or subtracted or controlled in power output. All system components must be put in models for accurate simulation. Modeling right to the last details becomes important. Since the WTG technology is fast developing, standards for modeling these are not yet in place. Take the case of a 4.5 MW TWG weighing 450 t developed by Enercons in Germany. The very size contributes to LVRT on small duration faults. This large piece of apparatus is not the last step in WTG development.

Accurate Hourly Weather Forecasts. Weather is not all that erratic. Weather behavior falls into a pattern—daily, seasonal, periodical, and geography specific. Excursions out of this pattern might be in a band that can be anticipated. This, along with daily weather forecast by meteorology departments, can help the system operator on unit commitments from the wind farm on the day-ahead schedule as well as on the daily schedule, balancing them fairly closely. (Please see Chapter 13.) The system operator need not commit too much capacity to reserves. In fact, although wind energy costs are next to nil, their operational costs go largely toward unit commitments, as Table 9-4 shows.

9.9 CAPACITY FACTOR OF A WTG

Effective load-carrying capacity (ELCC) is defined as the amount of additional load that can be reliably served. In determining this capacity, two factors enter into the picture.

A load served during peak hours has the highest going market rate. If there is a LOLP during this period, cost of power is taken into for determining the capacity factor of a WTG. This period will not necessarily always coincide with the period of delivery at the rated output of a WTG. Thus, WTG capacity during the period of LOLP counts against its rated capacity for planning purposes, unit commitments, and so on.

Capacity value of a WTG has been shown to range approximately from 10% to 40% of the wind plant rated capacity (Table 9-4; [1]).

9.10 CAPACITY CREDIT CONSIDERATIONS

How do the capacity considerations affect wind energy installations? In many countries today, major utilities, mostly based on thermal generation, have to buy "green" energy such as wind energy at prices stipulated by the energy regulator. Typically, the basis for these prices is fixed on the following factors [10]:

- Capacity factor
- Operations and maintenance cost reductions for the utilities
- Electrical loss reduction for the utilities
- Environmental benefit

Hourly capacity factors are determined. The average of top 50% of the load hours (on both sides of the peak hours) is close to the estimation of the capacity factor for wind energy converters (WECs). In the deterministic method, historical data on wind is collected and the capacity factor is calculated. In the probabilistic method, this is calculated from the wind data statistics and forecasts. This gives too high a capacity. Thus, the capacity factor is important for WEC's since it directly affects their earnings. The capacity factor is also important from the grid operational point of view, for daily planning as well as for hourly operations. The capacity factor is important for future planners of the grid system. Probabilistic methods might be more useful here.

9.11 CAPACITY FACTOR FOR WECs IN A HYBRID SYSTEM

WECs are often combined with other distributed generators to ride over the unpredictability, again determined by the deterministic method. To that extent, the capacity credit goes up, particularly when the hybrid contributes energy around the peak load hours.

9.12 WIND PENETRATION LIMIT

The wind penetration is defined as the extent to which wind power can be added to a power system without compromising its operational reliability. Operational reliability in turn is defined in terms of a system's dynamic response, requiring the system frequency to stay within the limits of protection for a set of wind-related events. This makes acceptable system dynamics the sole criteria for increasing wind penetration [11].

Wind penetration simply means the extent to which wind power can be added to a power system without affecting its operational reliability. This reliability depends on the extent of reserves, external to the wind capacity, which are available to the system, internally or externally.

As per Mullane [12], the grid rule for Irish Electric Systems requires that the primary reserve in a system should be 75% of the single largest feed of the system. This should be available within 5 to 15 seconds subsequent to the event of the feed going out.

Again, if the load factor of a system is 60% and the average wind plant capacity is 40%, then the penetration has been worked out by Smith as 30% [1]. Average wind penetrations has been between 20% to 30%.

9.13 WIND POWER PROPORTION

Wind power proportion (WPP) is calculated by Soder [13] as a percentage as

$$\text{WPP} = \text{maximal wind power}/(\text{lowest consumption} + \text{possible exchange})$$

Gotland, an island off Sweden, provides an example of a set minimum and maximum power consumption, based on all types of generation, at 45 and 160 MW. During lean consumption periods, 180 MW of surplus power is exported. It has a maximum wind power capacity of 90 MW. Thus,

$$\begin{aligned}\text{WPP} &= 90/(45 + 180) \times 100 \\ &= 40\%\end{aligned}$$

9.14 WIND INTEGRATION COST IN UNITED STATES

See Table 9-4 for this data.

9.15 WIND ENERGY FARMS

With rising contribution by wind power to electricity systems, individual WTGs or clusters thereof are giving up their place in the energy scenario to wind energy farms. These farms have a few noteworthy features. They are generally located away from load centers. Connecting them into the grid system requires almost dedicated transmission arms. Whereas WTGs can be planned, installed, and put in operation within a short time span, connecting transmission links require long-time planning and execution. Existing transmission lines might not have the capacity to carry the wind farm power, as happened in the case of an East European customer who contracted for wind energy from a Dutch supplier [14]. An irregular flow on the direct line could also affect load flows and voltage regulation on the neighboring lines. This necessitates central control from the system operator over the operation and contribution by the wind energy farm.

9.16 PROMOTING GROWTH OF WIND ELECTRICITY

How we can increase wind energy penetration and at what level of operation can we make it economical? The first issue involves operational reliability. The second issue involves dimensions and integration into the existing operations.

Operational reliability involves relativity between connected wind-energy-derived capacities to total system capacity before the wind connection. In a contingency, the largest component of the wind capacity might trip. There must be sufficient reserve capacity to cover this loss. Failure of a wind capacity member may happen very fast. For making up the deficit, the ramping-up rate of the balance system should be high. Ramping rate of the balance system is a critical factor in the allowable safe wind penetration. When a new WTG is added, some grids put a condition on its ramping rate. This ramping-up rate is also important.

Holding down captive reserves on account of wind energy in the system increases the cost of wind energy and absence of it retards its growth.

The solution to reducing these reserves lies in accessing a larger resource base and consolidating balancing areas through use of dynamic scheduling. Accessing larger re-

Table 9-4. Wind integration costs in the United States

Date	Study	Wind capacity penetration (%)	Regulation cost ($/MWh)	Load following cost ($/MWh)	Unit commitment cost ($/MWh)	Gas supply cost ($/MWh)	Total operating cost impact ($/MWh)
May 03 [9]	Xcel-UWIG	3.5	0	0.41	1.44	na	1.85
Sep 04 [7]	Xcel-MNDOC	15	0.23	na	4.37	na	4.60
June 06 [10]	CA RPS Multi-year Analysis	40.45	na	na	na	na	
June 03 [11]	We Energies	4	1.12	0.09	0.69	na	1.90
June 03 [11]	We Energies	29	1.02	0.15	1.75	na	2.92
May 2005 [12]	PacifiCorp	20	0	1.6	3.0	na	4.6
April 06 [13]	Xcel-PSCo	10	0.20	na	2.26	1.26	3.72
April 06 [13]	Xcel-PSCo	15	0.20	na	3.32	1.45	4.97

Source: [1].

sources is achieved by interconnecting into neighboring resources, such as Denmark to North Germany and Denmark to Sweden, and into large grids, as in the proposed European grid and interstate connection in the United States.

The case of Denmark is illustrative. With a 300 kV AC connection into the North German system, it could depend on reactive power supply in case of its grid fault. It also exchanges power with Germany. It also has an HVDC connection to Sweden, which has high hydroelectric production. It could export wind energy whenever its production was high and domestic consumption was low. Hydroelectric power is very flexible.

Both these factors have helped the remarkable growth of wind power in Denmark.

Strengthening transmission systems also helps. There is an execution gap between a wind farm project and strengthening or adding a new transmission line. Transmission support systems such as FACTs devices, capacitor banks, and so on can effectively help. The economics will have to be worked out.

Transmission tariffs and regulations need to be favorably biased. Thermal generators connected to a transmission system are not allowed to feed in excess energy beyond an allotted quota. This restriction needs be relaxed in the case of wind energy [1].

Transmission systems are allowed to sell fixed transmission rights (FTRs) over their lines. These are physical rights and can end in underutilization. Hunt has proposed financial transmission rights to correct this. Please see Chapter 15 of this book on this subject. If this change is made, it will promote wind energy supply.

Developing day-ahead and hour-ahead marketing can cut down unit allocation costs and promote growth of wind energy. An example of NETA Systems in the United Kingdom can be found in Chapter 14.

Dynamic scheduling by the system operator helps. This has been discussed previously.

Models are also important in today's dynamic scheduling when incorporating wind energy.

9.17 MAINTENANCE OF WTG

Problems applying to modeling also apply to standardizing maintenance. Manufacturers treat failure data as proprietary and are not very cooperative in disclosing this. Ribrant [15] has surveyed failures in the Baltic area countries and finds that WTGs have the highest exposure to mechanical stresses. Loads are also high and of a fluctuating nature. So far, maintenance is of a corrective type. Downtime, loss of production, and repair costs are high. Problems are aggravated in offshore installations. Preventive maintenance is desirable. Geared WTGs operate through a weak shaft with torsion stresses. Quite often, there is mechanical resonance in the system, similar to electrical resonance. At this mechanical resonance frequency (about 1–2 Hz), there is shaft relaxation at shaft faults, leading to failures [15]. Including mechanical characteristic in modeling is essential.

9.18 UNFCCC AND WIND ENERGY

There is continuous pressure from the environmental groups to reduce atmospheric pollution, of which the second largest offender is thermal electricity production. Originally, it was intended that by a given target date electricity production from renewable sources would be 10% of total electricity production. This has since been revised.

Most countries are heading toward this 10% target. This is being helped by technical improvements in WTGs, immense advances in weather forecasting, which helps in dynamic scheduling, and last but not the least, by an abundant supply of wind energy at "nil" cost.

Wind generated energy has a drawback in its uncertain nature. Efforts are underway to tie wind electricity generation with hydroelectric plants. Please refer to Chapter 2 for details on this.

The uncertainty of wind in wind electricity generation prevents it from providing ancillary services like load-following frequency control to the grid. We are talking of a stage at which wind electricity is one of the major components of power electricity systems. This is being aided by introducing FACTS devices between wind generation and a tie-up to a power grid [15].

Still, wind energy has a very bright future.

REFERENCES

1. J. C. Smith, M. R. Milligan, E. A. DeMeo, and B. Parsons, Utility Wind Integration and Operating Impact State of the Art, *IEEE Transactions on Power Systems,* Volume 22, Issue 3, Aug. 2007, pp. 900–908.
2. V. Akhmatov and P. B. Eriksen, A Large Wind Power System in Almost Island Operation—A Danish Case Study, *IEEE Transactions on Power Systems,* Volume 22, Issue 3,Aug. 2007, pp. 937–943.
3. P. Fairley, The Greening of GE [Alternative Energy], *IEEE Spectrum,* Volume 42, Issue 7, July 2005, pp. 28–33.

4. H. De Battista and R. J. Mantz, Dynamical Variable Structure Controller for Power Regulation of Wind Energy Conversion Systems, *IEEE Transactions on Energy Conversion,* Volume 19, Issue 4, December 2004, pp. 756–763.
5. C. Carrillo, A. E. Feijoo, J. Cidras, and J. Gonzalez, Power Fluctuations in an Isolated Wind Plant, *IEEE Transactions on Energy Conversion,* Volume 19, Issue 1, March 2004, pp.217–221.
6. A. Grauers, Efficiency of Three Wind Energy Generator Systems, *IEEE Transactions on Energy Conversion,* Volume 11, Issue 3, September 1996, pp. 650–657.
7. A. Causebrook, D. J. Atkinson, and A. G. Jack, Fault Ride-Through of Large Wind Farms Using Series Dynamic Braking Resistors (March 2007), *IEEE Transactions on Power Systems,* Volume 22, Issue 3, Aug. 2007, pp. 966–975.
8. Y. Coughlan, P. Smith, A. Mullane, and M. O'Malley, Wind Turbine Modelling for Power System Stability Analysis—A System Operator Perspective, *IEEE Transactions on Power Systems,* Volume 22, Issue 3, Aug. 2007, pp. 929–936.
9. I. Erlich, J. Kretschmann, and J. Fortmann, Modeling of Wind Turbines Based on Doubly-Fed Induction Generators for Power System Stability Studies, *IEEE Transactions on Power Systems,* Volume 22, Issue 3, Aug. 2007, pp. 909–919.
10. R. M. G. Castro and L. A. F. M. Ferreira, A Comparison Between Chronological and Probabilistic Methods to Estimate Wind Power Capacity Credit, *IEEE Transactions on Power Systems,* Volume 16, Issue 4, Nov. 2001, pp. 904–909.
11. L. Pei, B. Hadi, K. Ping-Kwan, G. F. Hamed, and O. Boon-Teck, Macromodel of Spatial Smoothing in Wind Farms, *IEEE Transactions on Energy Conversion,* Volume 22, Issue 1, March 2007, pp. 119–128.
12. A. Mullane, G. Lightbody, and R. Yacamini, Wind-Turbine Fault Ride-through Enhancement, *IEEE Transactions on Power Systems,* Volume 20, Issue 4, Nov. 2005, pp. 1929–1937.
13. L. Soder, L. Hofmann, A. Orths, H. Holttinen, Y. Wan, and A. Tuohy, Experience From Wind Integration in Some High Penetration Areas, *IEEE Transactions on Energy Conversion,* Volume 22, Issue 1, March 2007, pp. 4–12.
14. L. Gestmar, L. Liljestrand, and H. Lendenmann, Wind Energy Powers-That-Be Successor Generation in Globalization, *IEEE Transactions on Energy Conversion,* Volume 22, Issue 1, March 2007, pp. 13–28.
15. J. Ribrant and L. M. Bertling, Survey of Failures in Wind Power Systems With Focus on Swedish Wind Power Plants During 1997–2005, *IEEE Transactions on Energy Conversion,* Volume 22, Issue 1, March 2007, pp. 167–173.

BIBLIOGRAPHY

Al-Hallaj, S., More than Enviro-Friendly: Renewable Energy is also Good for the Bottom Line, *Power and Energy Magazine,* Volume 2, Issue 3, May-Jun 2004, pp. 16–22.

Andersson, D., A. Petersson, E. Agneholm, and D. Karlsson, Kriegers Flak 640 MW Off-Shore Wind Power Grid Connection—A Real Project Case Study, *IEEE Transactions on Energy Conversion,* Volume 22, Issue 1, March 2007, pp. 79–85.

Bathurst, G. N., J. Weatherill, and G. Strbac, Trading Wind Generation in Short Term Energy

Markets, *IEEE Transactions on Power Systems,* Volume 17, Issue 3, Aug. 2002, pp. 782–789.

Billinton, R. and B. Guang, Generating Capacity Adequacy Associated with Wind Energy, *IEEE Transactions on Energy Conversion,* Volume 19, Issue 3, Sept. 2004, pp. 641–646.

Billinton,R., C. Hua, and R. Ghajar, A Sequential Simulation Technique for Adequacy Evaluation of Generating Systems Including Wind Energy, *IEEE Transactions on Energy Conversion,* Volume 11, Issue 4, December 1996, pp. 728-734.

Bresesti, P., W. L. Kling, R. L. Hendriks, and R. Vailati, HVDC Connection of Offshore Wind Farms to the Transmission System, *IEEE Transactions on Energy Conversion,* Volume 22, Issue 1, March 2007, pp. 37–43.

Castro, R. M. G. and L. A. F. M. Ferreira, A Comparison Between Chronological and Probabilistic Methods to Estimate Wind Power Capacity Credit, *IEEE Transactions on Power Systems,* Volume 16, Issue 4, Nov. 2001, pp. 904–909.

Castronuovo, E. D. and J. A. P. Lopes, On the Optimization of the Daily Operation of a Wind-Hydro Power Plant, *IEEE Transactions on Power Systems,* Volume 19, Issue 3, Aug. 2004, pp. 1599–1606.

Chen, Z. and E. Spooner, Grid Power Quality with Variable Speed Wind Turbines, *IEEE Transactions on Energy Conversion,* Volume 16, Issue 2, June 2001, pp. 148–154.

Chompoo-inwai, C., C. Yingvivatanapong, K. Methaprayoon, and L. Wei-Jen, Reactive Compensation Techniques to Improve the Ride-through Capability of Wind Turbine During Disturbance, *IEEE Transactions on Industrial Applications,* Volume 41, Issue 3, May-June 2005, pp. 666–672.

de Almeida, R. G. and J. A. P. Lopes, Participation of Doubly Fed Induction Wind Generators in System Frequency Regulation, *IEEE Transactions on Power Systems,* Volume 22, Issue 3, Aug. 2007, pp. 944–950.

Doherty, R., H. Outhred, M. O'Malley, Establishing the Role That Wind Generation May Have in Future Generation Portfolios, *IEEE Transactions on Power Systems,* pp. 1415–1422.

Fairley, P., Steady as She Blows [Wind Power, Energy Storage], *IEEE Spectrum,* Volume 40, Issue 8, Aug. 2003, pp. 35–39.

Feijoo, A. and Cidras, Analysis of Mechanical Power Fluctuations in Asynchronous WECs, *IEEE Transactions on Energy Conversion,* Volume 14, Issue 3, September 1999, pp.284–291.

Glushakow, B. Effective Lightning Protection For Wind Turbine Generators, *IEEE Transactionson Energy Conversion,* Volume 22, Issue 1, March 2007, pp. 214–222.

Hurt, S., *Making Competition Work in Electricity,* Wiley, 2002.

Jayadev, J., Harnessing the Wind, *IEEE Spectrum,* Volume 32, Issue 11, Nov. 1995, pp.78–83.

Lie, X., Y. Liangzhong, and C. Sasse, Grid Integration of Large DFIG-Based Wind Farms Using VSC Transmission, *IEEE Transactions on Power Systems,* Volume 22, Issue 3, Aug. 2007, pp. 976–984.

Lubosny, Z. and J. W. Bialek, Supervisory Control of a Wind Farm, *IEEE Transactions on Power Systems,* Volume 22, Issue 3, Aug. 2007, pp. 985–994.

Mandelbaum, R., Reap the Wild Wind [Offshore Wind Farm], *IEEE Spectrum,* Volume 39,Issue 10, Oct. 2002, pp. 34–39.

Moreno, C. V., H. A. Duarte, and J. U. Garcia, Propagation of Flicker in Electric Power Networks Due to Wind Energy Conversions Systems, *IEEE Transactions on Energy Conversion,* Volume 17, Issue 2, June 2002, pp. 267–272.

Morren, J. and S. W. H. de Haan, Short-Circuit Current of Wind Turbines With Doubly Fed Induction Generator, *IEEE Transactions on Energy Conversion,* Volume 22, Issue 1, March 2007, pp. 174–180.

Muljadi, E., C. P. Butterfield, B. Parsons, and A. Ellis, Effect of Variable Speed Wind Turbine Generator on Stability of a Weak Grid, *IEEE Transactions on Energy Conversion,* Volume 22, Issue 1, March 2007, pp. 29–36.

Mullane A. and M. O'Malley, The Inertial Response of Induction-Machine-Based Wind Turbines, *IEEE Transactions on Power Systems,* Volume 20, Issue 3, Aug. 2005, pp.1496–1503.

Negra, N. B., O. Holmstrom, B. Bak-Jensen, and P. Sorensen, Aspects of Relevance in Offshore Wind Farm Reliability Assessment, *IEEE Transactions on Energy Conversion,* Volume 22, Issue 1, March 2007, pp. 159–166.

Nilsson, J. and L. Bertling, Maintenance Management of Wind Power Systems Using Condition Monitoring Systems—Life Cycle Cost Analysis for Two Case Studies, *IEEE Transactions on Energy Conversion,* Volume 22, Issue 1, March 2007, pp. 223–229.

O'Malley, M. and J. J. Sanchez-Gasca, Guest Editorial: Special Section on Wind Energy, *IEEE Transactions on Power Systems,* Volume 22, Issue 3, Aug. 2007, pp. 899–899.

Olken, M., Powerful Forces Wind, Energy Policy, and the Magazine—from the Editor, Power and Energy Magazine, Volume 3, Issue 6, Nov.-December 2005, pp. 4–6.

Palanichamy C. and N. S. Babu, Day-Night Weather-Based Economic Power Dispatch, *IEEE Transactions on Power Systems,* Volume 17, Issue 2, May 2002, pp. 469–475.

Papathanassiou, S. A. and F. Santjer, Power-Quality Measurements in an Autonomous Island Grid with High Wind Penetration, *IEEE Transactions on Power Delivery, June 2006, pp. 218–224.*

Piwko, R., D. Osborn, R. Gramlich, G. Jordan, D. Hawkins, and K. Porter, Wind Energy Delivery Issues, Power and Energy Magazine, Volume 3, Issue 6, Nov.-December 2005, pp. 47–56.

Pradhan, A. K. and G. Joos, Adaptive Distance Relay Setting for Lines Connecting Wind Farms, *IEEE Transactions on Energy Conversion,* Volume 22, Issue 1, March 2007, pp. 206–213.

Roberts, B. W., D. H. Shepard, K. Caldeira, M. E. Cannon, D. G. Eccles, A. J. Grenier, and J. F. Freidin, Harnessing High-Altitude Wind Power, *IEEE Transactions on Energy Conversion,* Volume 22, Issue 1, March 2007, pp. 136–144.

Rodriguez, J. M., J. L. Fernandez, D. Beato, R. Iturbe, J. Usaola, P. Ledesma, and J. R. Wilhelmi, Incidence on Power System Dynamics of High Penetration of Fixed Speed and Doubly Fed Wind Energy Systems. Study of the Spanish Case, *IEEE Transactions on Power Systems,* Volume 17, Issue 4, Nov. 2002, pp. 1089–1095.

Sorensen, P., N. A. Cutululis, A. Vigueras-Rodriguez, L. E. Jensen, J. Hjerrild, M. H. Donovan, and H. Madsen, Power Fluctuations From Large Wind Farms, *IEEE Transactions on Power Systems,* Volume 22, Issue 3, Aug. 2007, pp. 958–965.

Sweet, W., Danish Wind Turbines Take Unfortunate Turn, *IEEE Spectrum,* Volume 41, Issue 11, Nov. 2004, pp. 30, 34.

Tentzerakis, S. T. and S. A. Papathanassiou, An Investigation of the Harmonic Emissions of Wind Turbines, *IEEE Transactions on Energy Conversion,* Volume 22, Issue 1, March 2007, pp. 150–158.

Thiringer, T., T. Petru, and S. Lundberg, Flicker Contribution from Wind Turbine Installations, *IEEE Transactions on Energy Conversion,* Volume 19, Issue 1, March 2004, pp. 157–163.

Ummels, B. C., M. Gibescu, E. Pelgrum, W. L. Kling, and A. J. Brand, Impacts of Wind Power on Thermal Generation Unit Commitment and Dispatch, *IEEE Transactions on Energy Conversion,* Volume 22, Issue 1, March 2007, pp. 44–51.

Weixing, L. and O. Boon-Teck, Optimal Acquisition and Aggregation of Offshore Wind Power by Multiterminal Voltage-Source HVDC, *IEEE Transactions on Power Delivery,* Volume 18, Issue 1, Jan 2003, pp. 201–206.

Zorpette, G., Technology 1993—Power and Energy, *IEEE Spectrum,* Volume 30, Issue 1, Jan. 1993, pp. 61–64.

10

PHOTOVOLTAIC ENERGY—SOLAR CELLS AND SOLAR POWER SYSTEMS

10.1 PHOTOVOLTAIC ENERGY—HOW IT WORKS

When a photovoltaic (PV) cell is exposed to the sun's thermal radiation, it absorbs the thermal energy and converts it directly into DC electrical energy. The size of the PV cell and the DC voltage and energy it can deliver are small. A number of these cells are mounted on a plate and connected in series and parallel. These plates together form a solar array. These arrays require sizeable exposed area and reasonably clear skies to deliver useable quantities of electrical energy.

Exploiting solar energy and converting it into an electrical form has been under intense development for a long time. It has several advantages.

10.2 ADVANTAGES OF PHOTOVOLTAIC ENERGY

First of all, this energy is free. It consumes no fuel. No fuel means no pollution. It is easily and universally available. Increases in solar radiation leading to increases in solar array output coincide with the increase in the load demand during a day. One advantage of solar energy is that its availability, except for cloud formation, is precise and definite. Scheduling and delivery come on the dot.

Copyright © 2011 the Institute of Electrical and Electronics Engineers, Inc.

With countries all over the world getting pollution conscious, solar energy is becoming the primary renewable energy source to be developed to replace the present fossil-fuel-based thermal generation of electricity. To that extent, many economic incentives are offered in most of the countries in the world for developing solar energy. To begin with, there are direct subsidies and tax incentives to support the high capital costs. Second, there are carbon trading certificates which are issued for reduction of so many tons of carbon dioxide from emission into the air. Thermal plants have restrictions on how much carbon dioxide gas they can put into the atmosphere. When they exceed this limit, they have to buy carbon trading certificates. These certificates can be traded internationally and are a great attraction for solar energy producers.

Being dispersible is a big advantage for solar energy generators. Locations far from the generating stations suffer losses and voltage drops when they draw electric power from utility grids. Supporting PV generation at these locations can help both the utility and the customer greatly.

PV is a "must" at locations where a grid has not been able to reach.

10.3 DISADVANTAGES OF PV ENERGY

PV energy has some substantial disadvantages as well. First of all, the basic solar cell is still very costly. Collecting viable energy requires sizeable infrastructure support in panels and space. The capital costs are quite high. In spite of free energy and next-to-nothing labor costs, the energy costs with reference to servicing the capital costs are much too high.

Another big disadvantage is the uncertainty in energy supply due to cloud effects. This uncertainty could occur within the routines of a day. It could occur seasonally also. The uncertainty has to be supplemented through costly storage batteries or through other types of generations or through the increased reserves in the utility system. All these raise the overall costs. Solar energy also has problems in interfacing with regular utility grids.

10.4 SOLAR THERMAL DENSITY—INSOLATION

Insolation is a measure of solar thermal density in terms of W/m^2 of the colleting area. For comparing the output of various solar cells, the reference insolation level generally accepted is 1000 W/m^2. This solar density varies as the day proceeds, being low in the morning and evening hours and high during the day. Figure 9-1 gives this value in average terms. It shows how much insolation is available for how many hours of a day. The insolation varies seasonally also. Figure 9-2 gives the values for insolation and ambient temperature in one of the world's extreme climate places, Kuwait. In the figure, the area above 30% insolation denotes major availability of solar insolation. On the hottest day in summer, in July, it occurs between 7:15 AM and 16:45 PM. On the coldest day in winter, in January, it occurs between 8:45 AM and 15:15 PM. As we move geographically to the north, these time windows will narrow down, and expand as we move south toward the equator.

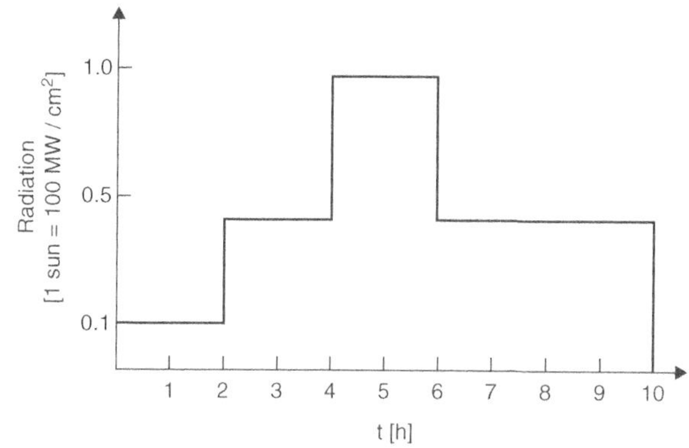

Figure 10-1. Daily insolation curve. (From [1], © IEEE 2001.)

10.5 OUTPUT OF A PV CELL

Figure 10-3 illustrates curren-versus-voltage characteristics of a PV cell. The curve P denotes the locus of points for various degrees of insolation at which we can get the maximum PV cell output.

10.6 VARIATION WITH AMBIENT TEMPERATURE

Figure 10-4 shows the output of a PV cell at varying ambient temperatures. Note that the output falls drastically as the ambient or PV cell temperature rises.

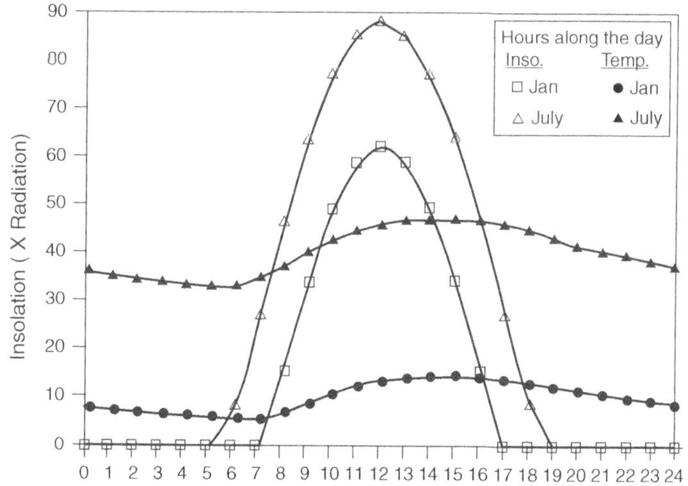

Figure 10-2. Hourly variation in solar insolation. (From [1], © IEEE 2001.)

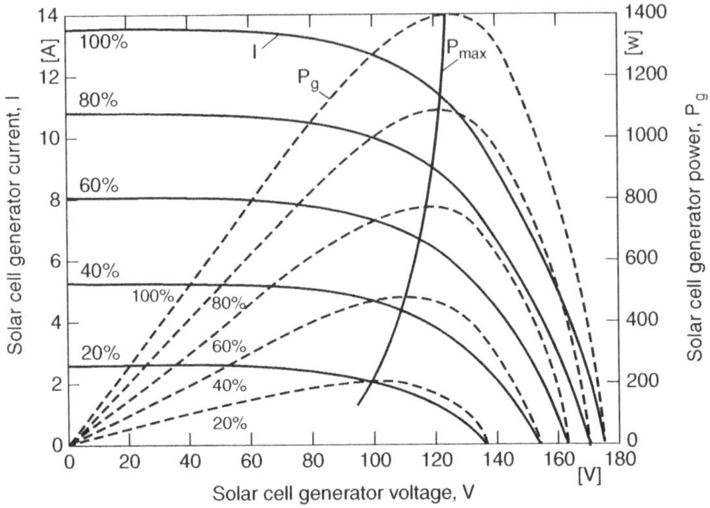

Figure 10-3. SCA voltage–current characteristics of a PV cell. (From [2], © IEEE 2001.)

10.7 VOLTAGE-VERSUS-CURRENT CHARACTERISTICS OF A SOLAR CELL

It may be seen that the power output of a PV cell is a complicated problem for analysis and control. The voltage created is a function of insolation. The current and power output depend on the properties of the silicon wafers commonly used in PV cells. Masoum [9] gives the following equation for the relation between voltage and the current:

$$v_{PV} = \frac{N}{\lambda} \ln\left(\frac{I_{sc} - i_{pv} + MI_0}{MI_0}\right) - \frac{N}{M} R_s i_{pv}$$

This is based on the basic diagram shown in Figure 10-5. Here,

v_{PV} = the output voltage of the PV cell at load terminals
i_{pv} = the output current of the PV cell at load terminals
i_{sc} = the cell short-circuit current, which is proportional to the insolation level
I_0 = the reverse saturation current
R_s = the series cell resistance
λ = a constant coefficient representing the cell material

Masoum gives the specifications shown in Table 10-1 for the solar panel he has experimented on. The voltage and power output with respect to current for two conditions of isolation as per computations based on the table and as per actual readings are given in Figure 10-6.

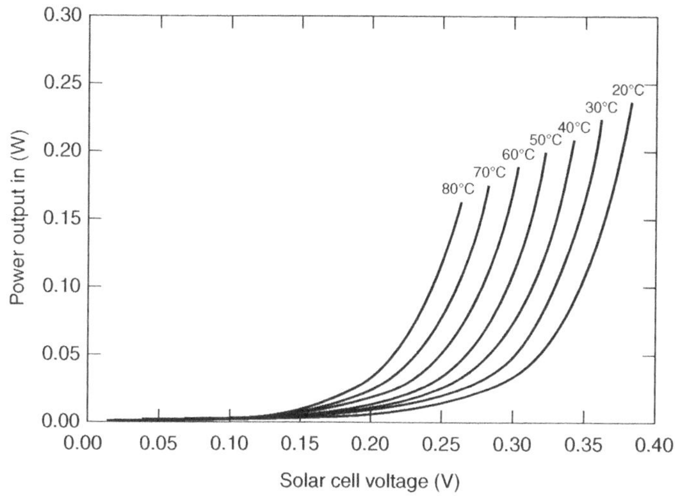

Figure 10-4. Effect of temperature on the maximum cell electrical power P_{emax} in terms of V_{max}. (From [2], © IEEE 2007.)

10.8 MATCHING THE PV WITH THE LOAD

10.8.1 Maximum Power Point Tracker (MPPT)

If we just connect a solar array and leave it at that, we might not get the best or maximum output from it. Output improves if we continuously correlate V and I. An MPPT enables us to do so. There are two important parameters for an MPPT. One is the i_{sc} or the short-circuit current of the cells. This is linearly proportional to the insolation. It varies per environmental conditions. It is periodically measured by an MPPT. The load is momentarily disconnected, the cell is shorted through a known resistance, and the i_{sc} value is read. Another important parameter is the V_{oc} or the open circuit voltage of the cell. It is logarithmically related to the insolation.

The actual cell voltage and currents are measured, referred to i_{sc}, and fed to the MPPT. The MPPT works out the maximum power point and gives switching signals to

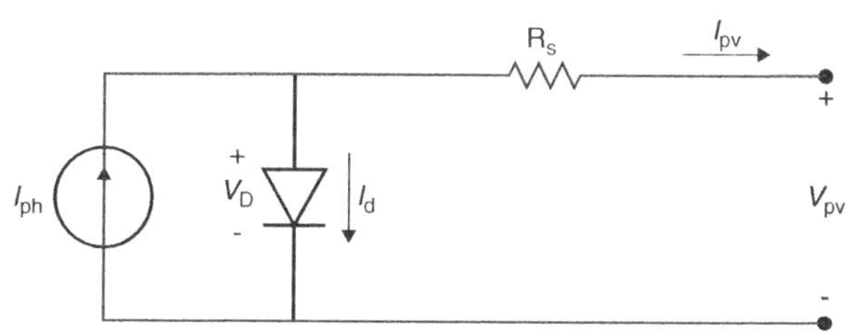

Figure 10-5. Equivalent circuit of PV solar cell. (From [3], © IEEE 2004.)

Table 10-1. Specifications of silicon solar panels manufactured by OFFC

Current temperature coefficient	$\alpha = 0.002086$	[A/°C]
Voltage temperature coefficient	$\beta = 0.0779$	[V/°C]
Reverse saturation surrent	$I_0 = 0.5 \times 10^{-4}$	[A]
Short-circuit cell current	$I_{ph} = I_{sc} = 2.926$	[A]
Cell resistance	$R_s = 0.0277$	[Ω]
Cell material coefficient	$\lambda = 0.049$	[1/V]

Source: [3].

the PWM system in the inverter. The inverter switches by modulating the pulse width control, the power input in the load. Note that in a normal electricity power generator, one controls the generator output by controlling the field excitation, energy input, inlet air control, and so on. Here, we pick up the best points across a PV cell for the maximum energy it is producing. The control is relative to the load and MPPTs are said to characterize the load. We do not control the voltage induced by isolation but pick up the most favorable value. In a voltage-reference-based VMPPT, the reference is made to the open-circuit voltage of the cell. This arrangement is suitable for resistive loads. Typically, it has increased the power output by 20%.

10.8.2 VMPPT and CMPPT

In a current-based MPPT (CMPPT), the reference is made to i_{sc}. This arrangement is suitable for a pump load. Here, the losses are slightly higher and the reported increase in pow-

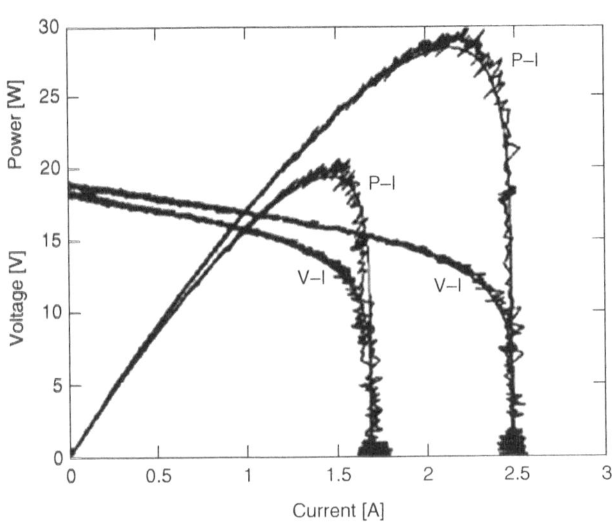

Figure 10-6. Computed (two conditions) and measured nonlinear V–I and P–I characteristics of one OFFC silicon solar panel (see Table 10-1). (From [3], © IEEE 2004.)

er output stands at 10%. The overall efficiency of an MPPT-based control system is fairly high—over 97%. The increase in output is not diminished by the losses in the MPPT.

A voltage-based MPPT (VMPPT) is more suitable for low-voltage, high-current loads such as battery chargers and low resistance levels. A CMPPT is more suitable for high-voltage, low-current loads such as DC motors on agricultural pumps.

10.9 OLD WORKING MODEL OF AN MPPT

A simple old-time MPPT based on saturating a transformer and controlling the current output is shown in Figure 10-7. In the figure, TVT is a time-variable transformer, for which the transformation ratio k is controlled and changed corresponding to the operating point of the load. As the name suggests, it is also time variable to enable it to keep in step with the insolation time variation. Thus, it gives maximum possible power-point operation.

Use of an MPPT depends on a variety of factors. If the seasonal temperature variation is high, it is preferable to have an MPPT. A load requiring constant current is bad for a PV system whose I–V and temperature characteristics make it difficult to comply with the requirements of constant current. An MPPT helps here. A solar array costs \$5000 to \$10,000 per watt. Corresponding MPPT costs will vary between \$150 and \$200 per watt. Its efficiency will be as high as 90%.

10.10 MAXIMIZING THE OUTPUT OF A SOLAR PANEL

10.10.1 By Orienting the Solar Panel

A low available solar input may be maximized by orienting the PV cells in the direction of the sun's rays. This is accomplished by rotating the panel around its axis during the day and rotating the axis itself as per season.

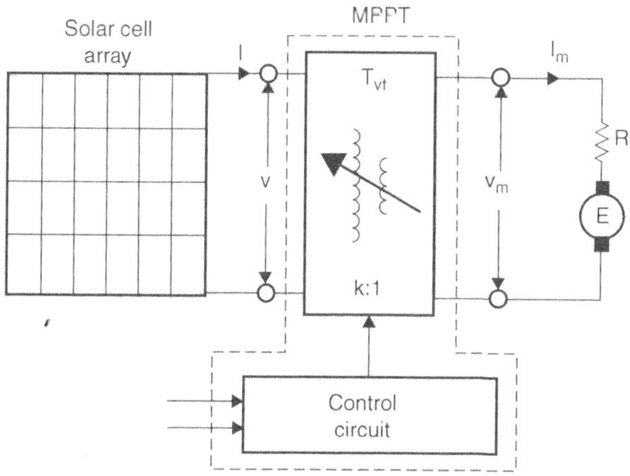

Figure 10-7. A solar cell system with an MPPT. (From [4], © IEEE 1993.)

The Carissa solar power plant rated at 6.5 MW installed by Pacific Gas and Electric Co. in the United States is reported to have incorporated both the above movements. It has also incorporated reflecting plates around the solar arrays, thus increasing the radiation collection. This plant is tied directly into a 110 kV electricity grid system.

Incorporating array orientation towards the sun's rays involves additional capital costs as well as some energy loss due to use of collected electrical energy for rotation. Both of these have to be balanced against the benefits in increased PV output and its value.

Rahman [5] has calculated that incorporating daily axis rotation is not viable on the basis of a cost-to-benefit ratio. Instead, axis orientation once a month is just as productive, with minimum costs involved.

An ingenious method reportedly uses air cylinders to rotate the panels [5]. These air cylinders develop air pressures as they get hot under sun's radiation and do not require external energy input for rotation.

10.10.2 By Water Cooling the Solar Panel Backs

Another installation has reported water cooling of the plates from the underside of array plates to keep the PV cell temperatures low [5]. This gives increased electrical outputs.

10.11 INTERFACE WITH A POWER SYSTEM

Figure 10-8 gives the workings of a complete system. The PV panel feeds into an MPPT. Output of the MPPT is supported by a paralleling device, in this case, an EDLC (electric double layer capacitor). This output goes to a power conditioning system, PCS. The PCS gets its control signals from a control system. There are a number of storage devices to complement the output from the PV panel, the EDLC or ECaSS being one of them. Table 10-2 compares the properties of an ECaSS with two different batteries.

10.12 POWER CONDITIONING SYSTEMS

The DC output of a PV system has limited direct uses, including residential panels for water heating and directly coupled agricultural pumps in a standalone system, for which the driving motors are rated at DC voltages.

In residential complexes, PV systems are used to substitute for grid energy when the sun's energy is available. PV supply parameters must be identical with the grid supply parameters for the same customer's appliances to work. In this case, a utility might use a PV system owned by it to reduce the power peak demand by air conditioners during the noon and afternoon periods of hot summer months. The utility might

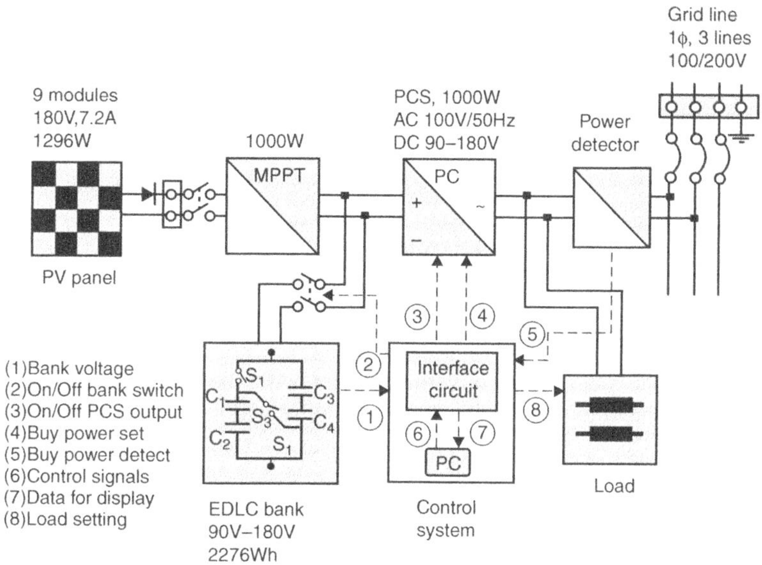

Figure 10-8. Block diagram of a solar power system. (From [6], © IEEE 1998.)

also want to use the PV system to support its distribution at far off points, both with active as well as with reactive power.

All these require an interface or a power conditioning system (PCS) that will match and tie up with the grid system in voltage, frequency, transient behavior, and so on.

The PCS converts DC into AC with parameters required at the utility connection point.

The PCS has three basic functions to perform:

1. It has to monitor and adjust so as to receive maximum power from the PV source and match its output with an MPPT. For this, the load characteristic represented by the PCS has to suit the output characteristic of the PV–MPPT system so that there is optimum power under all conditions of insolation.
2. It must transform DC into AC and hold on to the quality of power on the AC side.
3. It must monitor the utility system and maintain the steady-state compatibility, as well as respond appropriately to utility-side dynamics.

In this way, it forms an interface with the utility. It is an important and critical link.

10.12.1 Quality Requirements of a PCS

- It must have as perfect a reliability as possible.
- It must have a high efficiency.

- It should not add substantially to the overall costs of a PV-energy system, which already has the handicap of high-cost solar cells.
- It should take care of harmonic distortion and improve power factors where feasible and necessary.
- The high-frequency switching of the inverter, especially at higher power handling, tends to create electromagnetic interference on the output side. This has to be checked and guarded against.
- Safety to utility personnel and equipment has to be the topmost priority.

10.12.2 Converting DC into AC

There are two basic methods of converting DC into AC: the line commutation method and the forced switching system, also known as pulse width modulation (PWM).

10.13 SUPER CAPACITORS AND STORAGE BATTERIES

Table 10-2 compares the performance of supercapacitors to two types of batteries.

10.14 NERC GUIDELINES FOR CONNECTING A PV SYSTEM TO A GRID

NERC guidelines require the PV system to be isolated from the utility at the instant the utility supply is switched off for any reason. Should this not happen, the PV system will enliven the utility system and endanger personnel and equipment on the utility side. This protection is dealt with in detail in a later chapter. One of the most secure methods to ensure that the interface is not alive is to deenergize the inverter control grids.

Another NERC guideline requires the system reserves to be adequate enough to hold the supply parameters within limits when the largest single contributing member goes out. This is where the irregularity of PV supplies clashes with NERC rules.

Table 10-2. Comparison between ECaSS and popular storage batteries

Parameters	EcaSS	Lead–Acid	NiCd
Power density	400	10–300	100–160
Energy density	6.5	40–45	45–53
Depth of discharge	95	50–70	80–100
Efficiency (%)	–97	–80	–75
Charging time	1 min	5–10 h	15 min–8 h
Life cycle	–3,000,000	300–1000	300–500
Oruce ($/kW-hr)	4	0.2–0.8	0.8–1.5

Source: [6].

10.15 PROBLEMS OF INTERFACING PV SYSTEMS WITH THE GRID

Effect on grid:

1. An ad hoc PV connection into a utility grid system can overload the distribution lines. The distribution lines normally keep on adding loads, but are not frequently replaced or strengthened. If a PV output also flows along these lines to a consumer, the lines will get overloaded. This has to be particularly watched on residential PV systems.
2. Single-phase residential PV systems can cause phase imbalance and shift the neutral to unsafe values.
3. Line commutation, usually on low-voltage systems, degrades the overall power factor and hurts the supply voltages. It also increases harmonic distortion.
4. If there is a fault, it will be doubly fed, both by the utility as well as by the PV systems. This could be dangerous.
5. Similarly, protective relays, voltage regulation with load tap changes, and so on can be affected by a newly connected PV system.

Overcoming all the above problems requires careful study, including load flows, before connecting to a PV system. However, there are more serious problems to be thought of when we think of switching over to 10% of the total supply under a PV energy system, as envisaged under the Kyoto proposal on renewable energies. These are:

1. Cloud effects. Fast ramping rates from the grid are required. Whenever a dark cloud passes over a PV array its output dips substantially. Let us assume that an array supplying 1 MW of power is blanked out by a cloud in 10 seconds. Then the utility should be able to pump in extra power at the rate of 1 MW/10 seconds. Instances have been reported where clouds have covered arrays at the rate of 15 clouds per minute. Conversely, when the cloud disappears, let us say in just 10 seconds, the utility will have to absorb power at the ratio of 1 MW/10 seconds. This reverse flow could raise bus voltages and result in uncontrolled and undesirable inter tie flow of current between adjacent feeders.
2. Loss-of-load possibility (LOLP). Suppose that the ramping up or down rate of energy flow is too high for the system capacity to absorb. The voltage dips momentarily to, say, 70% of the rated system voltage. Many of the magnetizing coils on switches and contactors will not be able to hold on the contactors, thus disconnecting the connected load. This loss-of-load possibility is strictly forbidden. On the other hand, if the lowest voltage limit is, say, 5% of the rated voltage and this is permitted to go down to, say, 7% then one can correspondingly raise the percentage penetration of the grid by a PV system (415 V tolerance of ± 5% is brought down to ± 7%).

10.16 PENETRATION PERCENTAGE BY A PV ENERGY SYSTEM INTO A UTILITY GRID

Load following involves taking care of short-term or long-term dips and highs by the load and maintaining the power quality. For this, the supply system must have adequate capacity. When we add the cloud effect problems created by a PV system, the load following becomes more demanding. It requires more of the system capacity to handle this than before. Alternatively, we hold down these extra capacity requirements to manage the power quality control by limiting the percentage penetration by a PV of the utility capacity at that point. Thus, percentage penetration by a PV system is limited by the reliability of its energy supply when viewed against the potential capacity of the grid to absorb the cloud-effect shocks. This reliability increases as we increase the area under a PV system. Average cloud effect will fall as the coverage area increases and PV arrays are dispersed widely. Jewell [7] gives the following typical figures:

$$
\begin{aligned}
A_{H0} &= 10\ \text{km}^2 \\
&= 100\ \text{km}^2 \\
&= 1000\ \text{km}^2 \\
\text{penetration} &= 6.3\%\ \text{by PV} \\
&= 18.1\% \\
&= 35.8\%
\end{aligned}
$$

10.17 PROGRESS IN APPLICATION OF PV ENERGY

As detailed earlier, capital costs for utilizing solar energy are high. Besides, solar energy is dispersed widely. It does not come in a concentrated form like wind energy and all other types of energies. There is a universal reckoning that solar energy could be one of our saviors in the long run. The progress in solar energy utilization may be attributed to necessity, institutional encouragement, and market forces.

Growth due to necessity is exemplified by agricultural water pumps—growers need not wait for the arrival of electricity distribution systems—and local electrification schemes for far off hamlets. Residential complexes have been encouraged to install PV panels, generally for home water heating, through inducements offered by local and government agencies. Multistory buildings with many computers, residential complexes, large commercial marketing installations, hospitals, and so on are increasingly taking a look at the financial advantages that can be gained particularly under the new electricity marketing systems. Other uses of PV systems are PV packages for power factor improvement and PV panels on the tops of automobiles. We will examine these in detail.

10.17.1 PV Cells and Agricultural Pumps

Selection of a DC Motor. Kolhe [8], in a series of experiments, has calculated that the solar panel and the DC motor must be selected with due consideration for matching. Morning solar intensities are low. The starting torque speed characteristics of the pump motor are generally steep. At noon when the solar intensities are high, the torque and speed could both be high (Figure 10-9). With normal tracking, the performance has been improved by over 20% (Figure 10-10).

An agricultural water pump has three components whose characteristics must match one another:

1. PV cells, output characteristics of which were given in Figures 10-1 and 10-2
2. A DC electric motor
3. A mechanical load driven by the motor

The types of DC motors available are series connected, separately excited, shunt connected, and permanent-magnet excited. The connections refer to the field winding, which produces the air gap flux.

The mechanical load is normally a centrifugal pump that requires fairly low starting torques. The starting torque has an important bearing in the Indian context.

The water pump population per hectare has gone up. With each pump trying to pump out maximum water, the subsoil water levels fall quickly to lower and lower levels. The starting torques and, consequently, the power used, goes up. If a larger than normal pump is used, which is often the case, then the supporting PV array system must also be large enough.

With a large population of water pumps, the subsoil water level recedes to lower strata in the summer season and the running and starting torque requirements increase. Here again, more often than not, larger pump sizes are used than otherwise required. The PV array sizes will have to match accordingly.

The starting current and starting torque with relation to their running values have important effects:

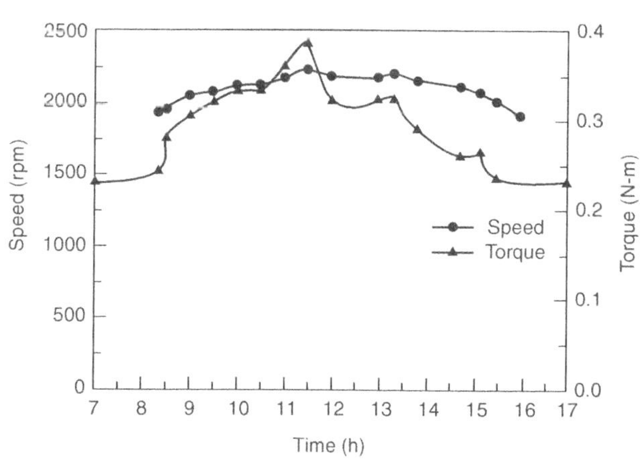

Figure 10-9. Speed and shaft torque with time for 5 m water head. (From [8], © IEEE 2004.)

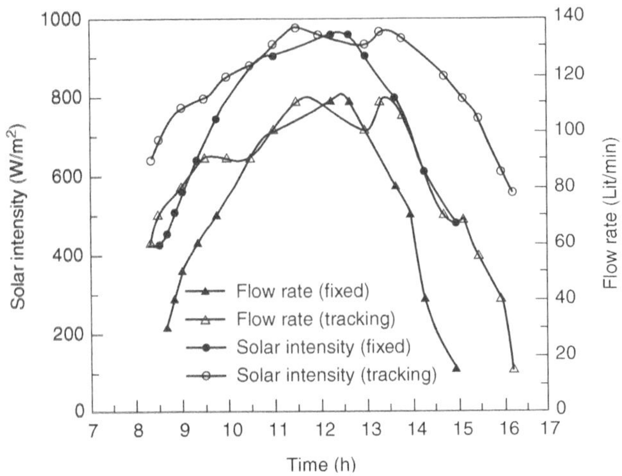

Figure 10-10. Solar intensity and flow rate for fixed and manual tracking modes. (From [8], © IEEE 2004.)

1. At the starting instant of a DC motor, the motor terminal voltage is equal to the voltage drop across the armature (and the field windings, in a series-type motor). As the armature picks up speed, it develops a back EMF. This reduces the current. With a PV cell, the starting current is limited to its short-circuit current at a given insolation. This implies two requirements. First, output from the PV cell should be drawn at the maximum power output point corresponding to the insolation level. An MPPT helps to achieve this. Second, the duration of these high starting currents, until the pump develops full speed, should be as low as possible. This requires as high a starting torque as possible. Long starting durations damage the armature windings.
2. The torque developed by a DC motor is proportional to the product of the air gap flux and the armature current. The starting torque should be high.

An MPPT Helps. Singer [4] has developed equations for the ratios of starting current to starting torques and starting currents to rated currents, with and without the use of an MPPT for various types of motors. These are given in Table 10-3.

Suitable DC Motor. Series and separately excited motors show better characteristics and are preferable. Shunt motors have the worst characteristics, followed by permanent-magnet motors. Again, performance with an MPPT is better than without it, justifying its additional costs and losses.

Operation of Solar Water Pumps. Solar water pumps cannot be switched on randomly. Starting a PV cell operated DC water pumps takes relatively longer time if started early in the morning, when insolation levels are low. It will develop an undesir-

Table 10-3. Starting torques and current ratios of DC motors rated at design insolation

	Permanent magnet	Series	Shunt	Separately excited (split)	Separately excited (unsplit)
	Starting to rated torque ratio				
Without MPPT	1.2	1.44	0.14	1.2	1.2
With MPPT	3.17	9.99	.97	8.33	26.35
Magnification m_T	2.64	6.94	6.94	6.94	21.96
	Starting to rated current ratio				
Without MPPT	1.2	1.2	1.2	1.2	1.2
With MPPT	3.17	3.17	3.17	3.17	3.17
Magnification m_T	2.64	2.64	2.64	2.64	2.64

Source: [4].

ably high starting torque if started at noon, when the insolation levels are high. High torques are also undesirable. They result in higher RPM and higher power output, leading to cavitations and formation of air bubbles in a pump. In such cases, the MPPT setting should be lowered.

Stand-alone Hybrid Power Stations for Remote Hamlets. These are covered in detail in Chapter 16.

Residential Complexes. The increasing cost of electricity and the increasing sizes of residential colonies and complexes are leading toward more solar energy than before, especially in Germany and Japan. First, technical problems due to the fluctuating nature of solar energy have to be overcome. This is dealt with extensively in Chapter 16 of this book, where detailed calculations for the storage needed to smooth out fluctuations are given.

There are financial advantages to solar systems under the new marketing schemes. Rahman [9] gives a detailed look into the possible financial gains in a Japanese system by proper sizing and operating a PV system and using this data and the metrological data for hourly buying power in the electricity market. There are substantial financial savings.

Optimizing Power Purchases from the Grid. The PV output energy is determined as that extra energy consumed during peak periods over an average daily energy. This is upgraded by allowing for efficiencies of MPPT, PCS, and so on. There is a difference between the PV output during the shortest and the longest days of the year. This difference is added to the PV rating. There is also a difference between peak load levels throughout the year. This difference is added. All this helps us to arrive at the total rating of the PV panel.

Statistical data (daily) on insolation levels and on cloud and other disturbing effects (rain, snow, etc.) is collected. With this data and forecast data on weather, next-day hourly forecasts of PV panel output are worked out. The load requirements are obtained from the Web. This helps to arrive at the quantum of electricity we need to bid for in the day-ahead electricity market. Note that day-ahead meteorological data is also available in Japan. The residence buys its electricity needs either at a flat rate or a nonflat rate, that is, at the ruling market rate in open competition. With all these preparations, it is shown that there are financial gains to the tune of 38.36% in buying hourly electricity on a nonflat-rate basis (Figure 10-11).

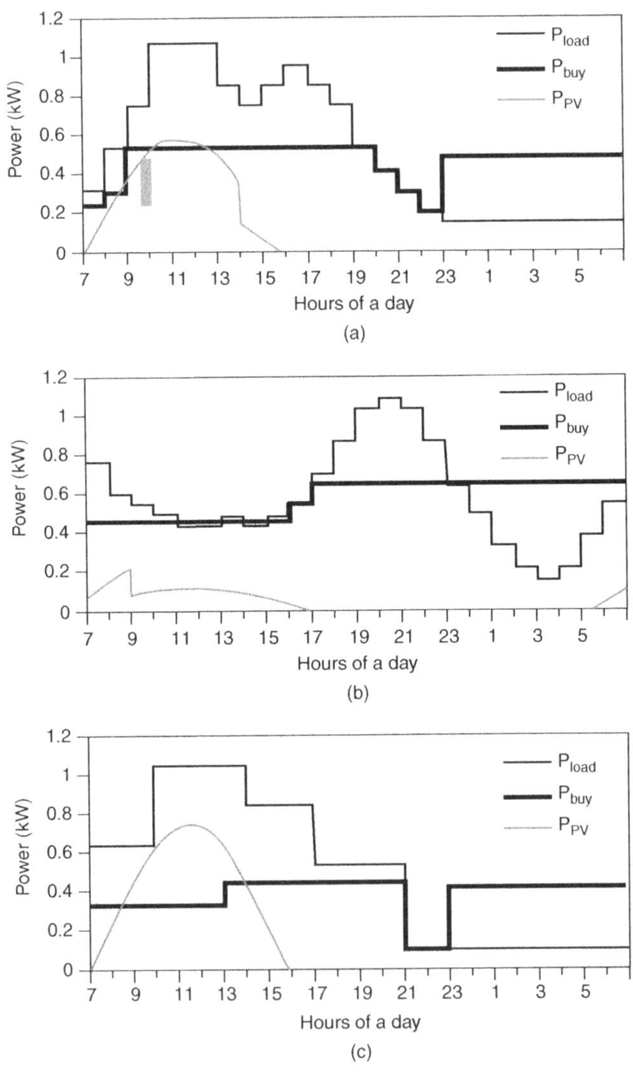

Figure 10-11. Simulated daily power patterns: (a) using Load-1, (b) using Load-2, and (c) using Load-3. (From [10], © IEEE 1992.)

With proper storage size and capacity, the residence can keep buying electricity during nonpeak hours, when it is cheap, and save on electricity purchases when different types of storage are available.

Note that the basic purpose of the PV application here is leveling of load around the year, taking into account weather and meteorological changes. This aims at attacking the costly portion of electricity buying while adhering to a common low price around the year. The purpose is not to compete with and replace grid energy totally, now or in future. This will have to wait until there are breakthroughs in solar cell technologies and as well threatening rises in fossil fuel (oil and gas). The Japanese example above shows a path that is leading to future healthy growth in utilizing solar energy.

Multistoried Buildings with a Large Number of Computers—Effect on Grid Power Factor and Harmonics. Multiple single-phase inverters are connected to the distribution system. These could be inverter interfaces for solar panels. There would be inverters as standbys along with storage batteries for computers within a building. The features of these inverters that may possibly affect power quality are power factor characteristics, harmonic distortion, DC injection, and islanding. In a building with many such inverters across a common step-down transformer from the grid supply, these characteristics will add up and pose a big problem for power quality. Let us study these effects individually.

Generally, the inverters are designed to operate at unity power factor (PF) at their rated output capacity. If the PF falls, the corresponding reactive currents add and can cause a voltage drop at the interface. Some inverters (PWMs) are programmed to provide their own reactive power and maintain a unity PF at even lower outputs. The best way to tackle this is to install suitable shunt capacitors at appropriate locations.

Harmonic distortion is a much more serious problem. Current harmonic distortion is important. IEEE 549 limits total current harmonic distortions to 5%. Figure 10-12 gives the spread of harmonics across an inverter interface. Table 10-4 tabulates these results. Two features may be noted: the second harmonic is sizeable and lower harmonics are predominant.

Within an inverter, a reference sine wave is either picked up from the grid supply or created within itself. The source of second harmonic is traced to the presence of the same in the reference sine wave that is, in the supply itself. Second harmonics along with other even harmonics may add to unwanted negative sequence currents, adversely affecting three-phase loads. There are stiff limitations on second harmonic current distortion. It should be less than 1%.

Note that the inductive reactance to the propagation of harmonics is low at lower frequencies and rises at higher frequencies. As a result, high-frequency harmonics attenuate rapidly and do not build up.

At the lower frequencies, except for the second harmonic, there is a possible phase shaft between successive/different waves, leading to an almost mass cancellation of the harmonic effects from different sources. However, Infield [11] reports that the phase shift between different inverter outputs of low harmonics is minor and tend to add. They also will not propagate due to cancellation by an aggregation effect, tet they can cause a nuisance between inverters.

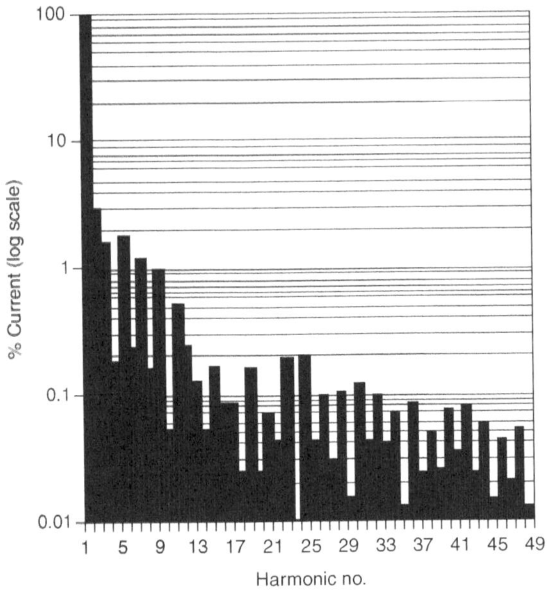

Figure 10-12. Harmonics from a single inverter at rated output. (From [12], © 1988.)

DC injection, if it accumulates and flows into the supply transformer, can distort its working. Its presence there is strictly prohibited. IEEE Std 1547/1710 on PV restricts DC injection to less than 0.5%. Many inverters use an isolating transformer for an interface with the grid. However, Infield has noted that under an aggregation effect, the aggregate DC components also tend to zero.

A PV Package for Power Factor Improvement. At the far end of a utility system, the voltages are low and line losses are high, because the loads are drawing high reactive power. A PV system with a PWM-type power conditioner is installed at this end. The power conditioner produces sufficient reactive power locally. Not only that, it makes up for the shortage of power as well, which a power factor improving shunt capacitor cannot do. Technically, this is a good proposal but financially it is not. As will be detailed below, capital costs of a PV system will be around US$10.65/W as against US$0.04/VA for a PF capacitor. The efficiency of a capacitor is also very high—99.5%.

Table 10-4. Harmonics up to the fifteenth order

Harmonic order	2	3	4	5	6	7	8	9	10	11	12	13	14	15
Percentage	3.01	1.62	0.18	1.80	0.21	1.13	0.17	0.96	0.05	0.51	0.22	0.12	0.04	0.16

Source: [12].

Typical installed costs of a PV system (Austin, Texas, capacity 300 kW) are:

PV module costs	\$5.61/W
Power conditioning system	\$0.78/W
Site preparation	\$0.30/W
Installation	\$0.99/W
Structural	\$0.90/W
Data acquisition system	\$0.36/W
Electrical	\$0.36/W
Administrative	\$0.40/W
Total	\$10.65/W

The above breakdown gives pre-1990 costs. These have come down substantially. The costs involved in other supporting items will also have come down.

A PV Panel on Top of an Automobile. In all the systems described above, the insolation does not change rapidly and frequently. This happens when we fit a PV panel on the top of an automobile in order to take advantage of the solar energy. The PV versus tme curve has a gradient. To deal with this, a widely adopted system called perturb and observe (P and O) is used. It senses consecutive signals and determines whether they are rising or falling. The control point moves accordingly. If the sensing time interval is too large, the system will oscillate and lose power. If the sensing interval is too small, there will be no oscillations, but the system will take a long time to reach the maximum power point. Logarithms have been developed to automatically adjust this sensing interval [13].

On solar panels on rooftops, the operation has long periods of steady operation, except under cloudy conditions when blocking of insolation will be very frequent and MPPTs must operate in the self-adjusting mode for the P and O mode.

REFERENCES

1. M. M. Saied, A. A. Hanafy, M. A. El-Gabaly, Y. A. Safar, M. G. Jaboori, K. A. Yamin, and A. M. Sharaf, Optimal Design Parameters for A PV Array Coupled to a DC Motor Via a DC-DC Transformer, *IEEE Transactions on Energy Conversion,* Volume 6, Issue 4, December 1991, pp. 593–598.
2. M. M. Saied, The Available Matching of Solar Arrays to DC Motors Having Both Constant and Series-Excited Field Components, *IEEE Transactions on Energy Conversion,* Volume 17, Issue 3, September 2002, pp. 301–305.
3. M. A. S. Masoum, H. Dehbonei, and E. F. Fuchs, Closure on "Theoretical and Experimental Analyses of Photovoltaic Systems with Voltage and Current-based Maximum Power Point Tracking," *IEEE Transactions on Energy Conversion,* Volume 19, Issue 3, September 2004, pp. 652–653.
4. S. Singer and J. Appelbaum, Starting Characteristics of Direct Current Motors Powered by

Solar Cells, *IEEE Transactions on Energy Conversion,* Volume 8, Issue 1, March 1993, pp. 47–53.

5. S. Rahman and G. Shrestha, Analysis of VISTA Photovoltaic Facility System Performance, *IEEE Transactions on Energy Conversion,* June 1990, pp. 245-251.
6. R. Chedid, H. Akiki, and S. Rahman, A Decision Support Technique for the Design of Hybrid Solar–Wind Power Systems, *IEEE Transactions on Energy Conversion,* Volume 13, Issue 1, March 1998, pp. 76–83.
7. W. T. Jewell and T. D. Unruh, Limits on Cloud-Induced Fluctuations in Photovoltaic Generation, *IEEE Transactions on Energy Conversion,* Volume 5, Number 2, June 1990, pp. 8-14.
8. M. Kolhe, J. C. Joshi, and D. P. Kothari, Performance Analysis of a Directly Coupled Photovoltaic Water-Pumping System, *IEEE Transactions on Energy Conversion: Photovoltaic Generation,* Volume 19, Issue 3, September 2004, pp. 613–618.
9. M. D. Rahman and H. Yamshiro, Novel Distributed Power Generation System of PV, Using Solar Energy Estimation, *IEEE Transactions on Energy Conversion,* Volume 22, Issue 2, June 2007, pp. 358-367
10. J. J. Bzura, Performance of Grid-Connected Photovoltaic Systems on Residences and Commercial Buildings in New England, *IEEE Transactions on Energy Conversion,* Volume 7, Issue 1, March 1992, pp. 79–82.
11. D. G. Infield, P. Onions, A. D. Simmons, and G. A. Smith, Power Quality from Multiple Grid-Connected Single-Phase Inverters, *IEEE Transactions on Power Delivery,* Volume 19, Issue 4, October 2004, pp. 1983–1989.
12. J. Stevens, The Issue of Harmonic Injection from Utility Integrated Photovoltaic Systems. II. Study Results, *IEEE Transactions on Energy Conversion,* Volume 3, Issue 3, September 1988, pp. 511–515.
13. Atmaram, G. H., B. Marion, and C. Herig, First Year Performance of a 15 kWp Amorphous Silicon Photovoltaic System, *IEEE Transactions on Energy Conversion,* Volume 5, Issue 2, June 1990, pp. 290–298.

BIBLIOGRAPHY

Aiello, M., A. Cataliotti, S. Favuzza, and G. Graditi, Theoretical and Experimental Comparison of Total Harmonic Distortion Factors for the Evaluation of Harmonic and Interharmonic Pollution of Grid-Connected Photovoltaic Systems, *IEEE Transactions on Power Delivery,* Volume 21, Issue 3, July 2006, pp. 1390–1397.

Alghuwainem, S. M., Matching of a DC Motor to a Photovoltaic Generator Using a Step-up Converter with a Current-Locked Loop, *IEEE Transactions on Power Delivery,* Volume 9, Issue 1, March 1994, pp. 192–198

Alghuwainem, S. M., Steady-state Performance of DC Motors Supplied from Photovoltaic Generators with Step-Up Converter, *IEEE Transactions on Power Delivery,* Volume 7, Issue 2, June 1992, pp. 267–272.

Appelbaum, J., Discussion of "Theoretical and Experimental Analyses of Photovoltaic Systems with Voltage and Current-Based Maximum Power Point Tracking," *IEEE Transactions on Energy Conversion,* Volume 19, Issue 3, September 2004, pp. 651–652.

Bailey, B. H., J. R. Doty III, R. Perez, R. Stewart, and J. E. Donegan, Evaluation of a Demand

Side Management Photovoltaic System, *IEEE Transactions on Energy Conversion,* Volume 8, Issue 4, December 1993, pp. 621–627.

Bzura, J. J., The New England Electric Photovoltaic Systems Research and Demonstration Project, *IEEE Transactions on Energy Conversion,* Volume 5, Issue 2, June 1990, pp. 284–289.

Casadei, D., G. Grandi, and C. Rossi, Single-phase Single-stage Photovoltaic Generation System Based on a Ripple Correlation Control Maximum Power Point Tracking, *IEEE Transactions on Energy Conversion,* Volume 21, Issue 2, June 2006, pp. 562–568.

Chen, Y-M. C.-H. Lee, and H.-C. Wu, Calculation of the Optimum Installation Angle for Fixed Solar-Cell Panels Based on the Genetic Algorithm and the Simulated-Annealing Method, *IEEE Transactions on Energy Conversion,* Volume 20, Issue 2, June 2005, pp. 467–473.

Esram, T. and P. L. Chapman, Comparison of Photovoltaic Array Maximum Power Point Tracking Techniques, *IEEE Transactions on Energy Conversion,* Volume 22, Issue 2, June 2007, pp. 439–449.

Jung, G-K., C.-C. Chang, and C.-L. Chen, Automatic Phase-Shift Method for Islanding Detection of Grid-Connected Photovoltaic Inverters, *IEEE Transactions on Energy Conversion,* Volume 18, Issue 1, March 2003, pp. 169–173.

Kang, F. S., S.-J. Park, S. E. Cho, C.-U. Kim and T. Ise, Volume Multilevel PWM Inverters Suitable for the Use of Stand-alone Photovoltaic Power Systems, *IEEE Transactions on Energy Conversion: Photovoltaic Generation,* Volume 20, Issue 4, December 2005, pp. 906–915.

Kellogg, W. D., M. H. Nehrir, G. Venkataramanan, and V. Gerez, Generation Unit Sizing and Cost Analysis for Stand-alone Wind, Photovoltaic, and Hybrid Wind/PV Systems, *IEEE Transactions on Energy Conversion: Photovoltaic Generation,* Volume 13, Issue 1, March 1998, pp. 70–75.

Korovesis, P. N., G. A. Vokas, I. F. Gonos, and F. V. Topalis, Influence of Large-scale Installation of Energy Saving Lamps on the Line Voltage Distortion of a Weak Network Supplied by Photovoltaic Station, *IEEE Transactions on Power Delivery,* Volume 19, Issue 4, October 2004, pp. 1787–1793.

Libo, W., Z. Zhengming, and L. Jianzheng, A Single-Stage Three-Phase Grid-Connected Photovoltaic System with Modified MPPT Method and Reactive Power Compensation, *IEEE Transactions on Energy Conversion,* Volume 22, Issue 4, December 2007, pp. 881–886.

Martynaitis, J., Discussion of "Theoretical and Experimental Analyses of Photovoltaic Systems with Voltage and Current-Based Maximum Power Point Tracking," *IEEE Transactions on Energy Conversion,* Volume 19, Issue 3, September 2004, p. 652.

Masoum, M. A. S., H. Dehbonei, and E. F. Fuchs, Theoretical and Experimental Analyses of Photovoltaic Systems with Voltage and Current-Based Maximum Power-Point Tracking, *IEEE Transactions on Energy Conversion,* Volume 17, Issue 4, December 2002, pp. 514–522.

Mohaddes, Optimal Pulse Width Modulated Statcom, Rural Network Control, *IEEE Transactions on Power Delivery,* T-PWRD, April 1999, pp 481–488.

Naik, R., N. Mohan, M. Rogers, and A. Bulawka, A Novel Grid Interface, Optimized for Utility-Scale Applications of Photovoltaic, Wind-Electric, and Fuel-Cell Systems, *IEEE Transactions on Power Delivery,* Volume 10, Issue 4, October 1995, pp. 1920–1926.

Oliva, A. R., J. C. Balda, D. W. McNabb, and R. D. Richardson, Power-Quality Monitoring of a PV Generator, *IEEE Transactions on Energy Conversion,* Volume 13, Issue 2, June 1998, pp. 188–193.

Roy, S., Optimal Planning for Utility Generation by Photovoltaic Sources Spread Across Multiple Sites, *IEEE Transactions on Energy Conversion: Photovoltaic Generation,* Volume 21, Issue 1, March 2006, pp. 181–186.

Sedghisigarchi, K. and A. Feliachi, Dynamic and Transient Analysis of Power Distribution Systems with Fuel Cells—Part I: Fuel-Cell Dynamic Model, *IEEE Transactions on Energy Conversion,* Volume 19, Issue 2, June 2004, pp. 423–428.

Sen, UPFC—Unified Power Flow Controller. Theory, Modeling and Applications, *IEEE Transactions on Power Delivery,* T-PWRD, October 1998, 1453–1464.

Senjyu, T., D. Hayashi, N. Urasaki, and T. Funabashi, Optimum Configuration for Renewable Generating Systems in Residence using Genetic Algorithm, *IEEE Transactions on Energy Conversion: Photovoltaic Generation,* Volume 21, Issue 2, June 2006, pp.459–466.

Shengyi, L. and R. A. Dougal, Dynamic Multiphysics Model for Solar Array, *IEEE Transactions on Energy Conversion,* Volume 17, Issue 2, Jun 2002, pp. 285–294.

Tan, Y. T., D. S. Kirschen, and N. Jenkins, A Model of PV Generation Suitable for Stability Analysis, *IEEE Transactions on Energy Conversion,* Volume 19, Issue 4, December 2004, pp. 748–755.

Varnham, A., A. M. Al-Ibrahim, G. S. Virk, and D. Azzi, Soft-Computing Model-Based Controllers for Increased Photovoltaic Plant Efficiencies, *IEEE Transactions on Energy Conversion,* Volume 22, Issue 4, December 2007, pp. 873–880.

Woyte, A., R. Belmans, and J. Nijs, Testing the Islanding Protection Function of Photovoltaic Inverters, *IEEE Transactions on Energy Conversion,* Volume 18, Issue 1, March 2003, pp. 157–162.

Woyte, A., V. Van Thong, R. Belmans, and J. Nijs, Voltage Fluctuations on Distribution Level Introduced by Photovoltaic Systems, *IEEE Transactions on Energy Conversion,* Volume 21, Issue 1, March 2006, pp. 202–209.

11

DIRECT CONVERSION INTO ELECTRICITY—FUEL CELLS

11.1 FUEL CELLS BYPASS INTERMEDIATE STEPS IN THE PRODUCTION OF ELECTRICAL ENERGY

When we burn a fuel, say, hydrogen or a hydrocarbon, we oxidize H_2 to H_2O and carbon to CO_2. The chemical action releases heat. The measure of this heat is termed the thermal capacity of the fuel. We use the heat to produce steam and use the steam to run a turbine. The turbine drives a generator and the generator finally gives us electrical energy as an output. At each stage of conversion, we lose energy by way of losses. Ultimately, energy used in the form of electrical energy is produced with low efficiency. We make this up by increasing the scale, say, by megameasures, so that the stage conversion efficiencies improve.

A fuel cell allows us to draw out energy in electrical form without resorting to two to three intermediate stages, that is, directly.

11.2 WORKING OF A FUEL CELL

A low-temperature, moderate-output fuel cell works as follows (see Figure 11-1):

1. A grooved porous holder plate at the entrance disperses H_2 uniformly.
2. A catalyst before the electrolyte breaks the H_2 into $H_2^+ + e^-$ electrons.

Copyright © 2011 the Institute of Electrical and Electronics Engineers, Inc.

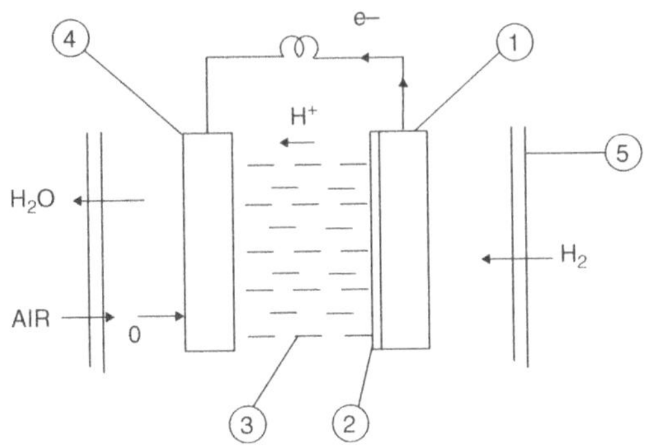

Figure 11-1. Basic fuel cell.

3. A proton energy membrane (PEM) is a polymer film that blocks passage of electrons but allows hydrogen protons to proceed through.
4. A porous graphite plate acts as an anode.
5. Electrons carried in from the external circuit to the anode react with H_2^+ protons and oxygen picked up from the air to form H_2O. Air is supplied externally at the anode.

Note the following. We have the following reactions:

At the cathode $H_2 \rightarrow H_2^+ + e^-$

At the anode $H_2^+ + \frac{1}{2}O + e^- \rightarrow H_2O$

Without the e^-, the chemical reaction of $H_2^+ + \frac{1}{2}O$ cannot take place. Only when electrons flow through the external circuit is the reaction complete and water formed. A portion of energy is liberated as heat at the anode and must be carried away. This is the portion that accounts for the losses in the fuel cell.

Net system efficiency is defined as the electrical energy output/thermal energy or the heat energy in the raw fuel feed amount.

11.3 A REFORMER FOR GETTING HYDROGEN FROM METHANE

Methane, CH_4, comes very close to being a copious supplier of hydrogen. Under high pressure and temperature steam, it has the reaction shown in Figure 11-2. A reformer splits $CH_4 + H_2O$ into $3H_2 + CO$. The CO further reacts with steam and produces CO_2 and H_2. Thus, CH_4 is fully utilized, the hydrogen portion giving electrical energy out-

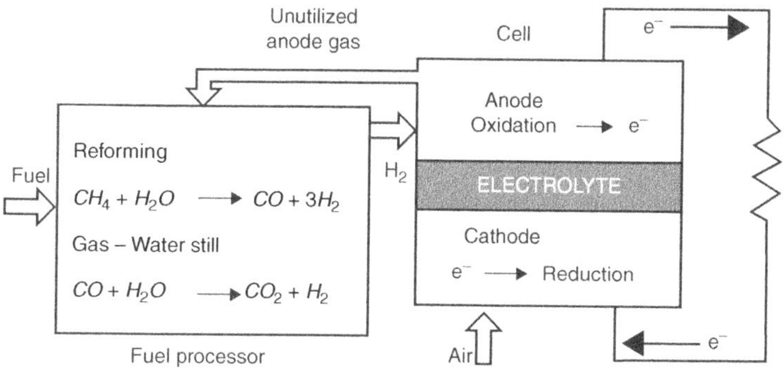

Figure 11-2. Autonomous fuel cell power plant. (From [1], © IEEE 2002.)

put and the C portion giving a thermal output. The heat generated is utilized in the reformer itself.

11.4 FUELS FOR A FUEL CELL

Liquefied natural gas (LNG) has generally the following composition by volume [2]:

Methane 89.3%
Ethane 5.92%
Propane 3.2%
Pentane 0.7%
Nitrogen 0.03%

Methanol is produced as an agricultural product by sugar factories in India. There is continuous pressure from the sugar lobby in the parliament on the Government for compelling petrol pumps to mix a certain percentage of methanol in petrol.

Methane, a liquid, is easy to store in a vehicle. Along with a small reformer and a controller, it forms an ideal fuel for a PEM fuel-cell driven automobile. This development is in full swing.

In place of methanol, regular petroleum is being tried for automobile applications. LNG can also be used in place of methanol. It should be desulfurized before passing it on to a reformer. Compared with hydrogen, methanol and LNG emit CO_2 into the atmosphere.

Energy is recovered from hydrogen and other components of LNG, mostly unburned carbohydrates, are burned fully. These other portions of LNG pass through the exhaust to a burner. This burning of carbon and other portions gives thermal energy that is used to heat the bath and incoming LNG to the bath temperature. In large-scale models, this energy is used for cogeneration. Hot air exhaust is compressed and fed to a gas turbine.

11.5 FUEL CELLS ON THE FOREFRONT OF DEVELOPMENT

11.5.1 Advantages of the PEM Fuel Cells

The PEM fuel cell has the following advantages:

1. Environmentally, it is the most desirable candidate, with low polluting gas emissions and danger of vast crude oil spills in the sea.
2. High fuel conversion efficiency, two to three times that of present-day internal combustion engines used in today's transport vehicles and passenger cars.
3. It has a fast start up. A motor car with a PEM fuel cell can meet all the requirements with respect to fast start. It can climb up an incline and run at a steady high speed. A multiple of these cells can be connected in a series–parallel combination to give a compact assembly.
4. Its operating temperature is low, 80°C, which makes it suitable for employing it in a car. Other high-temperature fuel cells, albeit with better efficiencies, cannot be used in a car.
5. Fuel cell operated electrical cars will cost less than other forms of propulsion. Maintenance costs are also low.
6. The PEM fuel cells have been developed to a performance level of 10 kA/m^2 surface area of the cell and to one kW per liter of its volume. DC voltages to the required value of the DC drive motor can be built up with cells in series. One can get the required kilowatts by connecting cells in parallel. Fuel cell packs with a rating of 30 kW for a station wagon and 200 kW for a passenger bus have already been demonstrated [3].
7. Costly and lossy invertors are not required in this application. If pure H_2 is used, the reformer/controller is also small and compact.
8. Since the package is simple, maintenance is much easier.

11.5.2 Disadvantages of PEM Fuel Cells

PEM fuel cells have today the following disadvantages:

1. Not universally available.
2. At present, the costs of PEM fuel cells, as well as those of hydrogen, are very high in comparison with the prices of petroleum products.
3. Energy density of hydrogen, a gas in which form it is available, is very poor. Therefore, it will have to be compressed at very high pressures for storage and handling (5000 to 12,000 psi). This requires many developments such as lightweight, high-strength materials capable of withstanding high pressures; standards on pressures to be used, and precautions to be taken; and standards on equipment and methods.
4. Quality of H_2 must be good. If other fuels are used for getting a supply of H_2, the process requires a reformer. The PEM electrodes are affected easily by CO and other gases.

5. The electrodes are given a coating of platinum, a rather costly material, for use in other types of fuel cells.
6. At its existing stage of development, it cannot compete in costs with petroleum products.

11.6 COMPARISON BETWEEN FUEL CELLS

A variety of fuel cells have been developed. Table 11-1 gives a comparison between these cells. The fastest to develop has been the proton exchange membrane type of fuel cell, called PEM FC, developed mainly for transportation vehicles.

11.7 TYPICAL CHARACTERISTICS OF VARIOUS FUEL CELLS

Table 11-2 sums up these characteristics.

11.8 DEVELOPMENTS IN FUEL CELLS

Major handicaps to the development of EM-fuel cells are the high costs and susceptibility to impurities. For large-scale deployment of fuel cells, we must have membranes and electrolytes that can override impurities and be useable with presently available energy-supplying fuels. LNG counts favorably. Coal gasification is being experiment-

Table 11-1. Comparing fuel cells

	Phosphoric acid	Alkaline	Proton-exchange membrane	Molten carbonate	Solid oxide
Electrolyte	H_3PO_4	KOH/H_2O	Polymer	Molten salt	Ceramic
Operating temperature	190°C	80–200°C	80°C	650°C	1000°C
Fuels	Hydrogen reformate	Hydrogen	Hydrogen reformate	Hydrogen reformate	Hydrogen, carbon monoxide reformate
Reforming	External	N/A	External	External, internal	External, internal
Oxidant	Oxygen, air	Oxygen	Oxygen, air	Carbon dioxide, oxygen, air	Oxygen, air
Efficiency, %	40–50	40–50	40–50	>60	>60
Scale	200 kW to 10 MW	100 kW to 20 MW	100 kW to 10 MW	>100 MW	>100 MW
Applications	Small utility	Aerospace	Motive, small utility	Utility	Utility

Source: [4].

Table 11-2. Typical characteristics of fuel cells

	PAFC*	AFC	MCFC	SOFC	SPFC	DMFC
Working temperature (°C)	150–210	60–100	600–700	900–1000	50–100	50–100
Power density (W/cm^2)	0.2–0.25	0.2–0.3	0.1–0.2	0.24–0.3	0.35–0.6	0.04–0.23
Projected life (kh)	40	10	40	40	40	10
Projected cost (US$/kW)	1000	200	1000	1500	200	200

*PAFC—phosphoric acid fuel cell.
AFC—alkaline fuel cell.
MCFC—molten carbonate fuel cell.
SOFC—solid oxide fuel cell.
SPFC—solid polymer fuel cell also known as proton exchange membrane fuel cell.
DMFC—direct methanol fuel cell.
Source: [5].

ed on keenly, since coal is still a major source of energy all over the world. The new fuel cells should be able to work with both these fuels. The development is shifting to higher operating temperatures, with emphasis on membranes and electrolytes suitable for high temperatures.

11.8.1 Molten Carbonate Fuel Cell

In this fuel cell (Figure 11-3 and Table 11-3), a solid carbonate is melted to 650°C and a gas rich in hydrogen is fed (LNG). The hydrogen reacts with the carbonate portion and liberates CO_2 and free electrons:

$$H_2 + CO_3 2e \rightarrow CO_2 + H_2O + 2e$$

CO_2 passes through the solid membrane. The membrane bars the passage of electrons. The electrons pass through the outer circuit to the cathode. Under a suitable catalyst,

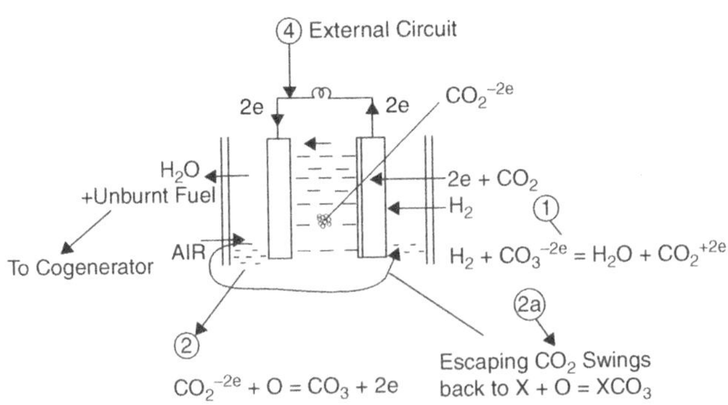

Figure 11-3. Molten carbonate fuel cell (MCFC)—high temperature and high output.

Table 11-3. Standard characteristics of a MCFC

Cell voltage	Observations
Operating conditions	
Current density	150 mA/cm^2
Temperature	650°C
Pressure	1 ata
Fuel composition	H_2: 72 CO_2: 18 Vol% H_2O: 0
Oxidant composition	O_2: 14.7 CO_2: 30 Vol% N_2: 55.3
Fuel utilization	80%
Oxidant utilization	30%

Source: [2].

CO_2 is turned back into $CO_3$2e. Under a process termed pressure swing qdsorption (PSA), CO_3 reacts with the metallic proton in the molten bath and forms metal carbonate. CO_2 is thus not lost from the system.

Sasaki's [2] system studies of MCFCs are summarized in Table 11-4. Here, in place of H_2, liquefied natural gas is used. Its constituents disassociate under hot steam. Methane supplies H_2 for the cell. LNG constituents are methane 89%, ethane 6%, other 6%.

Table 11-4. Conditions of system studies of MCFCs

	IIR-CB-AR	IIR-PSA-AR
Steam-to-natural gas ratio and feed	4.3	5.7
Burner excess air ratio	1.2	—
CO_2 recovery ratio	—	85%
Fresh air utilization	60%	
Anode gas recycle ratio	90%	
Operating pressure	1 atm	
Inlet temp of stack	600%	
Outlet temp of stack	650%	
Power conditioner efficiency	96%	
Parasite power	3% to net output power	
Heat release	Negligible	

Source: [2].

11.8.2 CO_2 Recycling under Pressure Swing Absorption

Figure 11-4 gives an overall sketch of this process. The heat generated in this process can be used elsewhere, as in a cogeneration process.

11.9 APPLICATIONS OF FUEL CELLS

11.9.1 Automobile Propulsion

We have dealt with this in detail earlier. This is the most important fuel cell application and is being developed on a commercial scale. However, presently there are quite a few serious roadblocks.

Lack of Widespread Servicing Facilities. Considerable development has to take place in this direction.

High Operating Pressures Required for Liquefying and Handling Natural Gas and Hydrogen. Hydrogen has very poor energy density. It has to be compressed to high pressure for storage and transport. Development of lightweight materials for such duty will be needed. BC Hydro, Canada, has reported the following developments:

1. A natural gas and hydrogen fueling station operating at 350 bars (5000 psi) pressure.
2. A natural gas and hydrogen fueling station operating at 700 bars.
3. An 875 bars (12,691 psi) lightweight composite tube trailer and satellite filing station that will deliver fuel at 700 bars.

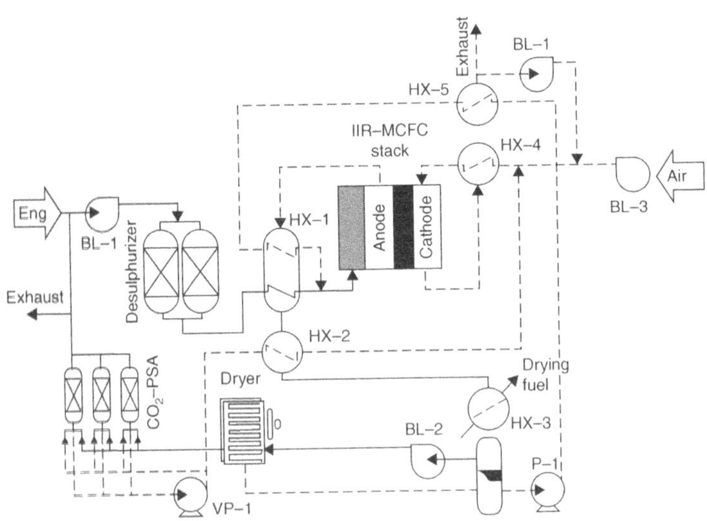

Figure 11-4. Configuration with CO_2 recycled by PSA. (From [2], © IEEE 1993.)

Competition from Cheaper Sources of Power. There are sources of large volumes of electrical energy that can be but are not produced, since the demand for electrical energy has very low troughs in certain periods of a day. Instead of backing down production, this energy can be utilized for production of hydrogen, thus helping the bottom line of financial operations of the generators. The typical examples are wind farms (as on sea islands), atomic generating stations, and hydro generators. There must be facilities to store and handle this hydrogen economically.

11.9.2 Residential Applications

A 5 kW capacity FC generator operating on LNG and employing a small reformer is being marketed in the United States. This would be an ideal addition to a household in India, where residents are continuously facing planned and unplanned electric power outages. Presently, inverters operating off a storage battery are being widely used. However, they have a very limited energy storage capacity.

LNG-based fuel cells can fit in nicely with the present-day widespread use of LNG in residences in India. Fuel cells have no bar on how long they operate at a time.

11.9.3 Electricity Utilities

Solid oxide fuel cell turbine generators (SOFCs) are favored for large-scale electrical energy production. They go a step further in improving net efficiency (Figure 11-5). Fuel is reformed to get a hydrogen rich supply. LNG or coal gasification products containing CO are also acceptable. The fuel goes to the anode. H^+ protons are separated by a high-temperature membrane. They travel to the cathode through the electrolyte. Here, they combine with oxygen-pumped in along with air and with the help of electrons traveling externally to form H_2O. Unutilized fuel and high-temperature (1000°C) exhaust, including steam from the cathode, go to a combustor that burns the unused gas. This high-temperature steam heats up compressed air that enters a gas turbine (see Chapter 4). The air under high pressure expands and drives the turbine blades. A generator driven by the turbine delivers electrical energy in AC form. This is rectified into DC. It is combined with the DC energy delivered by the solid oxide fuel cell (major portion) and then goes to the inverter. The inverter feeds the energy to the grid.

Overall fuel efficiencies are as high as 70%. Exhaust from the turbine is reused to heat the fuel being fed to the reformer as well as the solid oxide fuel cell, which improves its efficiency at a higher temperature. Table 11-5 gives the typical operating performance.

11.10 AN SOFC–GAS TURBINE SYSTEM

It will be noted that all the fuel cells require hydrogen as the main fuel to convert fuel energy into electrical energy directly. This gives a high to very high fuel conversion efficiency with respect to the low fossil fuel energies at which we are operating our major electrical energy producing systems today. More than that, with hydrogen, we are getting our environments many times cleaner, a major concern. This takes us to a

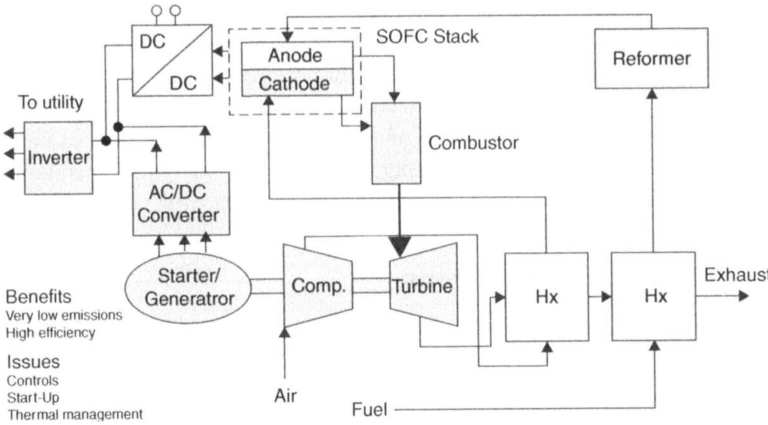

Figure 11-5. SOFC–gas turbine hybrid system. (From [6], © IEEE 2005.)

hydrogen economy of tomorrow for which great efforts are now being made. We will come back to this later.

11.10.1 Special Advantages

Molten carbonate type and solid oxide type fuel cell batteries are ideal sources for generating electrical energy on large scales. They have fairly high efficiencies and the following advantages:

Table 11-5. Typical performance for hybrid SOFC–gas turbine system for a 500 kW output

Stack and reformer	Value	Compressor and turbine	Value
Stack power, kW	441	Compressor inlet temperature C	15
Stack area, cm^2	400,000	Compressor outlet temperature C	168
Stack power density, cm^2	1.1	Compressor inlet pressure kPa	101.23
Stack voltage	0.74	Compressor outlet pressure kPa	323
Stack temperature °C	800	Compressor pressure ratio	3.19
Stack pressure (atm)	3.0	Mass flow through compressor g/s	609
Stack electrical efficiency %	50	Compressor power kW	93.65
Air utilization %	35	Turbine inlet temperature °C	1000
Fuel utilization %	85	Turbine outlet temperature °C	783
Reformer temperature °C	677	Turbine inlet pressure kPa	288.9
		Turbine expansion ratio	2.74
		Mass flow through turbine g/s	623
Stack power/generator power kW	500	Turbine power kW	161.96
Total system efficiency %	68.6	Generator power (available) kW	58

Source: [6].

1. They can be located anywhere, giving an advantage of voltage control and loss reduction on widespread distribution systems. Here, they compete with IC-engine-based generators.
2. Compared with the rest of the system of electricity generation, they have the fastest response to electrical transients. This response is governed mainly by the electronic control system (bandwidth of response, rate of rise/fall of voltage, etc.) and on the mechanical side by the inertia of the reformer-controller mechanism. This compares very favorably with the inertia of large rotating parts on other competitive systems.
3. Unlike other renewable energy sources, they enjoy full capacity credit in investment planning, in claiming higher tariffs on this account, and in load following.
4. They are quiet and scalable.
5. A natural gas supply arrangement can easily be made to operate fuel cell generators as distributed generators since LNG has reached many customers universally and supply depots for LNG cylinders are well spread.

11.11 EFFICIENCIES OF VARIOUS SYSTEMS IN THERMAL POWER GENERATION TECHNOLOGIES

Figure 11-6 gives an idea of electrical efficiencies for various types of fuel cell systems, the power range in which they are presently spread, and the zone of the power system in which they are favorably placed, for example, distribution, decentralized cogeneration, and large power stations.

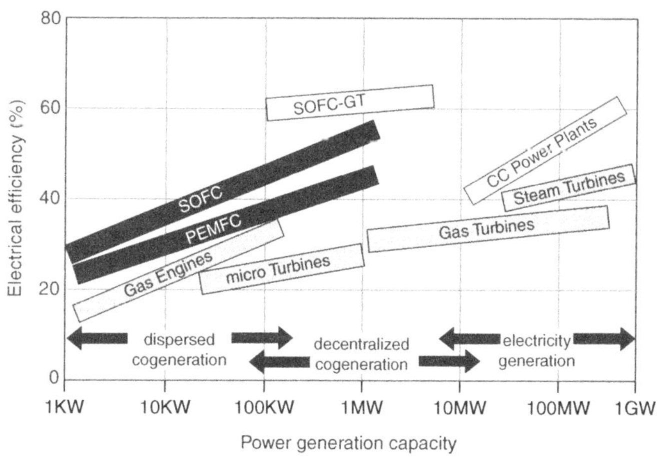

Figure 11-6. Efficiencies of various power generation technologies. (From [5], © IEEE 1993.)

REFERENCES

1. C. J. Hatziadoniu, A. A. Lobo, F. Pourboghrat, and M. Daneshdoost, A Simplified Dynamic Model of Grid-Connected Fuel-Cell Generators, *IEEE Transactions on Power Delivery*, Volume 17, Issue 2, April 2002, pp. 467–473.
2. A. Sasaki, S. Matsumoto, M. Fujitsuka, T. Shinoki, T. Tanaka, and J. Ohtsuki, CO2 Recovery in Molten Carbonate Fuel Cell System by Pressure Swing Adsorption, *IEEE Transactions on Energy Conversion*, Volume 8, Issue 1, March 1993, pp. 26–32.
3. H. B. Puttgen, P. R. MacGregor, and F. C. Lambert, Distributed Generation: Semantic Hype or the Dawn of a New Era? *IEEE Power and Energy Magazine*, Volume 1, Issue 1, January–February 2003, pp. 22–29.
4. T. Gilchrist, Fuel Cells to the Fore [Electric Vehicles], *IEEE Spectrum*, Volume 35, Issue 11, November 1998, pp. 35–40.
5. A. T-Raissi and D. L. Block, Hydrogen: Automotive Fuel of the Future, *IEEE Power and Energy Magazine*, Volume 2, Issue 6, Nov.-December 2004, pp. 40–45.
6. K. Rajashekara, Hybrid Fuel-Cell Strategies for Clean Power Generation, *IEEE Transactions on Industrial Applications*, Volume 41, Issue 3, May–June 2005, pp. 682–689.

BIBLIOGRAPHY

Aki, H., S. Yamamoto, Y. Ishikawa, J. Kondoh, T. Maeda, H. Yamaguchi, A. Murata, and I. Ishii, Operational Strategies of Networked Fuel Cells in Residential Homes, *IEEE Transactions on Power Systems*, pp. 1405–1414.

Andrews, C. J. and S. A. Weiner, Visions of a Hydrogen Future, *IEEE Power and Energy Magazine*, Volume 2, Issue 2, March–April 2004, pp. 26–34.

Azmy, A. M. and I. Erlich, Online Optimal Management of Pemfuel Cells Using Neural Networks, *IEEE Transactions on Power Delivery*, Volume 20, Issue 2, Part 1, April 2005, pp. 1051–1058.

Ball, G. J., D. C. Blackburn Jr., W. Gish, E. Gulachenski, E. A. Guro, E. W. Kalkstein, A. Pivec, P. W. Powell, F. Soudi, J. W. Stevens, M. Yalla, R. Zavadil, J. T. Emery, and R. Hill, Static Power Converters of 500 Kw or Less Serving as the Relay Interface Package for Nonconventional Generators, *IEEE Transactions on Power Delivery*, Volume 9, Issue 3, July 1994, pp. 1325–1331.

Brown, J. T., Solid Oxide Fuel Cell Technology, *IEEE Transactions on Energy Conversion*, Volume 3, Issue 2, June 1988, pp. 193–198.

El-Sharkh, M. Y., A. Rahman, M. S. Alam, A. A. Sakla, P. C. Byrne, and T. Thomas, Analysis of Active and Reactive Power Control of a Stand-Alone PEM Fuel Cell Power Plant, *IEEE Transactions on Power Systems*, Volume 19, Issue 4, November 2004, pp. 2022–2028.

El-Sharkh, M. Y., M. Tanrioven, A. Rahman, and M. S. Alam, A Study of Cost-Optimized Operation of a Grid-Parallel PEM Fuel Cell Power Plant, *IEEE Transactions on Power Systems*, pp. 1104–1114.

Gurney, J. H., Building a Case for the Hydrogen Economy, *IEEE Power and Energy Magazine*, Volume 2, Issue 2, March–April 2004, pp. 35–39.

Johnson, R. M., M. K. Donnelly, and K. E. Marcotte, A Pulse-Amplitude-Synthesis-and-Control

(PASC) Consolidation/Inversion System for Faraday Connected MHD Generators, *IEEE Transactions on Energy Conversion,* Volume 8, Issue 3, September 1993, pp. 448–454.

Khallat, M. A.,and S. Rahman, A Model for Capacity Credit Evaluation of Grid-Connected Photovoltaic Systems with Fuel Cell Support, *IEEE Transactions on Power Systems,* IEEE Transactions on Power Systems, Volume 3, Issue 3, Aug. 1988, pp. 1270–1276.

Kishinevsky, Y. and S. Zelingher, Coming Clean with Fuel Cells, *IEEE Power and Energy Magazine,* Volume 1, Issue 6, November–December 2003, pp. 20–25.

Li, Y.H., S. S. Choi, and S. Rajakaruna, An Analysis of the Control and Operation of a Solid Oxide Fuel-Cell Power Plant in an Isolated System, *IEEE Transactions on Energy Conversion,* Volume 20, Issue 2, June 2005, pp. 381–387.

Matsumoto, S., A. Sasaki, H. Urushibata, and T. Tanaka, Performance Model of Molten Carbonate Fuel Cell, *IEEE Transactions on Energy Conversion,* Volume 5, Issue 2, June 1990, pp. 252–258.

Monai, T., I. Takano, H. Nishikawa, and Y. Sawada, A Collaborative Operation Method Between New Energy-Type Dispersed Power Supply and EDLC, *IEEE Transactions on Energy Conversion,* Volume 19, Issue 3, September 2004, pp. 590–598.

Naidu, M., T. W. Nehl, S. Gopalakrishnan, and L. Wurth, Keeping Cool While Saving Space and Money: A Semi-Integrated, Sensorless PM Brushless Drive for a 42-V Automotive HVAC Compressor, *IEEE Industrial Applications Magazine,* Volume 11, Issue 4, July–August 2005, pp. 20–28.

Naik, R., N. Mohan, M. Rogers, and A. Bulawka, A Novel Grid Interface, Optimized for Utility-Scale Applications of Photovoltaic, Wind-Electric, and Fuel-Cell Systems, *IEEE Transactions on Power Delivery,* Volume 10, Issue 4, October 1995, pp. 1920–1926.

Rahman, S. and K. S. Tam, A Feasibility Study of Photovoltaic-Fuel Cell Hybrid Energy System, *IEEE Transactions on Energy Conversion,* Volume 3, Issue 1, March 1988, pp. 50–55.

Rau, N. S, and W. Yih-Heui, Optimum Location of Resources in Distributed Planning, *IEEE Transactions on Power Systems,* Volume 9, Issue 4, November 1994, pp. 2014–2020.

Riezenman, M. J., Fuel Cells for the Long Haul, Batteries for the Spurts [Electric Vehicles], *IEEE Spectrum,* Volume 38, Issue 1, January 2001, pp. 95–97.

Sedghisigarchi, K. and A. Feliachi, Dynamic and Transient Analysis of Power Distribution Systems with Fuel Cells—Part I: Fuel-Cell Dynamic Model, *IEEE Transactions on Energy Conversion,* Volume 19, Issue 3, June 2004, pp. 423–428.

Staunton, R. H., J. B. Berry, and C. A. Dunn, Compatibility Study of Fuel-Cell Protective Relaying and the Local Distribution System, *IEEE Transactions on Power Delivery,* Volume 20, Issue 3, July 2005, pp. 1825–1829.

Tam, K. S. and S. Rahman, System Performance Improvement Provided by a Power Conditioning Subsystem for Central Station Photovoltaic-Fuel Cell Power Plant, *IEEE Transactions on Energy Conversion,* Volume 3, Issue 1, March 1988, pp. 64–70.

Von Jouanne, A., I. Husain, A. Wallace, and A. Yokochi, Gone with the Wind: Innovative Hydrogen/Fuel Cell Electric Vehicle Infrastructure Based on Wind Energy Sources, *IEEE Industrial Applications Magazine,* Volume 11, Issue 4, July–August 2005, pp. 12–19.

Wongsaichua, W., L. Wei-Jen, S. Oraintara, C. Kwan, and F. Zhang, Integrated High-Speed Intelligent Utility Tie Unit for Disbursed/Renewable Generation Facilities, *IEEE Transactions on Industrial Applications,* Volume 41, Issue 2, March–April 2005, pp. 507–513.

12

HYBRID SYSTEMS

12.1 COUPLING OF ENERGY SOURCES

In earlier chapters, we have discussed various sources of energy, both conventional and renewable or nonconventional. The renewable energy sources, particularly photovoltaic and wind energy, are characterized by their uncertainties in the time frame as well as in magnitude. The best way to get around these uncertainties would be to couple them with other steady and controllable energy supplies. The coupling partner, such as storage batteries, IC-engine-driven generators, fuel cells, and so on, make up for the uncertainties and enable us to utilize the almost free energy available from the renewable.

12.2 WHAT EXACTLY ARE HYBRIDS?

Technical Committee 69 of the ICE defines a hybrid vehicle as "one in which propulsion energy during a specified operational mission is available from two or more kinds or types of energy stores, sources or converters. At least one store or converter must be on board" [1].

Electricity Power Generation. By Digambar M. Tagare

Copyright © 2011 the Institute of Electrical and Electronics Engineers, Inc.

12.2.1 Where Hybrids Can be Effective

There are many places where conventional electric supplies cannot become viable economically or physically. The best bet in such cases is to exploit the natural resources available and convert them into electrical energy. This leads to stand-alone hybrid systems in a multiplicity of combinations.

Today's production of electrical energy is substantially based on fossil fuels. Costs of these materials are going up almost linearly every year. At some point, it pays to partly offset the consumption of these conventional and costly energies with hybrids based mainly on freely available natural resources. Hybrids are making a sizeable dent in energy consumption because they save fuel.

The world is becoming environmentally cautious. Apart from CO_2, fossil fuels emit a good deal of SO_x and NO_x, which are directly harmful to mankind. Renewable energy sources are available in plenty, though not always in a concentrated form. They offer a chance to reduce pollution. As per U.S. statistics, automobiles are the second highest users of fossil-fuel-based products, the highest being the electrical utilities. The automobile industry has become a first-priority target for using fuel cells and shifting ultimately to use of hydrogen as an energy source. This will be coupled with an internal combustion (IC) engine.

We also have an interesting Mexican case [2] in which the State Forest Department denied permission for cutting a swath through their trees to lay down a transmission system. The village thus cut off was served with hybrids for energy supply.

12.3 STAND-ALONE HYBRID POWER SYSTEMS

There are numerous types of hybrids serving as stand-alone systems. Each of these systems has peculiar characteristics and engineering challenges, apart from economical advantages. We will study a few of these hybrids.

12.3.1 Options for A Rural Electric Supply—Case of A Remote Mexican Village

Vera has published an interesting comparative report on various alternatives possible for supplying electric power to a remote Mexican community [3]. The requirements were:

1. Suitability for a remote location with respect to logistics for fuel supply, installation, maintenance, and operations and infrastructure available locally.
2. Utility-grade quality power supply: reliability, capacity for tackling surges, faults, overloads, and so on.
3. Plant reliability: degree of modularity, spares, mean time to failure data, components failure modes, and remedies.
4. Plant operation and maintenance: unattended operation and maintenance with locally available capacity, safety, and minimum costs.

For this remote hamlet, he worked with the following data:

Number of adjoining villages to be covered—10

Electricity supply for cooling and heating for all, lighting for all, limited refrigeration, particularly for medical stores.

Type of distribution—central.

Power quality—as near to grid power quality as possible

The basic equipment was rated as:

PV arrays—4.3 kWp with respect to plane

PV arrays to share 32.2% of required electricity

Storage batteries—two to three days capacity

Inverter—2.5 kW

DG Set—17.3 kW

12.3.2 Six Alternatives with Advantages and Disadvantages in a Mexican Case Study

Relative costs of various options have worked out as below:

1. Grid supply—life cycle 25 years, cost of energy $3.026/kW-hr. Not feasible.
2. DG set only—installed cost $ 990,625, cost of electricity US$1.30/kW-hr. Advantages—lower installation cost, independent of climatic conditions, mature system. Disadvantages—if designed for peak load, average utilization only 50% of rated value; rising fuel prices; environmentally objectionable, low modality; systems failures are total.
3. Central PV systems plus batteries installed—costs $290,701, life cycle costs $0.891 kW-hr. Disadvantages—large committed area required, high capital costs, climate dependent, low battery life, battery disposal is a problem, trained maintenance is required.
4. Distributed PV plus battery—one PV system per home. Installed capital costs $346,950, annual operating costs $7213 (lowest), life cycle costs $1.28/kW-hr. Advantages—households responsible for maintenance, failure constrained to single homes. Disadvantages—no load sharing or leveling, battery disposal is a problem.
5. Wind energy plus battery plus inverter—wind generator of 5 kW rating gives 200 kW-hr. It requires +5 days battery plus battery charger plus inverter. Installed capacity costs $498,631, life cycle costs $ 1.49/kW-hr. Advantages—nonpolluting, no limit on available energy. Disadvantages—wind irregularity, with space between the wind turbine generator (WTG) $\geqq$ 10 × diameter of WTG. It is high.
6. Distributed DGs. Lowest capital cost $85,400, life cycle costs $3.30/kW-hr. If combined with PV, this cost reduces fuel costs by 24%. Disadvantages—environmental and noise problems, maintenance is difficult and costly.

Eventually, they settled for the following [3]:

PV array rating 12.39 kW, tilt 15
DG 48 kW, 60 Hz
Wind turbines—2, 10 kW each
Battery 250 kW-hr, 1.6 days to 88% DOD

Battery life plays an important part in the costing of the system. Charging/discharging with reference to the battery voltage is erroneous, as shown in Figure 12-1.

12.4 RENEWABLE SOURCE USE OF ENERGY IN MEXICO—SAN ANTONIO AQUA BENDITA

Reference must be made to the current fed into or drawn out, and also to the operating temperature, the life of battery, and so on. A properly used battery can last up to 8 years. In a badly operated case, it can be as low as 2 years. This plays an important part in saving fuel.

Bagen [4] makes a few further points on operation of these stand-alone systems containing WTGs, PVs, batteries, and DGs. A double DG is recommended—one for the base load and the other for a variable load. The base load DG should run continuously. Under light load, it should be used to charge the batteries. Intermittent and partly loaded DGs have low fuel efficiencies. Their maintenance costs are high. Frequent starting also creates mechanical problems. Failure to correctly start a DG reduces the reliability of the system. With excess wind/solar, power it is advisable to cut the renewable portions rather than lower the output of the base load generator.

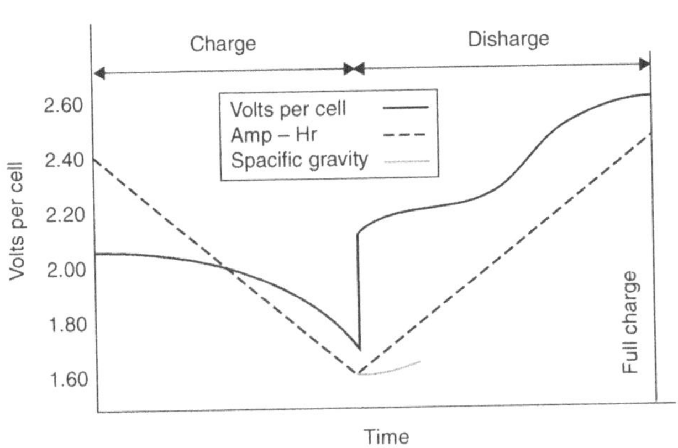

Figure 12-1. Battery life versus battery charge. (From [5], © IEEE 1994.)

12.5 SOME DEFINITIONS

12.5.1 Loss Possibility to Supply Power (LPSP)

This can be defined as the long-term average fraction of the load that is not supplied by the stand-alone system. Let us assume that at a given instance the load demands more energy than that generated by the PV and wind together. The charged battery will step in and shoulder the extra demand. However, quite soon, the battery will discharge down to an unhealthy charge level. Some portion of the load will have to be disconnected to prevent damage to the battery.

LPSP zero means all the load can be served by the hybrid. LPSP = 1 means that all the load will never be supplied.

LPSP for a given period *T* is the sum (loss of power × hours of loss) / sum (load for the period × hours of supply).

12.5.2 Battery Capacity

This is the amount of energy that can be extracted and not stored. Battery charge efficiency is equal to the round-trip efficiency (charge and discharge). Battery discharge efficiency is taken to be one.

12.5.3 Inverter Rating

This is expressed is in terms of peak load demand that it can serve. Inverter efficiency is a function of the ratio of load/inverter ratings.

12.5.4 Functions of a Battery Controller

When battery is overcharged and PV and wind are high, that battery must be cut. When the battery is near depletion and reaches it lowest permitted storage, it should cut off from the load. Battery life is greatest when it is kept near 100% of the capacity or returned to that stage quickly after a partial or deep discharge. Both MPPT and battery controller consume negligible energy by themselves.

12.5.5 Storage Batteries are Important in PV/Wind and Storage Battery Stand-Alone Hybrid Systems

There are four important players in the stand-alone system:

1. The PV arrays, which have a steady energy output over a given period. If a load goes up or comes down, PV arrays by themselves cannot level this load. They require assistance from the batteries.
2. The wind generators have, for all practical purposes, an infinite energy source behind them. They do not need much assistance for leveling the loads.

3. Storage batteries are an important player in leveling the load. They have a limitation on how much energy (charge) they can hold, as well as to what extent they can discharge or deliver energy. Their capital costs are high. They have low lives. Their life cycle costs are high.
4. The load will vary as per customer's requirements. Under power quality requirements, maximum loadless period allowed is one day in ten years.

12.6 COST BALANCE BETWEEN PV CELLS AND STORAGE BATTERIES

Thus, we are left with two variables: the number and costs of PV arrays and the number and costs of storage batteries, as against a minimum loss of load to serve the customer.

For a given hybrid system consisting of PV arrays, storage battery, and wind turbine generator, a relation can be established between the numbers and the costs of the PV arrays and those of storage batteries, with reference to the LPSP that is acceptable.

Borowy [6] explains this as follows. First of all, historical statistical data, preferably on an hourly basis, has to be procured for irradiance and wind velocities. The data can be converted into electrical energy from both of these sources. The depth of discharge and their capacities give the available energies from the batteries.

For the next part, similar statistical data on the past hourly loads has to be obtained. The hourly load data is also influenced by the climatic conditions.

Next we draw out a balance between energy available from all the three energy sources, including the batteries and the load. We arrive at the LPSP, the load that cannot be served.

Thus, for a targeted LPSP we have a number of combinations of PV arrays and storage batteries and the load requirement. The LPSP can be calculated. Against a targeted LPSP, we have a number of combinations of PV arrays and batteries. The relation between the two is logarithmic, as shown in Figure 12-2.

If we take the per-unit cost of PY arrays as x, and that for batteries as y, then the same graph gives us a ratio of unit costs in terms of x/y. This ratio is determined by the costing the PV arrays and of the batteries. With this ratio in hand, we draw a tangent to the above curve and arrive at cost-wise optimum numbers of PV arrays matching the number of batteries for an acceptable LPSP.

The optimum PV array selection discussed in the above applies to stand-alone hybrid systems. When a hybrid system incorporating PV arrays and batteries works as a complimentary system, say, in a residential supply, the criteria for selecting PV array size shifts to economic considerations. Please refer to Chapter nine.

12.6.1 Other Hybrid Illustrations

PV arrays backed by a storage battery have been used in residential systems, mainly with the purpose of saving on electricity bills. PV arrays with battery backing have also been used in agricultural pumps extensively. A system of PV arrays, wind turbine generator, and backup batteries gives a steady and reliable supply and is being used

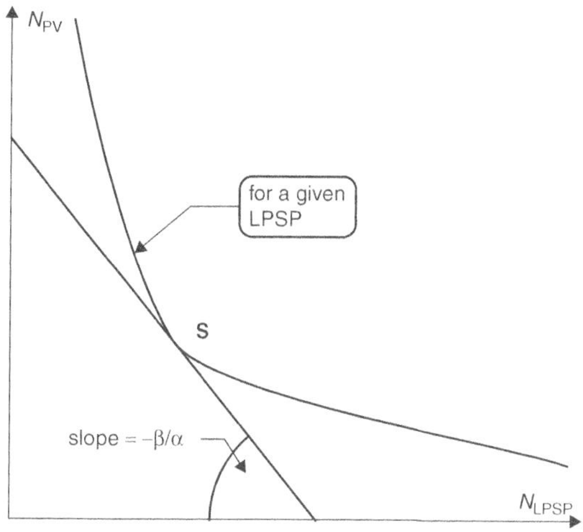

Figure 12-2. Plot of number of PV modules verses number of batteries for a given LPSP. (From [6], © IEEE 1996.)

widely. The solar insolation and wind usually arrive at complimentary times. The storage batteries ensure steadiness. A small DG set may also be added. The set comes into action when the battery is depleted to a certain low level of voltage. In this way, the diesel is used economically. Additionally, the DG set may also be used to meet the peak demand. A typical set has following ratings [3]:

Battery	120 V DC, 2084 A hrs
PV Arrays	12.39 kWp
Diesel generator set	48 kW
Rectifier	180 A, 200 V
Wind turbine	20 kW, 2 units
Energy sharing	PV, 33%; wind, 24.7%; diesel, 37.9%

12.7 HYBRIDS INCORPORATING FUEL CELLS

Fuel cells compete with batteries in their response to quick starting, and shutoff is almost instantaneous. Their response to transients is also instantaneous.

12.7.1 PV–Fuel Cell Hybrids for a Spaceship

Fuel cell hybrids development were first employed as a low-weight energy supplier and storage device employing recyclable materials in spaceship applications. Figure 12-3 gives a schematic layout. In this particular application, component designs have to be based on optimizing their use so that the power is available for maximum time.

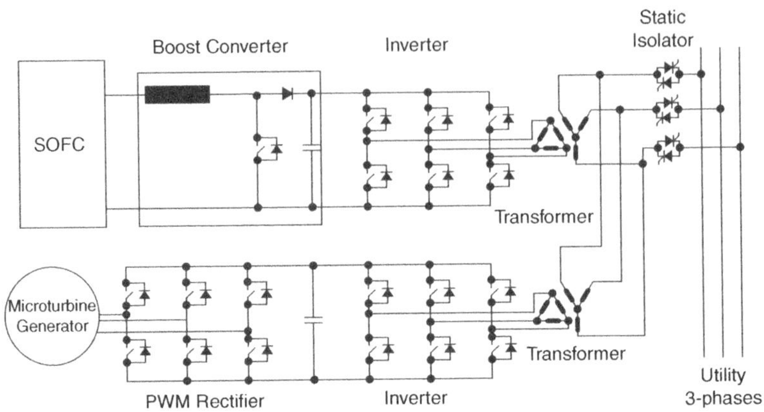

Figure 12-3. Hybrid photovoltaic fuel cell system for a spaceship application. From [7], © IEEE 2005.)

For more details on hybrids incorporating fuel cells based on hydrogen as well as on solid oxides, please refer to Chapter 11.

12.7.2 Diesel Generator–Wind Energy Hybrids

Wind energy is available in large quantities. This makes it a compatible source for supply of electrical energy in competition with large conventional energy sources employed by electrical utilities.

12.8 MIDSEA HYBRIDS

A striking example of this type of hybrid is given by the electric power supply to several islands in Greece. The Greek islands are vastly dispersed and are supplied individually, mostly by diesel generators. Typical system ratings are Kea Island, 2300 kW and Skyros, 26,300 kW.

It is profitable to employ wind energy on an as and when it is available basis to cut down on fuel consumption by diesel generators. However, there are certain limitations on how much wind energy we can employ with respect to the energy available from established sources such as diesel generators, as in a stand-alone hybrid case. Let us study this further.

12.9 WORKINGS OF A WTG AND DIESEL GENERATOR

12.9.1 Starting Of WTGs

Most of the wind generators have an induction machine for a generator. The induction machine draws reactive power to start and to run. The reactive power drawn is proportional to the slip speed, that is, the difference between the synchronous speed for which the induction machine is designed and its running speed. The first requirement of this hybrid is that a diesel generator must always be running for supplying reactive power to the WTG be. Secondly, it is preferable to let the WTG run by a reasonably strong wind to as near its rated speed as possible and then connect it to the supply system. The slip is then small. The reactive power drawn is low and the sudden dip in voltage could be held at a low value. In case of more than one WTG, subsequent machines can be switched in consecutively (one after another) in a similar fashion.

12.9.2 A Case of Low Wind

Now consider a case in which the wind is low and the WTG is running close to its idling position. A heavy load, say, a large water pump, is switched in. The diesel generator (DG) is supplying the power. With a sudden large load demand, its fuel controller opens up the fuel valves, thus running it at a slightly higher speed. The frequency moves up. The higher frequency tries to run in the induction generator at a higher speed, thus converting it into a loaded induction motor rather than a generator supplying power. The system will stall unless the DG set has a certain minimum power capacity or unless the automatic load controller on the output bus starts disconnecting some other load.

12.9.3 A Case of Wind Gust

Consider a case in which there is a strong wind or a sudden gust and the WTG capacity is fairly large in relation with the DG capacity. The load cannot absorb all the energy. The extra power at synchronous frequency will try to run the generator attached to the diesel as a synchronous motor now trying to drive the diesel engine. In this case, an automatic output damping will dump the extra wind energy into specially designed and connected damping resistances. Here again, there is a limitation on the size of WTG with respect to the size of diesel generator.

Let us now consider how much energy we can usefully draw from a WTG and under what guiding principles.

12.9.4 In a Hybrid System, Can We Draw Energy Wholly from WG?

Suppose we draw the wind energy on an as available basis, which means with no programming in advance in the case of a utility connected system. When the wind energy is almost near the load demand, we have to idle the diesel sets or shut them down and restart them. This lowers the life of the DG sets and is not desirable.

In the case of a utility, the problem is further accentuated. A utility has some basic high-capacity thermal generators. In the first place, their high inertia does not lend to quick output variations. In the second place, they are delivering the cheapest cost electricity, which one cannot forego if one is buying or compensating for WTG energy produced by independent producers.

Denny [8] has made an interesting third point about when thermal stations are backed down; they produce relatively more nitrogen oxide pollutants than when they are operating normally. See Figure 12-4. This runs contrary to the basic aim of reducing pollution by using renewable energy.

12.9.5 An Irish Rule on Permissible Wind Penetration

In case of Ireland, Denney [8] reports that pit coal is a natural source of energy and certain minimum energy production has to be based on the use of this as per legislative stipulations.

12.10 WIND ENERGY PENETRATION LIMIT

A limit for penetration by wind energy as a percentage of the load demand emerges whether it is a larger load as in case of the Greek islands or a small stand-alone WTG/DG hybrid in the form of a pit-coal-based utility supply on an island. The maximum wind energy penetration limit allowed in Ireland is in the neighborhood of 40%.

Wind energy is usually drawn on a metrological forecast-method basis whereby it is possible to schedule in advance, with sustainable success, cooperation of all the connected energy supplying sources.

12.11 WIND POWER–FUEL CELL HYBRIDS

These have been commented upon in earlier chapters. They have a high capacity potential. Figure 12-5 gives a typical layout.

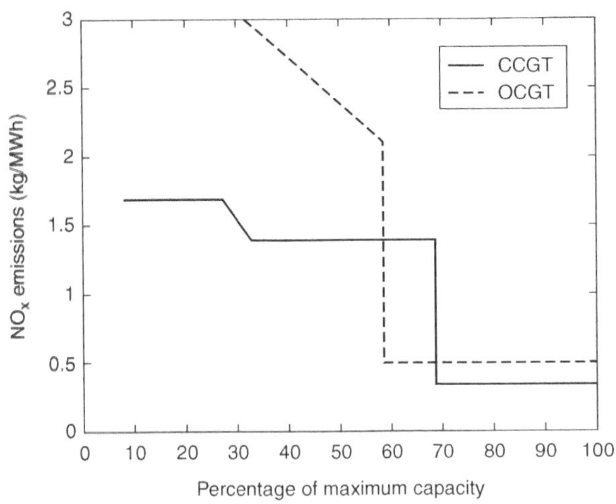

Figure 12-4. Typical NO emissions from a CCGT and an OCGT. (From [8], © IEEE 2006.)

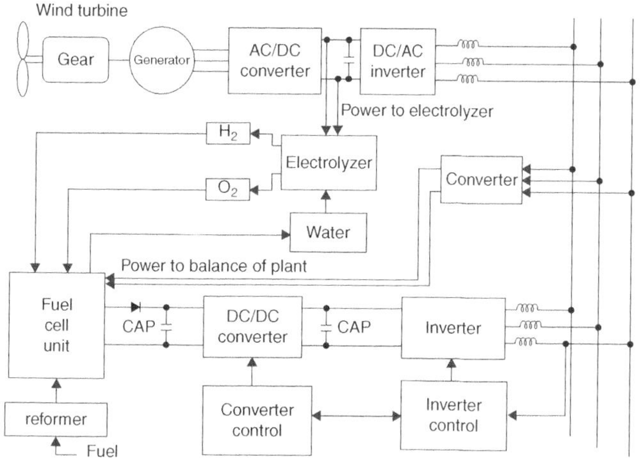

Figure 12-5. Wind–fuel cell hybrid power station. (From [9], © IEEE 2005.)

12.12 INTERFACING NONCONVENTIONAL ENERGY SOURCES WITH UTILITY SYSTEMS–STATIC POWER CONTROLLERS (SPCS)

Excepting wind turbine generators, all the other energy generators, or storage devices produce a DC voltage and be converted into matching and synchronizing systems compatible with the utility AC system to which they are to be connected. Static power converters incorporating appropriate protections are employed for the interface.

The major components of these static power converters are:

1. DC source interface including DC side fuses; switches; filters, if any; and energy the storage capacitors. It also includes transducers that monitor the converters of the DC supply.
2. The SPC control section is the brain of whole system. The controls are microprocessor based. This allows multiple functions to be incorporated in the interface on both sides of the interface.

For example, maximum power point tracking, in which the current drawn out from solar energy is adjusted so that the power output follows a maximum power yield point, can be incorporated in the microprocessor system.

12.13 PROTECTIVE CONTROLS BETWEEN A UTILITY AND A NEWCOMER

12.13.1 Routine Controls

For allowing a new supplier into the system, the utility requires control on the following parameters of the incomer's system:

- Overvoltage protection
- Undervoltage protection
- Under/over frequency protection
- Voltage controlled/voltage compensated overcurrent
- Islanding detection
- Out-of-phase reclosing
- Battery/DC understorage
- Synchronous check
- Voltage check
- Reverse power check

The battery check is for the purpose of ensuring that various relays can operate irrespective of any happenings on both sides of the interface. It also ensures that the intertie will operate upon receiving a signal and positively separate the two systems. In the eventuality of SPC and microprocessor system failures, an alternate system (say, manual or an independent relay control) should be able to trip the intertie breaker with the help of the DC supply from the station battery.

Another check on power flow from a SPC to the utility is to prevent flow of direct current and harmonics into the system. The utilities check these parameters at various levels of power injection from the RES (renewable energy sources).

12.13.2 Specific Controls

Utilities require specifically the following protections, if necessary as a separate package, so that the functions enumerated earlier can be performed.

- Open the connections for phase and ground faults within its protective zones
- Prevent over load of any utility facilities
- Open when voltage is out of tolerance
- Open when frequency is out of tolerance
- Open before any reclosing of live protective devices
- Prevent energizing a dead distribution system
- Open when SPC is operating as an Island

The interface protection must operate correctly during the following four modes of operation:

1. Nonfault with utility source connected
2. Nonfault with utility source disconnected
3. Fault with utility source connected
4. Fault with utility source disconnected

It is generally possible to incorporate all the above in the microprocessor programming. In addition to these it should be able to sense the RKVA requirement on the utility side and supply adequate reactive power within its operative capacity.

The third component of the SPC system consists of the semiconductors that act as high-frequency power switches and convert DC into AC. Two methods of communications, namely, line communication and self-communication, have been discussed earlier in detail.

The AC system interface includes interconnection components such as contactors, circuit breakers, isolating or step-up transformers where necessary, and filtering, metering, and power factor correction systems.

REFERENCES

1. V. Wouk, Hybrids: Then and Now, *IEEE Spectrum,* Volume 32, Issue 7, July 1995, pp. 16–21.
2. F. Valenciaga, P. F. Puleston, and P. E. Battaiotto, Power Control of a Solar/Wind Generation System without Wind Measurement: A Passivity/Sliding Mode Approach, *IEEE Transactions on Energy Conversion,* Volume 18, Issue 4, December 2003, pp. 501–507.
3. J. G. Vera, Options for Rural Electrification in Mexico, *IEEE Transactions on Energy Conversion,* Volume 7, Issue 3, September 1992, pp. 426-433
4. Bagen, I. and R. Billinton, Evaluation of Different Operating Strategies in Small Stand-Alone Power Systems, *IEEE Transactions on Energy Conversion,* Volume 20, Issue 3, September 2005, pp. 654–660.
5. J. Gutierrez-Vera, Use of Renewable Sources of Energy in Mexico Case: San Antonio Agua Bendita, *IEEE Transactions on Energy Conversion,* Volume 9, Issue 3, September 1994, pp. 442–450.
6. B. S. Borowy and Z. M. Salameh, Methodology for Optimally Sizing the Combination of a Battery Bank and PV Array in a Wind/PV Hybrid System, *IEEE Transactions on Energy Conversion,* Volume 11, Issue 2, June 1996, pp. 367–375.
7. Rajashekara, K., Hybrid Fuel-Cell Strategies for Clean Power Generation, *IEEE Transactions on Industry Applications,* Volume 41, Issue 3, May–June 2005, pp. 682–689.
8. E. Denny, and M. O'Malley, Wind Generation, Power System Operation, and Emissions Reduction, *IEEE Transactions on Power Delivery,* February 2006, pp. 341–347
9. F. Crescimbini, A. Di Napoli, L. Solero, and F. Caricchi, Compact Permanent-Magnet Generator for Hybrid Vehicle Applications, *IEEE Transactions on Industry Applications,* Volume 41, Issue 5, September–October 2005, pp. 1168–1177.

BIBIOGRAPHY

Ball, G. J., D. C. Blackburn Jr., W. Gish, E. Gulachenski, E. A. Guro, E. W. Kalkstein, A. Pivec, P. W. Powell, F. Soudi, J. W. Stevens, M. Yalla, R. Zavadil, J. T. Emery, and R. Hill, Static Power Converters of 500 Kw or Less Serving as the Relay Interface Package for Nonconventional Generators, *IEEE Transactions on Power Delivery,* Volume 9, Issue 3, July 1994, pp. 1325–1331.

Bansal, R. C., T. S. Bhatti, and D. P. Kothari, Bibliography on the Application of Induction Generators in Nonconventional Energy Systems, *IEEE Transactions on Energy Conversion,* Volume 18, Issue 3, Sept. 2003, pp. 433–439.

Bates, B., Getting a Ford HEV on the Road, *IEEE Spectrum,* Volume 32, Issue 7, July 1995, pp. 22–25.

Bonanno, F., A. Consoli, S. Lombardo, and A. Raciti, A Logistical Model for Performance Evaluations of Hybrid Generation Systems, *IEEE Transactions on Industry Applications,* Volume 34, Issue 6, November–December 1998, pp. 1397–1403.

Borowy, B. S. and Z. M. Salameh, Optimum Photovoltaic Array Size for a Hybrid Wind/PV System, *IEEE Transactions on Energy Conversion,* Volume 9, Issue 3, September 1994, pp. 482–488.

Cadirei, I. and M. Ermis, Performance Evaluation of a Wind Driven DOIG using a Hybrid Model, *IEEE Transactions on Energy Conversion,* Volume 13, Issue 2, June 1998, pp. 148–155.

Chedid, R. and S. Rahman, Unit Sizing and Control of Hybrid Wind-Solar Power Systems, *IEEE Transactions on Energy Conversion,* Volume 12, Issue 1, March 1997, pp. 79–85.

Chedid, R., H. Akiki, and S. Rahman, A Decision Support Technique for the Design of Hybrid Solar-Wind Power Systems, *IEEE Transactions on Energy Conversion,* Volume 13, Issue 1, March 1998, pp. 76–83.

Daniel, S. A. and N. Ammasai-Gounden, A Novel Hybrid Isolated Generating System Based on PV Fed Inverter-Assisted Wind-Driven Induction Generators, *IEEE Transactions on Energy Conversion,* Volume 19, Issue 2, June 2004, pp. 416–422.

Dokopoulos, P. S., A. C. Saramourtsis and A. G. Bakirtzis, Prediction and Evaluation of the Performance of Wind-Diesel Energy Systems, *IEEE Transactions on Energy Conversion,* Volume 11, No. 2, pp. 385-393, June 1996.

Ermis, M., Z. Cakir, I. Cadirci, G. Zenginobuz, and H. Tezcan, Self-Excitation of Induction Motors Compensated by Permanently Connected Capacitors and Recommendations for IEEE Std. 141-1993, *IEEE Transactions on Industry Applications,* Volume 39, Issue 2, March–April 2003, pp. 313–324.

Giraud, F. and Z. M. Salameh, Steady-State Performance of a Grid-Connected Rooftop Hybrid Wind-Photovoltaic Power System with Battery Storage, *IEEE Transactions on Energy Conversion,* Volume 16, Issue 1, March 2001, pp. 1–7.

Hatziadoniu, C. J., A. A. Lobo, F. Pourboghrat, and M. Daneshdoost, A Simplified Dynamic Model of Grid-Connected Fuel-Cell Generators, *IEEE Transactions on Power Delivery,* Volume 17, Issue 2, April 2002, pp. 467–473.

Hermance, D. and S. Sasaki, Hybrid Electric Vehicles Take to the Streets, *IEEE Spectrum,* Volume 35, Issue 11, November 1998, pp. 48–52.

Hybrid vehicles are worth it! (paper discussion and authors' reply), *IEEE Spectrum,* Volume 38, Issue 5, May 2001, pp. 79–84.

Jones, W. D., New York Wants More Hybrid Buses on the City's Streets, *IEEE Spectrum,* Volume 37, Issue 9, Sept. 2000, pp. 36–36.

Kellogg, W. D., M. H. Nehrir, and G. Venkataramanan, and V. Gerez, Generation Unit Sizing and Cost Analysis for Stand-Alone Wind, Photovoltaic, and Hybrid Wind/PV Systems, *IEEE Transactions on Energy Conversion,* Volume 13, Issue 1, March 1998, pp. 70–75.

King, R.D., K. B. Haefner, L. Salasoo, and R. A. Koegl, Hybrid Electric Transit Bus Pollutes Less, Conserves Fuel, *IEEE Spectrum,* Volume 32, Issue 7, July 1995, pp. 26–31.

Kyoungsoo, R. and S. Rahman, Two-loop Controller for Maximizing Performance of a Grid-

Connected Photovoltaic-Fuel Cell Hybrid Power Plant, *IEEE Transactions on Energy Conversion,* Volume 13, Issue 3, September 1998, pp. 276–281.

Lave, L. D., and H. L. Maclean, Are Hybrid Vehicles Worth It? *IEEE Spectrum,* Volume 38, Issue 3, March 2001, pp. 47–50.

Muljadi, E. and T. A. Lipo, Series Compensated PWM Inverter with Battery Supply Applied to an Isolated Induction Generator, *IEEE Transactions on Industry Applications,* Volume 30, Issue 4, July–August 1994, pp. 1073–1082.

E. Muljadi and H. E. McKenna, Power Quality Issues in a Hybrid Power System, *IEEE Transactions on Industry Applications,* Volume 38, Issue 3, May–June 2002, pp. 803–809.

Naik, R., N. Mohan, M. Rogers, A. Bulawka, A Novel Grid Interface, Optimized for Utility-Scale Applications of Photovoltaic,

Wind-Electric, and Fuel-Cell Systems, *IEEE Transactions on Power Delivery,* Volume 10, Issue 4, October 1995, pp. 1920–1926.

Phillips, D., Transportation (Technology 1998 Analysis and Forecast), *IEEE Spectrum,* Volume 35, Issue 1, January 1998, pp. 84–89.

Rahman, S. and K. S. Tam, A Feasibility Study of Photovoltaic-Fuel Cell Hybrid Energy System, *IEEE Transactions on Energy Conversion,* Volume 3, Issue 1, March 1988, pp. 50–55.

Rau, N. S., and B. Tayor, A Central Inventory of Storage and Other Technologies to Defer Distribution Upgrades-Optimization and Economics, *IEEE Transactions on Power Delivery,* Volume 13, Issue 1, January 1998, pp. 194–202.

Riezenman, M. J., Fuel Cells for the Long Haul, Batteries for the Spurts (Electric Vehicles), *IEEE Spectrum,* Volume 38, Issue 1, January 2001, pp. 95–97.

Riezenman, M. J., Engineering the EV Future, *IEEE Spectrum,* Volume 35, Issue 11, November 1998, pp. 18–20.

Senjyu, T. T. Nakaji, K. Uezato, and T. Funabashi, A Hybrid Power System Using Alternative Energy Facilities in Isolated Island, *IEEE Transactions on Energy Conversion,* Volume 20, Issue 2, June 2005, pp. 406–414.

Stempel, R. C., S. R. Ovshinsky, P. R. Gifford, D. A. Corrigan, Nickel Metal Hydride: Ready to Serve, *IEEE Spectrum,* Volume 35, Issue 11, November 1998, pp. 29–34.

Thiringer, T., T. Petru, and C. Liljegren, Power Quality Impact of a Sea Located Hybrid Wind Park, *IEEE Transactions on Energy Conversion,* Volume 16, Issue 2, June 2001, pp. 123–127.

Uhlen, K., B. A. Foss, and O. B. Gjosaeter, Robust Control and Analysis of a Wind-Diesel Hybrid Power Plant, *IEEE Transactions on Energy Conversion,* Volume 9, Issue 4, December 1994, pp. 701–708.

Valenciaga, F. and P. F. Puleston, Supervisor Control for a Stand-Alone Hybrid Generation System Using Wind and Photovoltaic Energy, *IEEE Transactions on Energy Conversion,* Volume 20, Issue 2, June 2005, pp. 398–405.

Zhu, T., S. R. Shaw, and S. B. Leeb, Transient Recognition Control for Hybrid Fuel Cell Systems, *IEEE Transactions on Energy Conversion,* Volume 21, Issue 1, March 2006, pp. 195–201.

13

COMBINED GENERATION—COGENERATION

13.1 DEFINITION AND SCOPE

Cogeneration is the generation of two types of energies within the same plant, generally thermal energy in the form of steam and electrical energy. Cogeneration extends from 25 MW up to 500 MW. It is connected into the transmission system network at voltages above 33 kV. This is where it differs from distributed generation, which is connected into distribution systems up to 33 kV. Distributed generation is not under the control of the regional load dispatcher, but cogeneration is.

Cogenerators have no franchised area and thus are not committed to serve any particular territory. Unlike utilities, they are not regulated. However, they have to accept some discipline and abide by the rules incorporated.

Cogeneration has been described as large-scale generation tied directly into EHV systems. This helps us to distinguish it from distributed generation. Historically, these large-scale generators connected to EHV belonged to old established industries like paper mills, oil refineries, and steel and aluminum plants. Going strictly by definition, cogeneration can include small-scale generation, which under new economic conditions makes it economical to produce both thermal and electrical energy at the distribution-level voltages.

Electricity Power Generation. By Digambar M. Tagare

Copyright © 2011 the Institute of Electrical and Electronics Engineers, Inc.

13.2 RISE OF COGENERATION

Large industrial plants using fossil fuel-based thermal energy or working with useable biomaterials like wood initially produced their own electrical requirements when there was a shortage of electrical energy and utilities were small. As the utilities expanded, both in their production as well as areas covered, they could supply electricity to large industrial consumers at affordable and economic prices. Industrial plants with cogeneration facilities went in for import of electricity from utilities. Still, they found that they had quite a considerable amount of thermal energy that could not be economically converted. However, with increase in sizes of industrial plants this picture also changed. Not only could industrial plants produce their own requirements affordably, they had surplus electrical energy to sell. With all the areas franchised to existing utilities, they had no option except to sell their surplus power to the utilities.

In the United States under government order PURPA of 1978, it was made compulsory for a utility to buy all the electric energy from a qualified cogenerator at prices fixed by the electricity regulator. The cogeneration expanded in scope and size [1].

Today all over the world where restructuring is going on, the generation of electric power is freed. The markets are open and there is competition to sell. This has pushed up cogeneration to larger sizes. Now they can sell electric power across the boundaries. Franchised areas do not exist anymore.

Cogeneration has one basic advantage over utility generation. Since they have alternate uses for energy, there are comparably fewer slowdown periods during which they have to back down production, unlike utilities. Thus, their utilization factors are better than those of utilities [2].

13.3 BASIC PURPOSE OF COGENERATION

Basic purpose of cogeneration is, concentrating on the main basic energy production, to utilize the spare capacities to produce alternate energy, add to the revenue and lower the overall operating costs.

13.4 THREE TYPES OF COGENERATORS

13.4.1 Primary Product—Steam

Some cogenerators produce steam as their primary product. For example, a hard coke plant requiring lots of steam for extracting petroleum liquids from hard coke, or a crude oil well where steam is required to lower crude oil viscosity for ease of pumping. Steam does this by heating up the raw crude. This is termed thermal enhanced oil recovery (TEOR). It is also required in a sugar factory for evaporating cane juice. Leftover steam capacity is diverted to electricity production, which is generally used for the internal load of the plant [3]. The cogenerator in this case is operating in a "bottoming cycle" mode.

13.4.2 Primary Product—Electricity

Cogeneration producers may have put up their plant to produce and sell electrical energy mainly. Extra steam is piped to supply power to industrial or residential establishments in the close vicinity. Here, the cogenerator is operating in a "topping cycle" mode.

13.4.3 Equal production—Steam and Electricity

Cogeneration producers may place equal value on both thermal production and electrical production. A typical example is in the Netherlands, a small European country in a cold region, with a high population/residential geographic density. Steam is piped around and sold, the same as electricity, which is produced and sold. Both energies are complementary for a consumer as well as for a producer [4].

13.5 ADVANTAGES OFFERED BY COGENERATION

Costs and sales prices of electricity are low. Because of their spread overheads on multiple products and because of input energy being used without waste, the cost of production and, resultantly, the selling price of electrical power is lower than that of utilities. For utilities, they are a cheap source of additional power. With open marketing, this benefit will now shift to the consumer, who can buy directly from a cogenerator.

Reliability, availability, and maintainability (RAM) of cogeneration have been found to be significantly better than that of utility generation. Unavailability could be of planned nature (EPOF) or an unplanned one (EUOF). These are determined by the frequency of outage, duration of outages, and capacity reduction. Total unavailability (EUF) of these two types leads us to the equivalent availability factor or EAF (equal to 1 minus total unavailability factor). Figure 14-1 and Table 14-1 bring out the sharp differences between EAF of utilities and those of cogeneration.

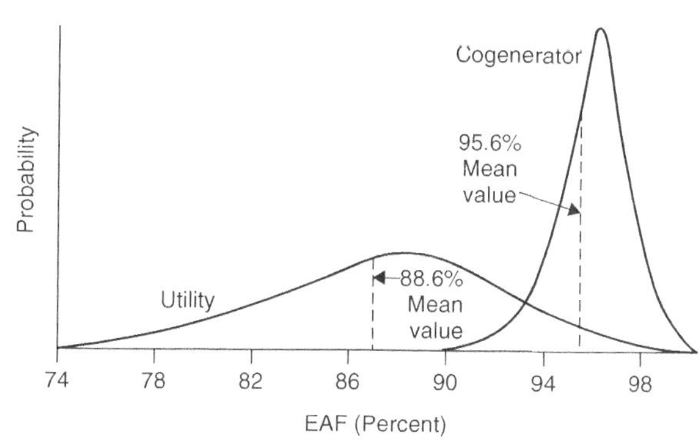

Figure 13-1. Comparison of utility and cogeneration EAF probabilities. (From [5], © IEEE 1991.)

Table 13-1. Performance comparison percentages

Operator	EUF Range*	EUF Mean	EUOF Mean	EPOF Mean
Utility	8–24	13.4	4.4	9.0
ogenerator	2–6	4.4	1.6	2.8

*There is a 90% confidence level that the values lie within this range.
Source: [5].

Record of past outages, availability of spares ahead of overhauling, adequate space, and facilities handling training of personal and skilled staff are the factors that affect (RAM) performance of a unit.

One basic factor that did not encourage the utilities to improve the availability factor was the fixed rate of return on their performance, which is fixed by the regulators.

Cogenerators can sell reserve capacity to a utility. Cogeneration capacity adds significantly to the operating reserves to be maintained by utility. However, their reliability is to the extent promised a day or two ahead by the cogenerators, particularly if the cogenerator is operating in the bottoming cycle mode.

13.6 PLANNING OF COGENERATION

There are three categories of cogenerators:

1. Old established industries that have had captive generation of their own, right from inception, to ensure uninterrupted supply to essential loads.
2. New high-tech industries in need of secure, reliable, and high-quality electric power supply.
3. Hotels, hospitals, commercial establishments, and so on who find it economical to maintain small cogeneration units of their own, akin to the uninterrupted power supply (UPS) system.

13.6.1 Planning by Old Established Cogenerating Units

The planning has to start from considerations of available, complementary, energy sources. These are:

- Spare process steam available. Type of industries: chemical industries, refineries, paper, textiles, sugar, and so on.
- Biological wastes—wood shavings from timber industries, rice bran, bagasse from sugar industries, and so on.
- Hot flue gases from metal smelters, metal refineries, and so on—aluminum, steel, blast furnaces, coke oven, batteries.

This availability dictates the quantum to add to the existing facilities.

There may be nonuniformity in the blocks added during the past growth period. Another characteristic of these industries is their scattered generation and in-plant distribution at various voltage levels and with presently working assets but at nonuniform ratings.

Planning for expansion will be on an incremental or modular basis. Adding assets at different levels needs careful attention to:

- Overloading of existing distribution facilities—conductor cables, transformers, and circuit breakers.
- Increased short-circuit levels, due to synchronization with grid supply.
- Earthing grid within the plant and possibility of high voltage points rising in the earthing.
- Relay settings.
- Scenario developing under islanding conditions.

In the United States, NERC has developed norms to be observed by cogeneration units for permission to be tied into a utility grid. These are given in Appendix 13-2. Planning has to include adherence to these norms.

PURPA legislation passed by the U.S. government in 1975 made it compulsory to buy power from cogenerating units. The cogenerator had to bring up total thermal efficiency, including all of the electrical and half of the thermal outputs, to 42.5% or more [1].

13.6.2 New High-Tech Industries

Power quality requirements for an IC manufacturer are very strict as to reliability, voltage, and frequency tolerances. A normal supply from a utility cannot meet these. They have to establish their own cogeneration facility, since thermal power is also required for these plants. Details of a particular plant in Taiwan are given in Appenix 13-1. A tie-up with the utility enables this high-quality power plant to exchange electricity. The exchange normally is done with economic considerations.

13.6.3 Small Establishments

Hotels, hospitals, and commercial establishments are facing increasing differential (time-of-use) rates for electric power. Costs of high-efficiency, small gas turbine generators are falling. Small establishments can find it profitable to go in for individual cogeneration of their own, as the following illustration from Japan will show [6].

The hospitals had a fairly high maximum demand but low average and annual energy consumption. The hotels had characteristically high morning and evening loads. The commercial establishments had a steady high load during working hours. All three consumers required thermal energy as well, for which they had gas connections. The following factors affected the design for an optimized solution for consuming electric power and gas in economic portions:

- Maximum demand charge. This had an impact on the size of the cogenerator.
- Ratio of on-peak charge to off-peak charge, termed the energy charge ratio.
- Rates for gas and electricity power.

Figures 13-2 and 13-3 show off peak kilowatt-hour share and the proportion of power costs to gas costs, both purchased from the same utility for the above three examples. The following may be noted:

1. It benefits the hotels and hospital to buy more electric power under off-peak hours and keep their cogeneration in reserve. This makes little difference to the office building.
2. Beyond an energy charge ratio of 5, the differences level out.
3. It benefits the utility to level out the peak demand by inducing the customers to use cogeneration during peak hours.

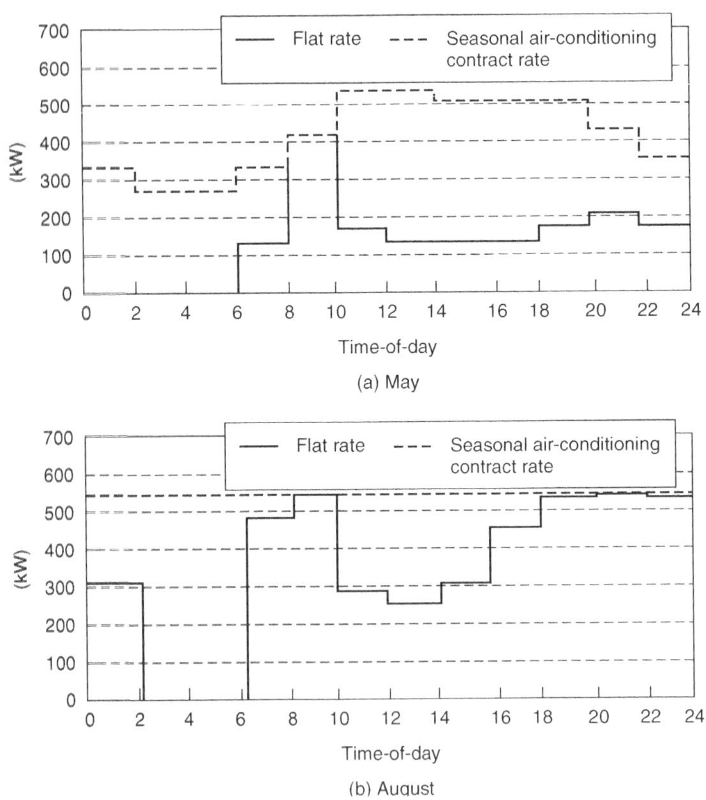

Figure 13-2. Effects of city gas rates on the optimal operational pattern of a gas-engine-driven generator. (From [7], © IEEE 1992.)

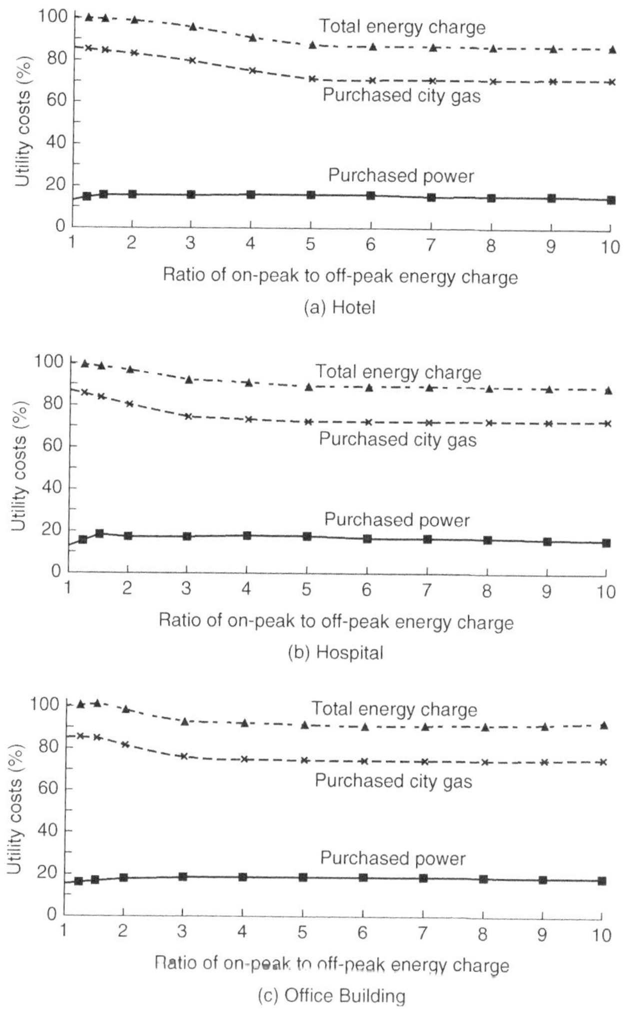

Figure 13-3. Effects of time-of-use rate structure on annual utility costs. (From [7], © IEEE 1992.)

As of March 1991, there were 616 such cogenerators producing a total of 274 MW in Japan.

13.7 ECONOMIC OBJECTIVES FOR A COGENERATOR

The two primary economic objectives for cogeneration facilities are optimization of fuel input for a given target and maximization of profits for electricity.

13.7.1 Optimization of Fuel Input

Tao [2] gives an algorithm for combined heat and power dispatch. Figure 13-4 gives a rough concept of optimization in this direction. By reducing the electrical power offtake slightly from point A to C, to the maximum increase in thermal output can be produced. Similarly, by reducing thermal offtake under C to D, more electrical power can be produced.

13.7.1 Profit Maximization under TOU Rates—An Illustration

Figures 13-5 and 13-6 show optimization of inputs and of exports of electricity, respectively, under a flat tariff and under a time-of-use (TOU) tariff. These figures pertain to an oil refinery in Mumbai. Tables 13-2 and 13-3 are self-explanatory. It pays to export power and to opt for a flat tariff in this case.

The cogeneration operator will have to be responsible for his own equipment. He cannot, by terms of his contract, hold the grid responsible for damage or losses on his side.

The contract between the cogenerator and the utility is for supply of KVA with payment being valued on the basis of kilowatt hours. The utility's greatest need is for KVAR and it will demand less of kilowatt hours and more KVAR-hrs. Unless KVAR-hrs are priced separately, this will be unfair to the cogenerator. His excitation system control has to be accurate and robust if he is going to supply large KVAR-hrs most of the time [11].

13.8 OPERATION OF COGENERATORS

13.8.1 Within Its Own Complex

A cogenerating plant has generally more than one generating unit, to suit their departmental requirements as well meet the electrical output goal. Ratings of these genera-

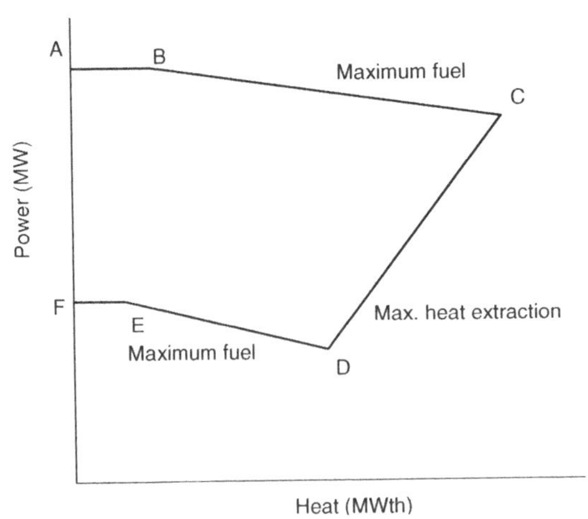

Figure 13-4. Heat–power feasible region for a cogeneration unit. (From [2], © IEEE 1994.)

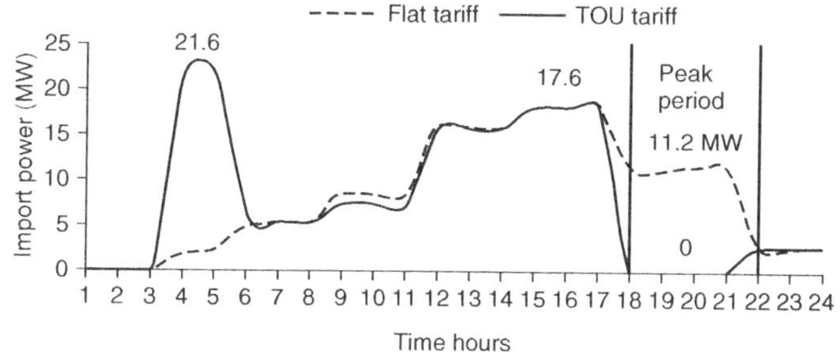

Figure 13-5. Import power to the grid under flat and TOU tariffs. (From [6], © IEEE 2003.)

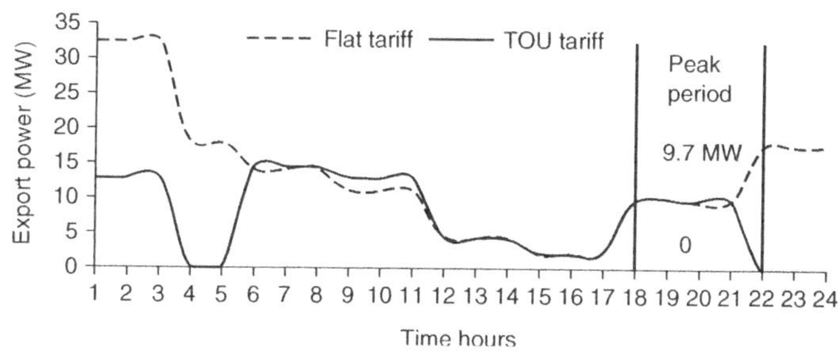

Figure 13-6. Export power from the grid under flat and TOU tariffs. (From [6], © IEEE 2003.)

tors will vary. The internal distribution voltages for a cogenerator will vary from plant to plant. The whole plant as such is tied into the utility system, generally at a utility substation or to a transmission line. The cogenerators have automatically controlled turbines so that they maintain a frequency equal to that of the utility whether they are receiving power or dispatching power, or whether they are islanded. Active power receipt or dispatch is controlled through the turbine power inputs. It is affected by, but not controlled by, slight frequency variations. Reactive power receipt or dispatch is controlled by the excitation current control of the cogenerator.

Table 13-2. Results of optimal response of cogeneration system (under no power export condition)

Tariff type	MD, MW	Average partial peak period coincident demand, MW	Average peak period coincident demand MW	Monthly import of energy from grid, GW-hr	Monthly electricity cost, millions of rupees	Total operating cost per month, millions of rupees
Flat	17.6	8.5	11.2	5.9	33.5	148.7
TOU	21.6	7.2	0	5.5	31.7	149.6

Source: [6].

Table 13-3. Results of optimal response of cogeneration system (under power export condition)

Tariff type	Average power export to grid, MW		Monthly energy export to grid, GW-hr	Monthly revenue from electricity export, millions of rupees	Total operating cost per month, millions of rupees
	Partial peak period	Peak period			
Flat	10.9	9.7	9.5	31.7	134.5
TOU	12.8	9.7	5.3	18.3	145.4

Source: [6].

13.8.2 As a Tie-Up between a Cogenerator and a Utility

The tie-up into a utility is governed by an agreement that establishes certain qualifying norms for the cogenerator. Qualifying norms drawn by the Northern Electricity Regional Council (NERC) in the United States, for such a tie-up are given in Appendix 1. Figure 13-7 gives a schematic of a tie-up between a cogenerator and a utility.

With the shortage of electrical energy supply in India and a large idle capacity in cogeneration, a large number of these cogenerators are expected to be tied into the existing grid system.

13.9 WORKING TOGETHER WITH COGENERATION

In a parallel-connected cogeneration utility system, a power supply of varying capacities is paired with a utility with very large capacities. A multitude of problems arise out of this combination, particularly with respect to the protection and operation of both the systems. Some of these prominent problems are pointed out in the following.

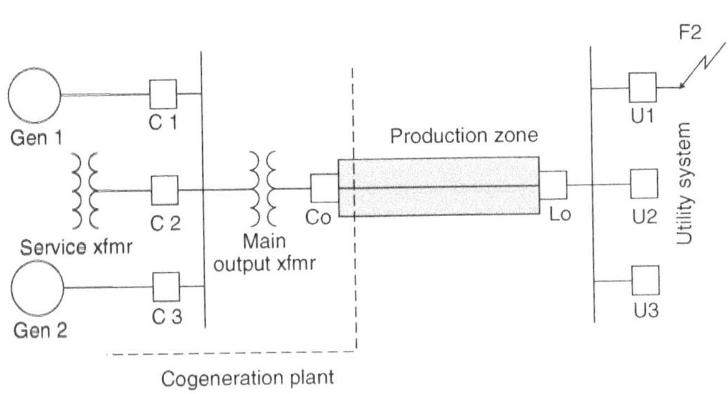

Figure 13-7. A tie-up between a cogenerator and a utility. (From [8], © IEEE 1997.)

13.9.1 Excitation Control of Cogenerators

When the KVAR demand of the network is not met, the system voltage tends to go down. However, there could be contingencies where this demand is of a sudden nature and beyond the capacity of the cogenerator or cogenerators operating in parallel. For example, a large induction motor (say, 8000 HP) is being started or a large block of load that was disconnected comes back following a recloser action on the main utility portion. In both these cases, the protection system has to clamp down on the excitation current of the cogenerator and limit its output KVAR. A practice followed is to allow the excitation to grow and hold it for up to 15 seconds and then clamp it down to a value corresponding to the terminal voltage while delivering full load current (Figure 13-8).

Should this 15 sec interval and KVAR supply be inadequate to reduce the starting KVAR demand of the large induction motor, the system voltage will continue to stay low, with other motors and systems cutting out and, finally, the large motor itself cutting out. Restoring power haphazardly could lead to a voltage collapse. The remedy would be disconnecting some of the system load while starting the large motor and reconnecting it later.

On the other hand, should the terminal voltage rise, the excitation control will try to reduce the excitation control. Should it rise abruptly, then it will trip the excitation control, simultaneously tripping the generator. This will happen if a large capacitor bank is switched on near a cogenerator. This will also happen if the load at the end of an EHV line trips off or an EHV line is switched in.

A sudden tripping of a cogenerator can lead to cascade tripping, with the overload and undervoltages coming on the remainder system. Following the capacitive current

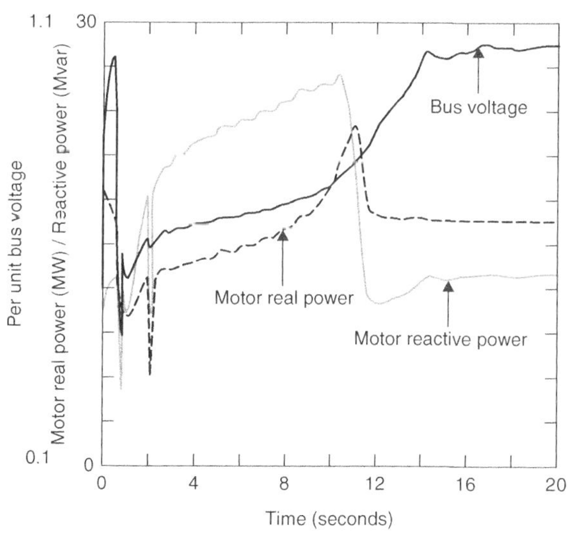

Figure 13-8. Dynamic simulation showing real and reactive power taken by a motor load during a low-voltage condition. (From [9], © IEEE 1997.)

shock, a small time delay is introduced in the generator trip, which counts on the residual magnetism of the generator to carry through until the capacitive shock subsides. This shock subsides most of the time in a few cycles.

Suppose that the cogeneration section is being supplied by a low-capacity generator. The load increases and a larger capacity generator is switched in. There will be a power swing between the two generators for a while. A similar situation will arise if a large load goes off on the utility side, which will try to suck in more power from the cogeneration section.

This sudden increase/decrease in load on the cogenerator increases the rotor swing angle. This in turn increases the excitation current. The field in the air gap around the rotor increases. This has a stabilizing effect on the rotor by damping down the rotor oscillations.

13.9.2 Short-Circuit Faults and Overcurrent

Large induction motors and synchronous generators react to a severe fault as follows. At the instant of a fault, the current supplied by a generator into the fault is the highest. It is controlled by all the electrical steady-state reactance from the generator to the point of fault in the system. However, this fault current from the generator falls rapidly, as shown in Figure 13-9.

As the fault current flows, the bus voltage falls down for 2–10 cycles, depending on the proximity of the fault to the bus. The induction motors in the system (acting as gen-

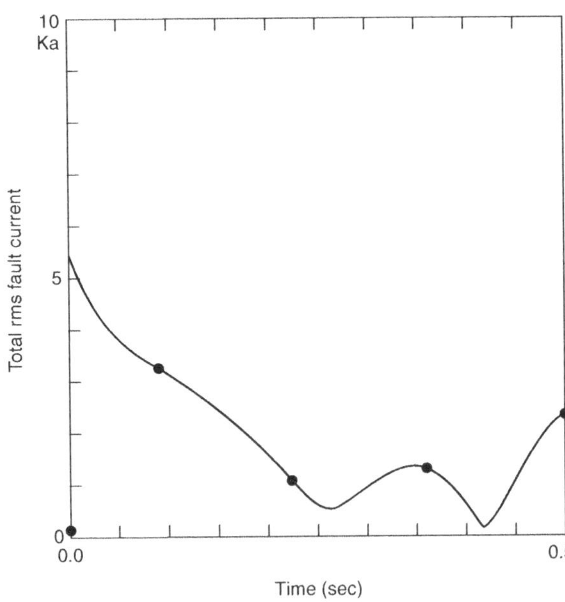

Figure 13-9. Decay in fault current must be known to set time-over-current relays properly. (From [10], © IEEE 1989.)

erators driven by the inertia of the load) deliver part of the reactive current, reducing the proportionate supply by the cogeneration. Then they start drawing heavy reactive current themselves, reducing the system voltage and reducing the available reactive current going into the fault. All this time, the synchronous motors in the cogeneration section are delivering large KVARs to keep up the bus voltage. After 1–2 seconds, these motors will slow down, drop out of synchronism, and trip. It may be noted that the large industry offering cogeneration has generally large-sized induction motors as well as a few synchronous motors for their own drives.

By this time, the voltage and speed of the cogeneration have dropped down and they are producing out-of-step voltage with the utility system. If the utility system returns after a reclosure, there is a heavy electrical and mechanical shock to the cogeneration windings, which will eat into its balance of life period. The cogeneration should trip under an out-of-step relay control if it has not already tripped under its overcurrent and/or reverse power relay control. For a system not to go out of step, the critical fault clearing times are given in Table 13-4.

13.9.3 Clearing Times for an Out-of-Step Relay Control

Let us consider a case in which the fault is on a feeder of the utility connected into its substation and not in the substation itself, and the utility fails to trip this feeder. The cogeneration has only a small generator whose overload current setting and the remote fault current almost match. Then the cogeneration will trip on its overcurrent, which would be a nuisance tripping. A better way out of this would be for an overcurrent relay or a reverse current relay in the tie line to trip before the relays on cogenerator trip. Figure 13-10 illustrates this.

Overcurrent relay has to detect all faults in its primary and back-up zone. It has to be fast and sensitive enough for its successful operation. As seen above, it must also be selective, which may not agree with its requirement of being fast and sensitive. Load flow diagrams for all types of permutations and combinations have to be worked out and selectivity will have to be incorporated accordingly.

13.9.4 Loss of Excitation Relay—Maloperations

Switching in large capacitor banks on EHV lines can cause transients. A sudden loss of load will also cause transients. The loss of load will go on increasing the voltage level.

Table 13-4. Utility fault type and clearing current time.

Fault type	Time in cycles
Three phase	8
Double line-to-ground	17
Single line-to-ground	80

Source: [11].

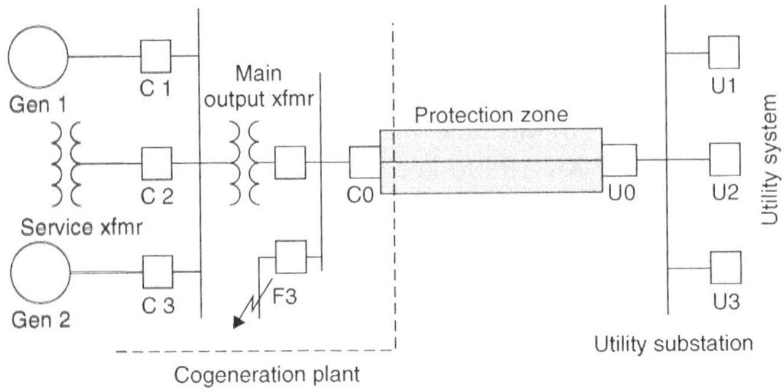

Figure 13-10. A fault in a feeder connecting a utility. (From [11], © IEEE 1997.)

All the transients will trip the loss of the excitation relays. These relays should be set so that they do not respond to these transients.

13.9.5 A Series Inductance in the Tie Line Works as A Stabilizer

If there is some inductive reactance in series with the tie line, this could prevent the violent reactions between a cogenerator and a utility and a destabilizing disturbance. It will also damp down power hunting between the two. If the cogenerator-to-distribution voltages are stepped up with a 34 kV–138 kV step-up transformer and the tie line operates at 138 kV, then the transformer impedance will help. A choke coil or an isolating transformer at transmission voltages is sometimes provided [12].

13.10 ISLANDING OF COGENERATION SECTION

If there is a large fault on the utility side due to sudden overloads or cut-offs, the tie line protection system separates the cogeneration section into a self-servicing island.

13.10.1 Sudden Overloads

Steps Leading to Tripping of Cogenerators. The cogeneration section has been importing power from the utility. This supply is cut off and the cogenerators are suddenly called upon to serve a large load of industry within this island. The voltage will fall and the frequency will start falling down. This will activate an underfrequency relay, which has a 2/3 step-time delay. A microprocessor-based system starts tripping connected loads on the cogeneration section selectively and rapidly. This will arrest the rate of fall of frequency and can restore it gradually. If this does not work, the cogenerator will trip.

13.10.2 Sudden Load Cut-Offs

Cogenerator Holds onto Frequency. The cogeneration section will have had a surplus supply capacity when it was exporting power to the utility, which has abruptly been cut off. The cogeneration voltage and frequency will both rise. The voltage and frequency relays will operate the control vanes on the steam-supply side of the turbine driving the cogenerator and reduce the steam flow. The steam control on the turbine side has two possible modes of operation:

1. A droop characteristic mode in which the turbine is allowed to lose frequency but recover rapidly within a few cycles.
2. An isochronous mode in which the turbine holds fast to the synchronous frequency, with only a minor drop in frequency. Here, as explained earlier, an impact load stresses the generator winding. Normally, cogeneration tied into a utility system will have droop characteristic modes of operation.

Reverse Power Flow in Generator. This can take place if the steam supply to the turbine coupled to its generator is cut off without tripping the connected cogenerator. The power from the mains will now flow into the generator and run it as a motor. It will draw large reactive power and lower the bus voltage. More than that, it will now drive the turbine and cause intense mechanical damage within the turbine. A reverse power flow relay should trip the generator should this happen. Suppose that a section containing a number of unevenly loaded cogenerators is islanded. The steam turbines with low ratings have low load settings and they are vulnerable to reverse power flow.

Proportionate Load Shedding through Rate of Change of Frequency. A cogeneration section importing a sizeable amount of electricity from the utility gets islanded due to a disturbance on the utility side and the tie line tripping out. The cogeneration section load will suddenly come on the cogenerators and they will slow down and have low terminal voltages and low frequencies. Sensitive and important load, within a few cycles to seconds, will trip off.

To make this transition smooth, without affecting the power quality, by maintaining frequency and voltages within the permissible range, it is necessary to shed preplanned unimportant loads rapidly.

If we make frequency a criterion for shedding loads step by step, we will lose precious time until the frequency reaches the marked level. Sensitive loads will start tripping out.

The rate of change of frequency, $d\text{Hz}/dt$, is a measure of the degree of power shortage. Therefore, load shedding is based on the rate of change of frequency. The faster this rate, the larger must be the load to be shed.

Planning and Executing Load Shedding

1. Decide on the priority use. How much energy in the form of steam and how much in the form of electricity for different overload contingencies is to be apportioned?
2. Shed electrical load that does not affect the overall processes in the plant and is easy to restart first and, if necessary, in simultaneous lots.
3. Electrical loads that are difficult to restart are shed last.

4. Intermediate shedding will be as per magnitude and importance in keeping the processes running.
5. Load shedding should be fast so that synchronous motors in the section and generators do not fall out of sync and trip.
6. All loads that can be switched off should be tripped before tripping the generators.
7. When the short fall in energy supply due to islanding is small, do not trip larger loads unnecessarily; large excess load tripping should be avoided. Loads suddenly cut off can cause transient disturbances.

For a critical high-speed response, it is desirable to carry out a number of transient stability simulations (accounting for transient response of generators as well as for the loads) to ensure that avoidable tripping causing such transients does not occur [4].

Reclosures. The majority of the electrical power circuits are protected with fuses. These fuses, selected as per their I^{2t} characteristic, blow due to a high current arising out of a fault and protect the supply side. Fuses have to be reinstalled. This involves time delay and lots of personnel time. To avoid this, reclosers are used, which open up a circuit against a fault but recluse, allowing for a temporary fault to clear by itself, or to be cut off by the action of the sectional fuse. A temporary fault, for example, could be caused by a tree branch brushing against a live line, birds, or lightening flashover. If the fault persists, it will reclose a second time and repeat, in most cases for a third time. If the fault still persists, the recloser locks itself out and has to be reset manually.

For a cogeneration system, depending on the tie line impedance, every reclosure has a system impact, stressing the generator coils electrically and mechanically. A triple reclosure will build in a lot of damage in a generator winding and reduce its life. Reclosures are best avoided in the vicinity of a paralleled cogeneration system or operation inr single phase. The trip and reclose cycle operates on all feeders that are multicircuit. This calls for three separate mechanism boxes, one for each phase.

Passive Reclosing. Should a recloser close on a fault that is not cleared, there will be heavy torque on the rotor shaft of the cogenerator. This will cause rotary oscillation, which will eventually die out under the influence of the air gap flux. A fuse goes out and, after an interval, the recloser closes again on the fault. If the first rotor vibrations have not died down before the second impact comes in, the shaft will be severely stressed. In such a case, the second recloser is delayed beyond its preset time interval, leading to passive reclosure.

13.11 ENVIRONMENTAL CONSIDERATIONS

Typical environmental issues were discussed in Chapters 6 to 8. However, there are some special features faced by an oil exploration company. They want permission to put up a cogeneration plant for electrical energy as well as thermal energy (steam).

The issues involved are:

SO_x—Control of sulfur in fuel

PM—Particulate materials

Fuel composition

Biological issues—The operations were not to damage protected animal and plant species.

Seismic considerations—All the building and structure designs have to be proved to withstand seismic levels prescribed for that area.

The following emission control techniques were to be maintained:

NO_x—Water or steam injection to lower the burning temperature. Selective catalytic reduction (SCR). Dry low NO_x. Catalytic combustion.

CO_2—Combustor design. Catalytic device.

UHC—Unburned hydrocarbons. Combustor design.

The actions to be undertaken were:

1. Permanent emission monitoring equipment on each stack for measuring SO_x, NO_2, and CO_2.
2. Limits on maximum fuel that can be burned daily.
3. Ensure adequate water flow for NO_x control.
4. Limits on sulfur content of fuel oil and natural gas used.
5. Establish minimum instrumentation requirements.
6. Vapor emission from oil well head casting, down which steam was injected for oil recovery, to be monitored.

13.12 COGENERATION IN BRAZIL

Brazil has provided a very interesting example of cogeneration. Beginning with the oil shock of 1973, it created tax incentives for producing ethanol by its vast sugar industry. With the second oil shock in 1979, followed by increase in interest rates it was paying for external borrowing, the costs as well as government pressure for replacing gasoline by ethanol increased. Cars were modified to suit ethanol consumption. Ethanol production kept on increasing.

With a cyclic slowdown in gasoline prices, ethanol consumption by cars fell. There was a glut of ethanol. By just adding an effluent boiler, ethanol producers diverted ethanol to electricity production. They produced electricity at competitive costs to such an extent that they are now competing in the annual electricity supply auctions against electricity from hydro and gas resources. Cogeneration has become a sizeable national electricity supplier [13].

APPENDIX 13-1 A TYPICAL COGENERATING SYSTEM FOR A HIGH-TECH, SCIENCE-BASED INDUSTRIAL PARK IN TAIWAN

This industrial park in Taiwan produces computers, semiconductor devices, and communication equipment. It has a large production of IC chips. All these plants have microprocessor-based drives and controls. Voltage sags and faults are just not acceptable. They have analyzed that 60% of the total plant interruptions were due to system faults and 40% due to voltage sags. Voltage sags were defined as voltages between 0.1 pu to 0.9 PU. Long-duration sags lasted for more than 1 minute and short-duration sags lasted between 0.5 cycles to 1 minute. Absolute power quality was essential.

Standby diesel generators had an unacceptably long start-up time. Uninterrupted power systems (UPS) were not acceptable, if one considers that the power requirements were on the order of 130 MW. Some of the plants required steam as well. Installing a cogenerating station with accurate controls was the answer. Figure 13-11 gives a schematic of the installation.

The cogeneration system consisted of two numbers of 46 MW gas turbine units and a 36.6 MW steam turbine unit. The tie connection was through 161 kV:161 kV transformers with onload tap changers to provide for voltage control as well. Gas turbines

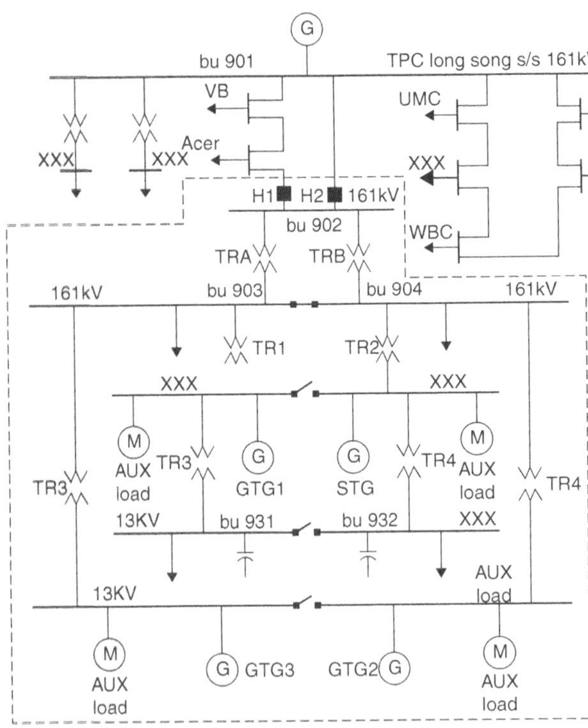

Figure 13.11. Online diagram of the power system under study. (From [9], © IEEE 1997.)

had advantages of quick start and fast response to disturbances. On the other hand, cost per kilowatt-hour was higher than that of the utility. The utility had low nighttime rates. The two gas turbines were closed during the night and power was purchased from the utility. Steam consumers were encouraged for nighttime use. An underfrequency relay and an overvoltage relay were used to trip the tie line.

A Load Shedding Scheme

This comes into play when the tie line is tripped. It is connected to the underfrequency relay system. The load shedding is done with respect to the rate of frequency decay due to overloading of the cogeneration after separation from the utility. The amount of load (P_{step}) to be interrupted is calculated as:

$$P_{\text{step}} = 2H/60\ mo \text{ pu}$$

where mo is the initial frequency decay rate and H is the equivalent inertia of the system:

$$H_{\text{sys}} = (H_1\text{MV}A_1 + H_2\text{MV}A_2 + \ldots\ldots + H_n\text{MV}A_n)/\text{MV}H_{\text{sys}}$$

The isolating transformers bring up the transient voltage to an acceptable value of 0.9 PU in less than a second as against more than 1 second without these. See Figure 13-12. The voltage dip stays at 0.3 pu as against below 0 pu without the isolating transformer.

The frequency response with load shedding improves very considerably, the frequency staying at and above the acceptable figure of 57.5 Hz and returning to 60 Hz in a maximum of 4 seconds. Without load shedding, the system is heading for a cascade

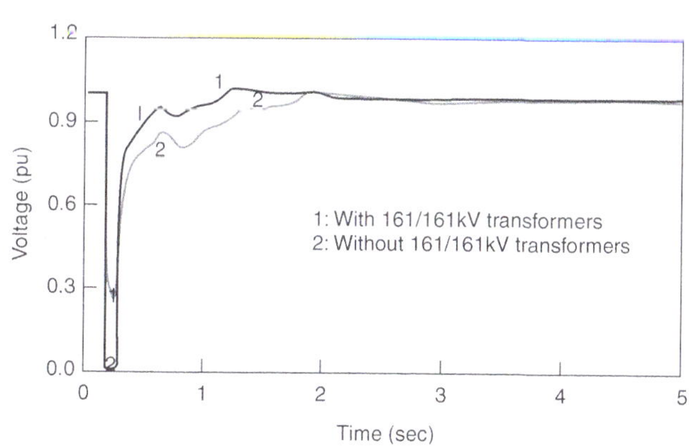

Figure 13-12. Voltage responses of bus 932 for a bolted ground fault at Long-Song substation. (From [9], © IEEE 1997.)

trip. See Figure 13-13, in which curves 1, 2, and 3 are typical cases in the system worked out for illustration purposes.

The mechanical controls on the turbine hold the mechanical power input steady, but at an increased value, without load shedding. With load shedding, this power reduces correspondingly, thus holding the rotor steady. Figures 13-14 and 13-15 show a model system that can provide top power quality to a super quality conscious IT park 24 hours a day.

APPENDIX 13-2 NERC DIRECTIVE

In 1992, the North American Electricity Reliability Council (NERC) issued a directive for integrating nonutility generation, with reliability considerations for integrating nonutility generators (NUGs) with bulk electric systems. The most important considerations for generation and transmission power system are:

1. Load following capacity of the NUG plant as per contract terms or any other agreed arrangement.
2. Generation schedule to enable the utility to accurately match the generation and load demand.
3. Fuel supply and storage capabilities:

new stocks must be ordered when ground stocks reach a predetermined level or order out stocks at regular time intervals.

4. Maintenance, routine or emergency, is a must for all components of the nonutility supplier.
5. Emergency availability of NUG facilities to be under the directive of the purchasing utility until the system returns to normal.

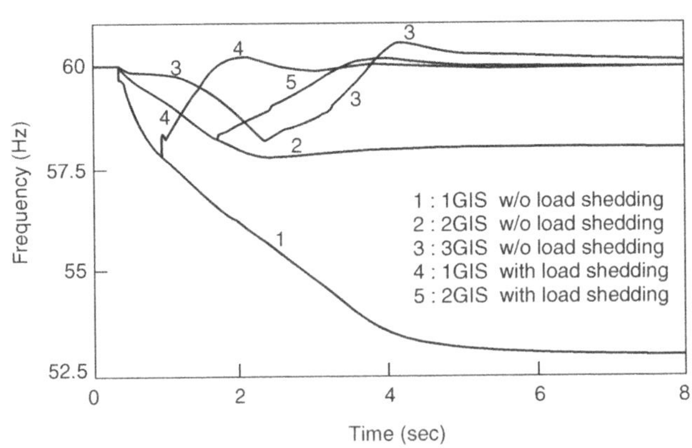

Figure 13-13. Frequency responses of the cogeneration system for the three operation modes with and without executing load shedding. (From [9], © IEEE 1997.)

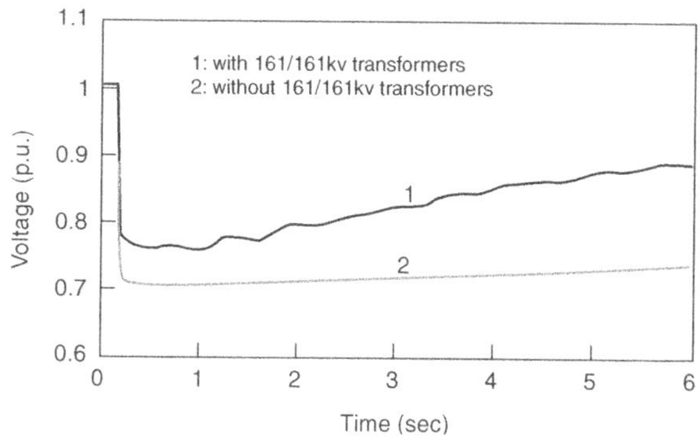

Figure 13-14. Voltage responses of bus 932 for a fault with ground impedance at Long-Song substation. (From [9], © IEEE 1997.)

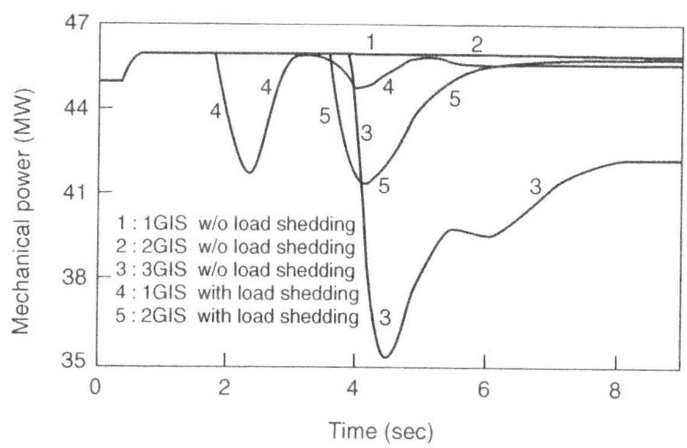

Figure 13-15. Mechanical power responses of the gas turbine generator for the three operation modes with and without executing load shedding. (From [9], © IEEE 1997.)

6. Restoration procedures of the NUG plant under standard conditions should be throughly written down operating instructions.

Cogeneration units may not be available at their agreed capacities at fixed times for all the days in a year.

Penalties for noncompliance are:

1. Utility to pay NUG when utility does not purchase the agreed quantity at the agreed time.
2. The above penalty will be cancelled if at any time utility buys more quantity than agreed to.
3. NUG pays the utility if it fails to supply the agreed quantity as per schedule.

APPENDIX 13-3 COMBINED POWER GENERATION AND CAPTIVE POWER PLANTS—A TYPICAL EXAMPLE

Background

Madhya Pradesh in India has a number of steel producing plants, a large aluminum producing plant, paper mills, and so on. The state is rich in minerals and forests. There is a sizeable production of combined power generation to the tune of 3000 MW, as reported by the end of 1991.The hot flue gases from the open hearth blast furnaces and coke ovens produce steam. Steam is also agriculture-based industries like paper mills. Steam engines based on this steam drive generators that produce electricity. The captive production is tied with grid supply from the state electricity board [43].

There is some negative background to grid supply. The grids are overloaded and work under low capacities. The voltages dip to as low as 169 kV across a 220 kV supply and 160 volts across a 230 V LT supply. The frequencies vary over a region of 48.5 to 52 Hz against a declared frequency of 50 Hz. Load shedding is unpredicted and frequent.

This has resulted in major customers opting out for captive power to cover essential operations that cannot accept supply interruptions. Nonessential operations are generally transferred, to grid supply.

The grid tariffs are generally high, being higher during peak hours. In a typical case, the BALCO aluminum company went in strongly for captive power, reducing their CD by 200 MW. BALCO is a committed, high-tension consumer and is also a regularly paying consumer. The 200 MW so released were then distributed to spread out agricultural load that was incurring higher distribution losses. The agricultural load was a loss-making alternative with defaulting revenue.

Problems in Cogeneration and Grid Interconnections

The interconnection brought out a number of interesting technical points. Problems on account of increased short-circuit levels were:

1. CB ratings within the plant have to be enlarged sufficiently to cover the increased fault level and total load as well as to cover possible load expansion over the next 10–15 years.
2. Synchronizer CB should be tested to withstand a high voltage at $2\sqrt{(2Vs)}$.
3. Earthing systems had to be adequate. In special cases, resistance earthing and changes in relay setting will have to be made.
4. On-load tap changers (OLTC) on power transformers should have a higher duty rating considering two way flows between the grid and the CPP.
5. Distance protection relays had to be adjusted in consideration of two way flows.
6. Metering CTs and summation CTs needed special attention and mutual agreements on CT ratios and on accuracies. These have the potential to raise disputes.

7. A fast autoreclosure on the grid side could create problems requiring auick load regulation within the plant.
8. A CPP with a double system connection (both 220 and 132 kV) can have a feedback power flowing, say from the 220 kV supply into 132 kV via the CPP's meter, along with the CPP feeding power from its own generator. This created problems in metering and billing accuracies.

Grid Discipline for the CPP

Wherever parallel operation is envisaged, grid discipline needs to be observed by CPP with regard to:

- Regulation of import/export of MW power during the specified time.
- MVAR reactive power sharing and maintenance of voltage level under steady-state conditions and after system disturbance.
- Draw of power at 0.9 power factor and supplying the grid with an appropriate power factor for voltage regulation.
- Planning and coordination of plant shutdown (PP with utility, to avoid load draw under system peak-load conditions) or generation scheduling.
- Emergency load regulation under forced outage at the CPP.
- Harmonic suppression at consumer premises.
- Communication facilities, with area load dispatch center, for MW load/RKVA regulation, as well as generation schedule/control of load flow over the tie line.
- Operating instruction for resumption after various contingency events, including loss of communication.
- Drawing up a protocol for shutdown, including line permits, ensuring no back-feed for periodic/annual maintenance, testing, and coordination.
- Instruction for extension of startup power after system collapse/black start.

APPENDIX 13-4 COGENERATION IN SUGAR MILLS IN INDIA

			MW Capacity					
Mill no.	Place	State	Condensing	Back pressure	Total	Bagasse*	Selling price, Rs./unit	Cost of generation Rs./unit
1	Mawana	UP	19	31	50	1.8/2.5	2.98	2.00
2	Titawai, Muzzafarpur	UP			40			
3	Dharmpur	UP			30		2.80	2.20
4	Balarampur, Chini	UP	19.55	5	24.55	1.7	2.89	
5	Simbhaoli	UP			13.9	3 to 4		
6	Chilwaria	UP	7		2.8/2.9			
7	Rajashree, Thevi	T Nadu			12		3.16	
	Villupuram				22	4	3.15	2.98

(*continued*)

Mill no.	Place	State	MW Capacity Condensing	Back pressure	Total	Bagasse*	Selling price, Rs./unit	Cost of generation Rs./unit
8	Ugar Sugar	Karanataka			4.4	1.8/2.8	3.32	2.60
9	Bammari Amman	T Nadu			20	2.45	3.01	2.57
					20+	2.3	3.52	2.07
					16	2.4	2.92	2.89
10	Dharni Sugar Chemical	T Nadu			15	2.33	2.6	2
11	KCP Sugar and Indl. Corp.	AP	12	3	15	3.5	2.79	2
12	Nancylamal	Nagaland			26			
13	Jammu and Kashmir	JandK			19	3.5	2.89	
14	Rana Sugar	Punjab			12			
Total					386.45			

Source: Power Line Magazine, 2006.
*Bagasse is chaff left over after crushing of cane.

REFERENCES

1. U. S. Department of Energy, The Changing Structure of the Electric Power Industry 2000: An Update. Energy Information Administration. DOE/EIA—056200D.
2. G. Tao, M. I. Henwood, and M. van Ooijen, An Algorithm for Combined Heat and Power Economic Dispatch, *IEEE Transactions on Power Systems,* Volume 9, Issue 3, August 1994, pp. 1778–1784.
3. J. R. Holmquist and R. W. Jones, Reducing the Effects of Lightning in Cogeneration Plants, *IEEE Industry Applications Magazine,* Volume 7, Issue 5, Sept.-Oct. 2001, pp. 20–28.
4. T. J. Hammons and A. G. Geddes, Assessment of Alternative Energy Sources fFor Generation of Electricity in the UK Following Privatization of the Electricity Supply Industry, *IEEE Transactions on Energy Conversion,* Volume 5, Issue 4, December 1990, pp. 609-615
5. M. J. Smith, Reliability, Availability, and Maintainability of Utility and Industrial Cogeneration Power Plants, *IEEE Transactions on Industry Applications,* Volume 27, Issue 4, July–August 1991, pp. 669–673.
6. S. Ashok and R. Banerjee, Optimal operation of industrial cogeneration for load management, *IEEE Transactions on Power Systems,* Volume 18, Issue 2, May 2003, pp. 931–937.
8. D. J. Love and N. Hashemi, Considerations for Ground Fault Protection in Medium-Voltage Industrial and Cogeneration Systems, *IEEE Transactions on Industry Applications,* Volume 24, Issue 4, July–August 1988, pp. 548–553.
9. C. T. Hsu, C. S. Chen, and J. K. Chen, The Load-Shedding Scheme Design for an Integrated Steelmaking Cogeneration Facility, *IEEE Transactions on Industry Applications,* Volume 33, Issue 3, May–June 1997, pp. 586–592.
10. H. K. Clark and J. W. Feltes, Industrial and Cogeneration Protection Problems Requiring Simulation, *IEEE Transactions on Industry Applications,* Volume 25, Issue 4, July–August 1989, pp. 766–775.
11. R. M. Rifaat, Considerations for Generator Ground-Fault Protection in Midsize Cogenera-

tion Plants, *IEEE Transactions on Industry Applications,* Volume 33, Issue 3, May–June1997, pp. 628–634.
12. C. Chao-Shun, K. Yu-Lung, and H. Cheng-Ting, Protective Relay Setting of the Tie Line Tripping and Load Shedding for the Industrial Power System, *IEEE Transactions on Industry Applications,* Volume 36, Issue 5, September–October 2000, pp. 1226–1234.
13. S. Mocarquer, Balance of Power, *IEEE Power and Energy Magazine,* September/October 2009, pp. 26–35.

BIBILOGRAPHY

Asano, H., S. Sagai, E. Imamura, K. Ito, and R. Yokoyama, Impacts of Time-of-Use Rates on the Optimal Sizing and Operation of Cogeneration Systems, *IEEE Transactions on Power Systems,* Volume 7, Issue 4, November 1992, pp. 1444–1450.

Bagen, B. R., Evaluation of Different Operating Strategies in Small Stand-Alone Power Systems, *IEEE Transactions on Energy Conversion,* Volume 20, Issue 3, September 2005, pp. 654–660.

Bagen, B. R., Incorporating Well-Being Considerations in Generating Systems Using Energy Storage, *IEEE Transactions on Energy Conversion,* Volume 20, Issue 1, March 2005, pp. 225–230.

Barton, J. P. and D. G. Infield, Energy Storage and its use with Intermittent Renewable Energy, *IEEE Transactions on Energy Conversion,* Volume 19, Issue 2, June 2004, pp. 441–448.

Bin-Kwie, C. and H. Chen-Chuan, Optimum Operation for a Back-Pressure Cogeneration System Under Time-of-Use Rates, *IEEE Transactions on Power Systems,* Volume 11, Issue 2, May 1996, pp. 1074–1082.

Cheng-Ting, H.,Cogeneration System Design for a High-Tech Science-Based Industrial Park, *IEEE Transactions on Industry Applications,* Volume 39, Issue 5, September–October 2003, pp. 1486–1492.

Daniel, S. A. and N. A. Gounden, A Novel Hybrid Isolated Generating System Based on PV Fed Inverter-Assisted Wind-Driven Induction Generators, *IEEE Transactions on Energy Conversion,* Volume 19, Issue 2, June 2004, pp. 416–422.

Dialynas, E.N., D. J. Papakammenos, and N. C. Koskolos, Integration of Nonutility Generating Facilities into the Reliability Assessment of Composite Generation and Transmission Power Systems, *IEEE Transactions on Power Systems,* Volume 12, Issue 1, Feb. 1997, pp. 464–470.

Dickin, J., R. A. Hanna, S. Randall, and C. Dedeurwaerder, Challenges Encountered when Expanding a World-Class Petrochemical Facility, *IEEE Transactions on Industry Applications,* Volume 37, Issue 4, July–August 2001, pp. 1109–1119.

Doughty, R. L., L. Gise, E. W. Kalkstein, R. D. Willoughby, Electrical Studies for an Industrial Gas Turbine Cogeneration Facility, *IEEE Transactions on Industry Applications,* Volume 25, Issue 4, July–August 1989, pp. 750–765.

Dwarkanath, A. R., Techno-Economic Operation of Captive Power Plants, *IEEE Power Conference,* Nagput, India, October 2000.

Knight, A. M. and G. E. Peters, Simple Wind Energy Controller for an Expanded Operating Range, *IEEE Transactions on Energy Conversion,* Volume 20, Issue 2, June 2005, pp. 459–466.

Korpas, M. and A. T. Holen, Operation Planning of Hydrogen Storage Connected to Wind Power Operating in a Power Market, *IEEE Transactions on Energy Conversion,* Volume 21, Issue 3, September 2006, pp. 742–749.

Kusko, Standby Electric Power Systems. Short Term, Long Term Energy Storage Methods, *IEEE Transactions on Energy Conversion, IAS,05,* pp. 2672–2678.

Mansour, M. and C. Davidson, A Unique Cogeneration Application in a Large Pulp Mill Electrical System, *IEEE Industry Applications Magazine,* Volume 3, Issue 5, Sept.-Oct. 1997, pp. 34–40.

Paine, D. M., Increasing the Electrical Output of a Cogeneration Plant, *IEEE Transactions on Industry Applications,* Volume 38, Issue 3, May–June 2002, pp. 726–735.

Palanichamy, C., N. S. Babu, and C. Nadarajan, Municipal Solid Waste Fueled Power Generation for India, *IEEE Transactions on Energy Conversion,* Volume 17, Issue 4, December 2002, pp. 556–563.

Powell, L. J., An Industrial View of Utility Cogeneration Protection Requirements, *IEEE Transactions on Industry Applications,* Volume 24, Issue 1, Part 1, Jan.-Feb. 1988, pp. 75–81.

Puttgen, H. B. and P. R. MacGregor, Optimum Scheduling Procedure for Cogenerating Small Power Producing Facilities, *IEEE Transactions on Power Systems,* Volume 4, Issue 3, August 1989, pp. 957–964.

Reddy, K. N. and V. Agarwal, Utility-Interactive Hybrid Distributed Generation Scheme With Compensation Feature, *IEEE Transactions on Energy Conversion,* Volume 22, Issue 3, September 2007, pp. 666–673.

Rifaat, R. M., Critical Considerations for Utility/Cogeneration Inter-Tie Protection Scheme Configuration, *IEEE Transactions on Industry Applications,* Volume 31, Issue 5, September–October 1995, pp. 973–977

Rooijers, F. J. and R. A. M. van Amerongen, Static Economic Dispatch for Co-Generation Systems, *IEEE Transactions on Power Systems,* Volume 9, Issue 3, August 1994, pp. 1392–1398.

Roy, S., Optimal Efficiency as a Design Criterion for Closed Loop Combined Cycle Industrial Cogeneration, *IEEE Transactions on Energy Conversion,* Volume 16, Issue 2, June 2001, pp. 155–164.

Short, T. A., J. J. Burke, and R. T. Mancao, Application of MOVs in the Distribution Environment, *IEEE Transactions on Power Delivery,* Volume 9, Issue 1, January 1994, pp. 293–305.

Swider, D. J., Compressed Air Energy Storage in an Electricity System with Significant Wind Power Generation, *IEEE Transactions on Energy Conversion,* Volume 22, Issue 1, March 2007, pp. 95–102.

Vardeman, R. D., Permitting a Large Cogeneration Project in California, *IEEE Transactions on Industry Applications,* Volume 24, Issue 6, Nov.-December 1988, pp. 1160–1167.

Venkatesh, B. N. and V. Chankong, Decision Models for Management of Cogeneration Plants, *IEEE Transactions on Power Systems,* Volume 10, Issue 3, August 1995, pp. 1250–1256.

Voorspools, K. R. and W. D. D'haeseleer, The Impact of the Implementation of Cogeneration in a Given Energetic Context, *IEEE Transactions on Energy Conversion,* Volume 18, Issue 1, March 2003, pp. 135–141.

Whipple, D. P. and F. J. Trefny, Current Electric System Operating Problems from a Cogenerator's Viewpoint, *IEEE Transactions on Power Systems,* Volume 4, Issue 3, August 1989, pp. 1037–1042.

Young-Moon, P., C. Myoen-Song, L. Jeong-Woo, K. Y. Lee, An Auxiliary LQG/LTR Robust

Controller Design for Cogeneration Plants, *IEEE Transactions on Energy Conversion,* Volume 11, Issue 2, June 1996, pp. 407–413.

Zhenhua, J., G. Lijun, R. A. Dougal, Adaptive Control Strategy for Active Power Sharing in Hybrid Fuel Cell/Battery Power Sources, *IEEE Transactions on Energy Conversion,* Volume 22, Issue 2, June 2007, pp. 507–515.

14

DISTRIBUTED GENERATION (DG) AND DISTRIBUTED RESOURCES (DR)

14.1 DEFINITION AND SCOPE

14.1.1 Definitions

1. Distributed generation (DG) is defined under IEEE Std 1547 as electricity generating facilities connected to an electric power system through a point of common coupling. In-sue [21] defines DGs as small-value generators up to 10 MVA. DGs are a subset of distributed resources (DR).
2. Distributed resources (DR) are resources of electric power that are not directly connected to bulk power transmission systems. DR includes both generators and energy storage facilities.
3. Point of common coupling (PCC) is the point at which a local electric power system (EPS) is connected into an area of EPS (covering number of local EPSs; local EPS may contain DR/DG).
4. Interconnection is the process of adding a DR to a unit area of EPS.
5. An island is a local area covered by electric supply supplied by a local EPS while disconnected from the rest of EPS.

14.1.2 Scope

This chapter covers utility interconnected power systems, based on both conventional energy sources as well as nonconventional energy sources with respect to their inter-

Copyright © 2011 the Institute of Electrical and Electronics Engineers, Inc.

connections. This chapter also covers stand-by resources of electric power, which are not connected into an ePS.

14.2 WHO ARE THE PLAYERS IN DISTRIBUTION GENERATION?

Presently the players are those utilizing renewable energy sources for generation of electricity. These are

- Natural-gas-based IC engines, turbines, and so on coupled to electricity generators
- Wind turbine generators wind farms
- Solar-energy-based PV technologies
- Combined generators producing electricity as an additional product
- Small hydro projects are entering this field worldwide, utilizing hydro energy whenever it is available.

There is a very large population of diesel-engine-driven electricity generators. They mainly serve as stand-by generators for critical users of electricity. Now and then, the question of admitting them as distribution generators crops up.

14.3 PROMINENT FEATURES OF DRs

DRs use the utility frequency for their reference. They cannot (at least when connected to a utility) have their own independent frequency. DRs cannot regulate utility voltage—boost it up on a utility feeder on their own. They impact the utility voltage system under the directions of the utility area controller only to share system RKVA. DRs are not under the direct control of the utility area operator.

The implications for the utility of the interconnections are:

1. Customers within the supply zone of a utility are covered under power quality parameters—voltage and frequency. Any deviations out of tolerance limits for the parameters are subject to financial damage claims by customers.
2. Safety precautions for personnel and property have to be compulsorily observed by the DR.

14.4 TYPES OF DGs

We can classify the DGs into three types, as per their generators (see Table 14-1):

1. Distributed generators based on induction motors turning into generators when driven above their synchronous speeds. They do not generate their own excitation current for the magnetic field but draw it from the circuit to which they are

Table 14-1. Characteristics of DG technologies

Technology	Internal Combustion (IC) Engine	Gas Turbine (GT)	Micro turbines (MT)	Fuel Cells (FC)	Electricity Grid
Costs					
Unit costs ($/kW)	300–900	300–1000	700–1100	2800–4700	Distribution service charge
OandM costs ($/kW-hr)	0.007–0.015	0.004–0.010	0.005–0.010	0.005–0.010	$26.17/mo and $3.387/kW
Other characteristics					Demand charges:
Fuel type	Natural gas	Natural gas	Natural gas	Natural Gas	June-Sep: $6.429/kW
Equipment life	20 Years	20 Years	10 Years	10 Years	Oct–May: $4.84/kW
Start-up time (cold start)	10 secs	10 mins	2–5 mins	< 0.1 Hr	kW-hr charges:
Electrical efficiency (HHV)	30–37%	22–37%	23–28%	30–46%	Peak: 3.155 c/kW-hr
Availability (%)	91.2–95.8	90–93.3	95	90	Off Peak: 0.524 c/kW-hr
Forced outage rate (FOR%)	4.7–6.1	2.1–6.5	0.5	0.5	0.1
Emission					
NO_x (lb/MW-hr)	4.7	1.15	0.44	0.3	2.54
SO_x (lb/MW-hr)	0.454	0.008	0.008	Negligible	5.83
PM-10 (lb/MW-hr)	0.78	0.08	0.09	Negligible	n/a

Source: [1].

delivering active power. As a result, they operate at poor power factors and have to be supported with capacitors. They pull themselves into synchronism with the synchronous frequency of the grid and hold onto it.

2. Distributed generators based on synchronous generators. These systems are self-sufficient in their requirement of excitation current as well as in holding onto the controlled frequency. Generally, larger DG sets based on this system are termed synchronous condensers.
3. Distributed generators based on inverter–converter interfaces picking up energy from renewable energy sources such as photocells, wind turbines, and fuel cells, have autonomous control of frequency and provide a variable supply of active and reactive power to the grid. Fuel cell technology is technically superior but costs have to come down.

14.4.1 Background

In the past, the rate of growth in the use of electricity was predictable. Forecasts and planning to cover the next, say, 10 year period could be comfortably made, executed, and found to be satisfactory. However, the world economic scenario (and with it the demand for electricity) is changing fast, beyond imagination.

14.5 PUSH FACTORS, STAY-PUT COSTS, AND INVESTMENT PROSPECTS FOR ELECTRICITY

A number of global forces have brought this about. Cell phones, personal computers, e-mails, and the Internet, along with satellite communications, have changed the life style and expectations of people all over the world. The technological continued progress in microprocessor chips and their applications have increased production capacities. The demand for electricity keeps on outstripping the renovated electricity power systems. The priority questions that arise are naturally:

1. What are the immediate and long-term options in investments available?
2. How far can we overload the existing facilities and what will be the effects of overloading on their life expectancies? With a universal use of modern gadgets, can we neglect the acute requirements on power quality?
3. What are the costs involved in not meeting the demand in time, in cost to the consumers, and losses to suppliers?

Let us study this in detail.

14.6 INVESTMENT OPTIONS

Investment options are governed by load growth and the cost of available alternatives.

14.6.1 Load Growth, Including Time Factor

There is continuous load growth in electricity consumption in the overall area under a distribution company (DISCO). However, some sectors are growing very fast, creating problems both in quality and reliability of electric supply and in losses. Generally, these sectors have a radial distribution, with a step-down transformer and a number of feeders radiating out. At the end of these feeders there are distribution transformers for clusters of customers.

A decision has to be taken on upgrading the distribution system, either by strengthening the existing facilities with additions or by locating a distributed generator at an advantageous point in this section.

All said and done, putting in a DG is a stopgap arrangement. In a competitive market, it helps us to postpone a risk in investment and maximize our profits with the available assets in hand. The decision requires a study of load characteristics of the past and expected rates of growth and forecasts of this growth. Figure 14-1 enables us to visualize this better.

The problem of low capacity arises with respect to the peak load. Beyond the peak load point, the utilities' costs and customer's costs shoot up. We have two options to solve this:

1. Make additions to substation and feeders. This implies high investment and high operating costs. It also involves long completion times.
2. Install a DG. This is low cost. If the load growth is slow, option 2 can carry us over a couple of years and this will do. Under option 2, incremental addition of DGs might help. If the growth rate is fast, option 1 is the best choice.

14.6.2 Costs of Available Alternatives—DG versus Substations

Table 14-2 covers a four-year forecast. Column 1 shows the calculation for substation expansion only. Column 2 shows results of adding DGs strategically and adding just

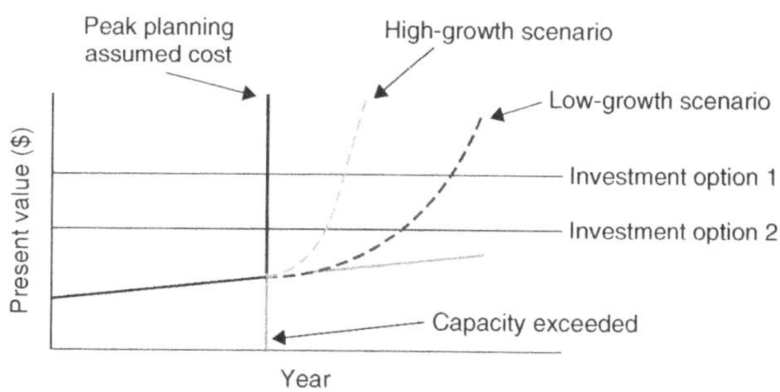

Figure 14-1. Comparison of traditional peak planning method with alternate cost minimization methods. (From [2], © IEEE 2001.)

Table 14-2. DG versus substation expansion

	Substation expansion	DG and substation expansion	DC and substation expansion (planner decision)
Total expansion cost (MS)	29.472	25.808	26.373
Total expansion cost (MW-hr)	76.74	74.013	74.434
Additional substation purchased power (MVA)	4.329, 10	0.7328, 0	0, 0
Total losses (MVA)	3.2296	1.6328	1.496
Total system demand (MVA)	54.3296	52.7328	52.5960
Expanding substation cost (MS)	0.4	0.2	0
DG capacity (MVA) and location	nil	3 (4 MVA and 1 MVA) @ buses 2, 6, 8	2 (4 MVA and 1 MVA) @ buses 2, 8 2 (3 MVA and 1 MVA) @ buses 4, 6
DG investment cost (MS)	nil	7.5	9
Losses cost (MS)	23.7247	1.2134	0
DG O&M costs (MS) nil	14.1912	14.89607	

Source: [3].

one substation transformer. Column 3 shows only DG additions. This option gives optional DG sites and designs and results in low total planning costs.

14.6.3 Costs of Overloading Existing Assets

The normal limits of loading a transformer or a feeder are 50% to 80% of the emergency limits. Emergency loading limits for lines are maximum continuous current rating (at the time of peak load) for one hour rating. For transformers, the emergency rating is 130% to 150% of the nameplate rating, within temperature rise limits over ambient temperature levels.

For aiming at an investment option for DG, due allowances have to be made for including customer costs/ and costs of unserved energy. Any energy served over and above the normal rating of an asset by overloading it shall be added to, in our calculations, as the cost of unserved energy over that period.

Costs of alternatives to DGs for handling overloads at substations result from:

Use of capacitors—cost for reference < $10/kW

Use of voltage regulators—cost for reference, < $10/kW

Modify feeders cost for reference—$20 to $100/kW

Add transformers costs for reference—$20 to $100/kW

Use DG cost for reference—$350 to $300/kW, depending on the type of DG

Let us now consider the costs involved in the loss of service to the customers or the cost of unserved services.

14.6.4 Costs of Unserved Energy

Peak planning costs imply costs incurred when the peak capacity is exceeded and load shedding is undertaken. There is unserved energy cost incurred by the customer as well as by the DISCO. Unserved energy cost by the customer results in loss of production/business by the customer. These could be covered by operating his standby DGs. Typical customer cost reported by Bae [4] are:

Residential load: $0.35 to $1 per kW-hr

Industrial and commercial load: $4.00 to $10.00 per kW-hr

Mixed residential load: $3.75 to $6.00 per kW-hr

On the utility side, one has to consider the cost of loss of business potential as well as threat of losing customers on account of frequent load shedding and poor quality of supply.

14.6.5 Interruption Costs

Table 14-3 lists these costs.

14.6.6 Line Losses Will Keep on Increasing with the Load

Strengthening the distribution lines is a costly and time-consuming exercise. Note that land permission can also get involved. The option is to use capacitors and/or DGs. Capacitors will deliver reactive power and improve the voltage profile. They cannot deliver energy. For that, DG sets are necessary. We are introducing a technical note here on sites for this option.

Table 14-3. Sector interruption cost (dollars per kilowatt)

	Interruption duration (minute or hour) and cost ($/kW)				
User sector	1 min	20 min	1 hour	4 hour	8 hour
Lager user	1.005	1.508	2.225	3.968	8.240
Industrial	1.625	3.868	9.085	25.16	55.808
Commercial	0.381	2.969	8.552	31.317	83.008
Agricultural	0.060	0.343	0.649	2.064	4.120
Residential	0.001	0.093	0.482	4.914	15.690
Government institutional	0.044	0.369	1.492	6.558	26.040
Office	4.778	9.878	21.065	68.830	119.16

Source: [4].

14.7 PLANNING SITES FOR A DG

Distributed generation improves the voltage profile and reduces the losses in a way identical to that of capacitors The DGs can be located in a way identical to that worked out for capacitors.

A long-accepted and theoretically proven rule of thumb known as two-thirds rule, developed for placement of shunt capacitors, requires that capacitors be located at two-thirds the distance of a feeder from its starting point, for a uniformly distributed load.

The same two-thirds rule DGs also applies to. This holds good for a uniformly distributed load. We can size the rating of the DG at two-thirds the capacity of feeder power (allocated) and locate it at two-thirds the distance from the starting point.

Table 14-4 shows reduction in power losses with respect to spread of load along a feeder and location of the DG. However, DGs will be most probably installed by independent operators who want these either for cogeneration or for their own consumption due to prolonged load shedding, and also with an idea of sale of energy to a DISCO. In that case, the siting will be of their choice and at their convenience.

Voltage support for a distribution line requires both active and reactive power. This voltage support is best obtained by matching the load and generator characteristics by

Table 14-4. Effect of adding DGs on power loss

Cases (Assuming that DG supplies all the loads in each case)	P^o_{loss} (Power loss before adding DG)	P_{loss} (Power loss after adding DG)	Percent of power loss reduction (%)	Optimal place x_o
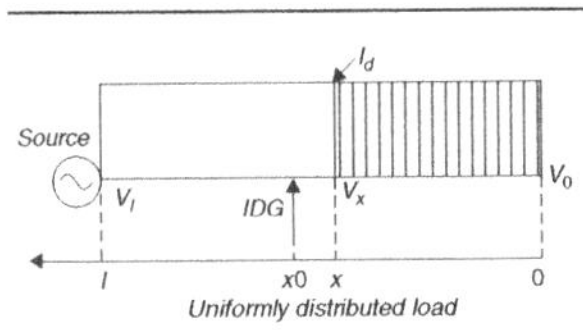	$I_d^2 R l^3/3$	$I_d^2 R l^3/12$	75%	$\frac{l}{2}$
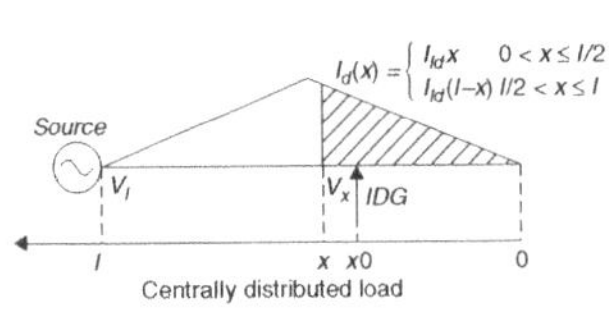	$\frac{23}{960} I_d^2 Rl$	$\frac{1}{320} I_d^2 R l^5$	87%	$\frac{l}{2}$
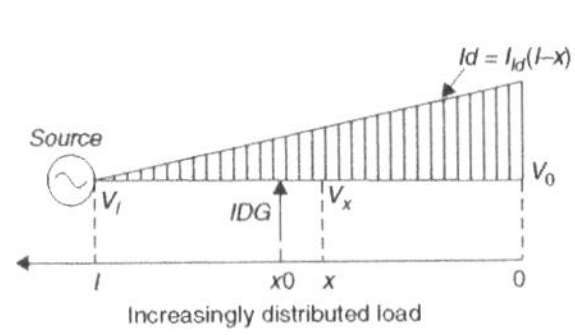	$0.1333 I_{ld}^2 R l^5$	$0.01555 I_{ld} R l^5$	88%	$(1-\sqrt{2}/2)l$

Source: [3].

working out a "voltage sensitivity factor" for a given system. The voltage sensitivity factor helps in correlating DG outputs with line and load parameters.

Let us consider a long rural distribution line with resistance R and reactance X. A DG is fixed at the far end to bring up the far end voltage within the tolerance limits specified. This voltage support is basically toward compensating the line drops due to both R and X. As such, it needs active power injection as well as reactive power injection.

The KVA injected by the DG is a function of the parameters G_g and BG_g of the generator and of R and X of the line. This can be determined as a ratio of KVA to be injected to KVA of the load. A voltage sensitivity factor helps us to arrive at this.

A generator is modeled mathematically as a passive element. When used for active power injection, it is termed G_g, a variable conductance; when used for reactive power injection, it is termed a variable susceptance, B_g:

$$\text{voltage sensitivity ratio} = (\Delta V_0/\Delta B_g)/(\Delta V_0/\Delta G_g)$$

where V_0 is the incremental rise in voltage. Voltage sensitivity can be calculated during DG commissioning (put the DG at the end of the line and increase its outputs incrementally and note the incremental rise in voltage) or analytically [5]. Figure 14-2 gives typical voltage sensitivity sets for a distributed load.

Several points may be noted from Figure 14-2. When $R > X$, the active power losses in line will be high and will have to be made up by injected active power by the DG. Were we to rely on voltage improvement through use of power capacitors only, this feature could not be realized. In that case, the capacitors would have to be supplemented by laying down an additional line or changing the line, a costly proposal in comparison with compensation through a DG.

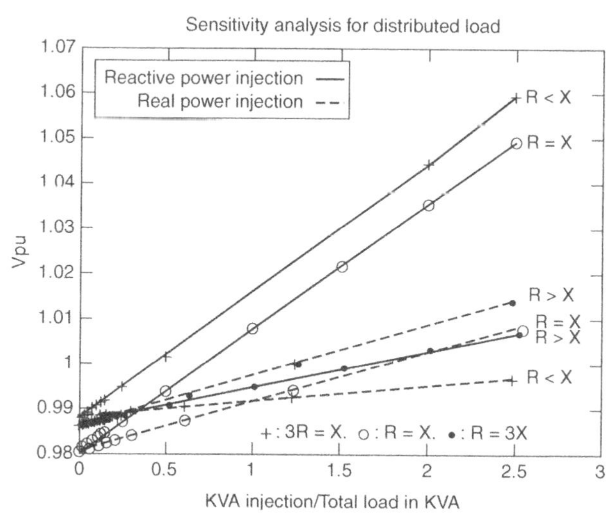

Figure 14-2. Voltage sensitivity for a distributed load. (From [5], © IEEE 2004.)

With $X > R$, reactive power will have to be injected by controlling the exciter winding of the generator. When generator is supplying reactive power mainly, its fuel compensation is very marginal. On most rural lines, the requirement for reactive power is high.

Capacitors can also perform this function, except that their output is not variable as it is with DGs as is required for power quality control.

14.7.1 Voltage Support for a Rural Line with Active and Reactive Power under Different Load Conditions

In the majority of rural lines under light load conditions, no DG support is required. Under moderate loads, mainly reactive power support is required. Under peak load conditions, both active and reactive power support are required (Figure 14-3). Please note that a very valuable system support during peak load is the active power that a DG injects into the distribution system in this period. There is a voltage regulator at 18.95 km in Figure 14-3 and two DG sets at 107.4 km and 69.5 km, respectively.

In these discussions, we are considering mainly thermal-energy-based portable generators.

14.8 OPERATION OF DGS IN AN ELECTRIC POWER SYSTEM

The presence of DGs in an EPS requires reconsideration of a few operational techniques. A DG helps to ride through a load fault. But relay settings need examination.

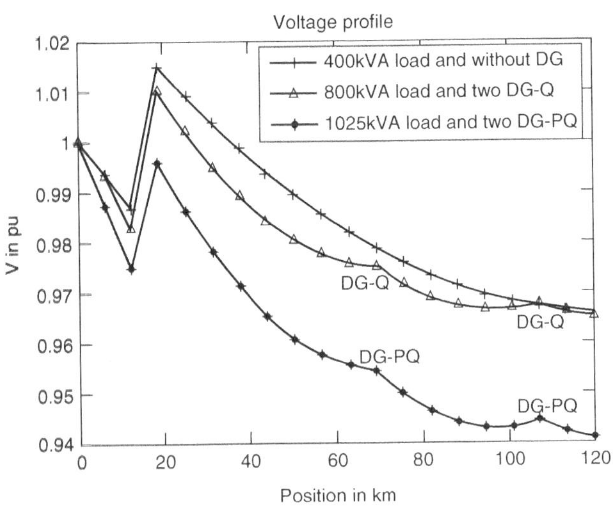

Figure 14-3. Voltage profile for low, medium, and heavy loads. (From [5], © IEEE 2004.)

14.8.1 A Ride Through a Voltage Dip

A fault drawing a short-circuit current causes deep voltage dips. The action of these voltage dips is controlled by interconnected fuses and autorecloser relay actions. Both the magnitude and duration of these dips count.

Figure 14-4 shows a representative system. Here, a generator *G*, in parallel with a DG, supplies a feeder carrying normal load and sensitive equipment, SE. The feeder develops a short and a heavy current flows via the connecting sections into the fault. The voltage level on the bus feeding both the feeder and the SE is roughly given by the fault current times the impedance, Z_f. Voltage level at the SE is slightly higher due to voltage drop in Z_{im} with current portion supplied by the DG. This statement will vary for varying positions of the faulted feeder and SE with respect to G_S and the DGs.

The circuit breaker on the DG is wired to trip within a fixed time after a sudden dip in the overall voltage due to a fault. Generally, it is 0.2 seconds or 10 cycles on a 50 Hz supply. The sensitive equipment has a voltage–time relation. As the voltage goes down, the time for which it can hold on (not trip) gets reduced. The Computer Business Equipment Manufacturers Association (CBEMA) has published sensitivity curves for sensitive equipment. These curves have become standard industry reference curves. Figure 14-5 shows these curves.

There are two possibilities open:

1. The protective fuse on the shorted feeder blows, say, within six cycles and the voltage at SE during this interval is such that the SE can hold on beyond six cycles at the reduced voltage level. There is a temporary dip and no equipment, save the feeder, is disconnected. In other words, the sensitive equipment has ridden through and the DG has not been disconnected.

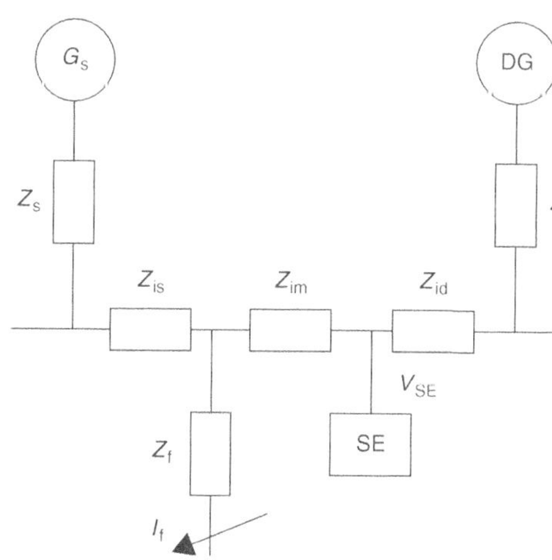

Figure 14-4. Generic circuit with DG for voltage sag studies. (From [6], © IEEE 2003.)

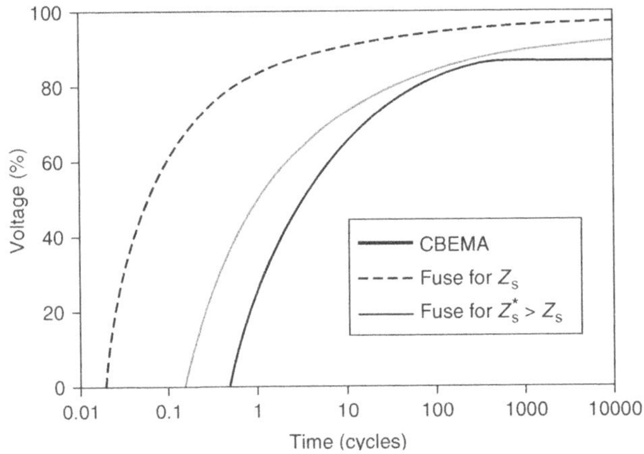

Figure 14-5. Change of fuse time–voltage characteristic due to DG disconnection. (From [6], © IEEE 2003.)

The DG has helped in this ride-through in two ways. First, it has added to the fuse current, making it blow quicker. In other words, it has changed the fuse characteristics by increasing fault current I_f. Second, it has given a marginal increase in voltage level at SE, enabling it to hold on a little longer.

2. The feeder fuse takes a longer time to blow. The DG is cut off after 0.2 seconds and the SE is cut off at the reduced voltage level after the ride-through period available to it is over. Then, the main circuit breaker across DGs trips. After a certain interval, the recloser on the feeder from G_s automatically switches on the circuit breaker. The circuit breaker will not trip again if the fault has been cleared. If the fault is still there, it will lock out the supply. It is very important for the DG not to be active in this case until after an interval the supply is firmly established.

There are many protective devices available: fuses, relays on protection equipment, circuit breakers, and so on. The major protective devices on electrical power systems are fuses. Ratings of fuses are graded so that a system section or sections are cut off sequentially and reclosures are built in so that the supply is not interrupted by temporary faults.

The system settings have to be carefully rechecked and readjusted when DGs are added, to take into account the increased fault current.

14.8.2 Small-Disturbance Stability of a DG

To maximize the small-disturbance stability of a DG, do not operate it at its full rated output. The majority of today's DGs have induction generators of various ratings tied into the system. The generators or their controls are not within the review of operation of the system controller. As a result, when a utility employs these, their outputs are almost steady and nearer to their rated capacity during the peak hours.

Assume that a DG is connected to a bus in the supply system. Figure 14-6 shows PV curves for the bus on which a DG is connected under three conditions [7]:

1. There is no induction generator in the system.
2. The DG puts in full output at 30 MW—curve A_1–B_1.
3. The DG puts in reduced output at 25 MWs—curve A_2–B_2.

If a disturbance (such as a short) requires more output from the bus, it will go into the unstable mode (curve A_1–B_1) and its voltage will drop down. As the voltage across the bus and, consequently, across the induction generator, drops down, its electrical torque is reduced and the rotor speeds up to make up for the loss in power delivery at reduced electrical torque. See Figure 14-7 for electrical torque versus speed at various voltages. The DG will speed up and lead to bus voltage collapse due to loss of synchronization. This leads to the conclusion that a fully loaded DG on a bus makes up for unstable operating conditions, whereas a small disturbance on a fully loaded bus can lead to voltage collapse. On the other hand, a DG with some margin in its output capacity reserve can form part of a stable operation during or after a small disturbance. See Figure 14-6, curve A2-B2.

14.8.3 Working of a Protective Fuse and a Backup Recloser Affected by The Presence of a DG

Fuse operation is fast. If current in the grid trip relay needs resetting, attending to each fault manually is a costly business, both for the supply company as well as for the cus-

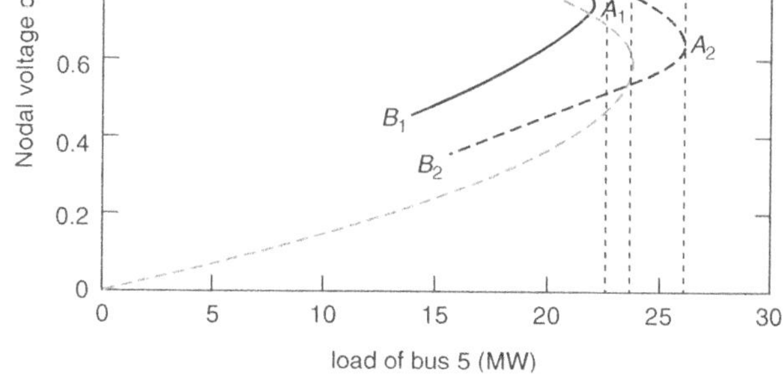

Figure 14-6. PV curves of Bus 5. (From [8], © IEEE 2005.)

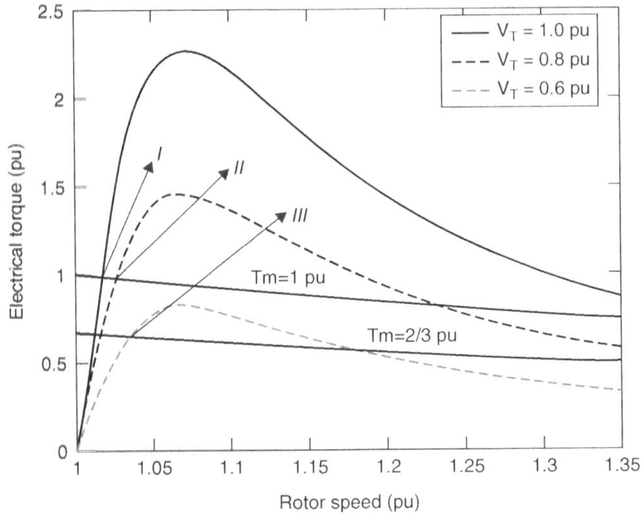

Figure 14-7. Electrical torque verses speed curves. (From [8], © IEEE 2005.)

tomers. Minor faults are cleared with fuses. The circuit outside these faults should reclose by means of an autorecloser switch. The sequence is as below:

1. A fuse blows and disconnects the fault.
2. An autorecloser opens up the circuit slightly after the blowing of the fuse.
3. After a set time, the autorecloser closes and and reestablishes the circuit.
4. If the fault still persists, the autorecloser will open and, after an interval, close a second time.
5. If the fault still persists, the autorecloser will lock out and wait for manual attention, including reset of the lock-out relay.

Line protection relays have to be reset manually.

With an interconnected DG, a fault close to the intertie will be fed by both the grid system as well as the DG system. Now suppose that we want to limit the fault current to 100 A, which the fault was drawing when there was no DG and the relay was set to actuate at 100 A on the grid side. The DG feeds, say, 40 A. Thus, the relay on the grid will have to be adjusted to trip at 60 A.

Here we take a slight detour and consider DGs from renewable energy sources, connected through an inverter, statcom, and so on. Fuse input current rises sharply under a dynamic inverter characteristic.

14.8.4 Correlation between a Fuse and a Trip Relay

An inverter-type DG has the following characteristics. The inverter has either a voltage control or a current control. This control operates within a small frequency band.

As a result, it has a small time constant, which gives it a rapid output response. This enables the inverter to respond instantly to dynamic system disturbances. The same characteristic now responds to a short circuit with its sudden dip in voltage by increasing its output sharply. Note that at the back side of an inverter we have a large energy-storage DC capacitor holding the input voltage to the inverter steady. Thus, the voltage output response of the inverter against a sudden short circuit is not dynamic. However, the pulse current output is dynamic and high.

The recloser is also subjected to high dynamic currents. Conditions for reclosing are explored in the following section.

14.8.5 Boost-up of Fault Current by an Inverter and its Effect on Reclosing

We now take a close look at the behavior of the fault current. The amplitude of the first sine wave of the fault current is very high. It stabilizes in the next 5–10 cycles to its steady-state value when the fuse is expected to blow. The inverter, with its dynamic and high-frequency characteristics, adds to the first cycle. This blows the fuse much earlier than what might be expected from its i^2R characteristics. The component to watch out for is the recloser. It is subjected to high-impulse stresses, both electrical as well as mechanical. If the summated inverter capacity of all DGs exceeds, say, 10% of the feeder capacity, the recloser short-circuit level will have to the considerably raised.

Figure 14-8 gives RMS fault current profiles of parallel DG and a recloser.

For stability, operate the D-Statcom at 0.95 lag and not near unity.

14.8.6 An Inductance Generator with a D-statcom

DGs with D-statcoms are ideal systems to operate. However, it is suggested that they be operated at a slightly lagging power factor, say, at 0.95. If they are operated at a leading power factor and the DG is suddenly cut off due to a fault elsewhere in the system, the system voltage will fall steeply due to a large fall in the supply of leading KVAR by the DG.

14.9 ISLANDING OF AN EPS SECTION FROM THE MAIN BODY

14.9.1 Disconnect on Islanding

An island contains a DG and is in operating condition on a cutoff from the mains. A local portion of an EPS including a DR gets disconnected. Assume that the DR is not disconnected but continues to supply energy into the connected line load. By its very nature, a DR keeps up its frequency and voltage with reference to the utility voltage and frequency. Without this reference, it will go outside the accepted tolerances. Another factor that will accelerate this imbalance is having no control on the load balancing against the supply capacity of the DR. A third factor is that the load will require a leading KVAR, which a normally working DR will not be able to provide. DRs work at unity power factor mainly. For a reconnection, the DR should have been disconnect-

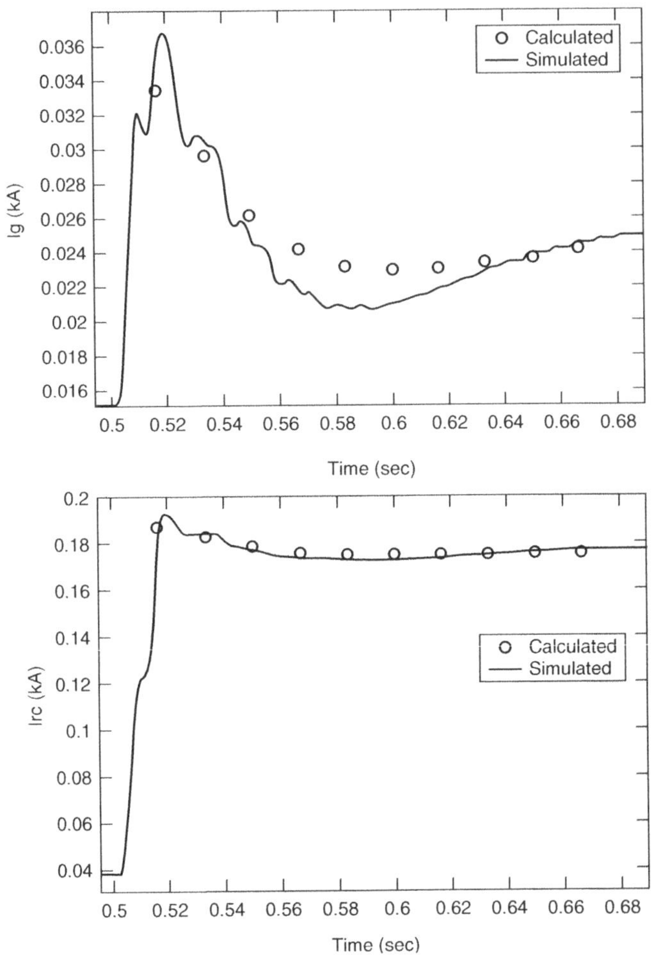

Figure 14-8. Fault currents profiles of parallel DG and recloser. (From [3], © IEEE 2005.)

ed for a minimum of 5 minutes, within which voltage and frequency will have been held steady and then revved up to near synchronization limits.

Considering all these factors it is essential that the DR be almost instantly disconnected upon detecting islanding conditions for it.

14.9.2 Vector Surge Relay (Out of Step)

Assume that a portion of the distributed system develops a sustained fault and a section containing a DR is cut off by a circuit breaker. It is imperative that all the DGs in the isolated or islanded section be switched off in a very short time, preferably within 260–300 msec.

There is a phase angle difference between the generator-induced voltage and the terminal voltage of a generator. Power is delivered at the terminals because of the phase angle difference. If there is a sudden change in the power delivered, due to islanding, this phase angle changes substantially. The vector sensor relay has a reference-generated EMF against which it compares the terminal voltage. For a given variation, it sends a trigger signal to shut off the generator. The VSR has quite a few additional safety and precautionary steps incorporated in it.

In addition to measuring the phase angle difference, the VS relay measures the cycle duration, which is changing continuously. The DG tends to run fast as it gets unloaded. The phase duration is now with respect to the reduced cycle—it is vert short. A VS relay is able to detect events within one cycle of the utility frequency.

14.9.3 Rate of Change of Frequency Relay

Depending on the residual load on the DR after islanding, the generator will run fast or slow and its output frequency will change. A rate of change of frequency (ROCOF) relay or *df/dt* relay is based on the rate of change in frequency. This gives it a fast detection ability. For a 60 Hz system, the ROCOF settings are between 0.10 to 1.20 Hz/s.

A ROCOF relay is based on a fixed 50/60 Hz cycle as a reference period. As a result, it is slow in comparison with VS relay. It measures "slow events" within a few cycles. Both relays are very sensitive and susceptible to false tripping on large magnitude signals. This causes dynamic system voltage problems but these can be offset by a suitable time delay as well as a frequency dip.

14.9.4 Built-in Protection for Inverter Systems

The gate signal control is derived from the utility line voltage. When this is out, an inverter ceases to operate and ceases to energize [9].

14.10 ALLOWABLE PENETRATION LEVELS BY DRs

Total energy capacity added to the utility impacts the system on:

- Steady-state voltages
- Short-circuit currents and transient angle stability of the generators
- Voltage stability and dynamic characteristics
- System security

The operating factors with respect to this penetration are:

- Minimum generation and maximum demand
- Maximum generation and maximum demand

- Maximum generation and minimum demand
- Largest capacity link tripping out

14.11 SYNCHRONOUS GENERATOR AS A DG WITH EXCITATION CONTROLS

Important features of such a system are:

- The first controlling vector is the voltage at the terminals. The limitations are maximum voltage at the bus and the difference between maximum and minimum voltages, which shall not exceed ± 5%.
- A second is the short-circuit current. If the SC MVA capacity goes up, the system costs will shoot up. For example, a circuit breaker (CB) with 630 MVA capacity will be costlier than one with 400 MVA short circuit (SC) capacity. The short circuit capacity plays a part in transient angle stability of the rotor of the DG.
- Should the excitation of this DG be controlled with reference to the power factor, the field excitation against low voltage, resulting from a short circuit, can cascade. Excitation control with voltage reference is preferable. Field excitation of the DG with reference to voltage also gives maximum possible DG output contribution.
- Dynamic voltage stability is improved with DGs. The P-V characteristics in Figure 14-9 show this. Diesel engine connected electricity generators are widely used as standby generators by various categories of consumers if a grid supply fails.

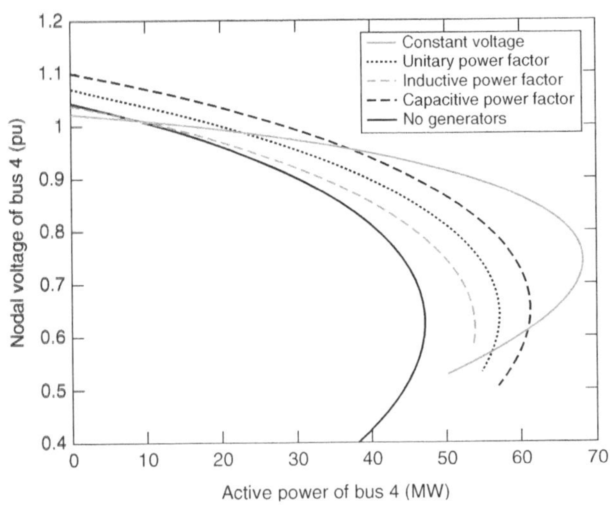

Figure 14-9. PV curves of bus 4. (From [8], © IEEE 2005.)

14.12 HOW CAN A DG EARN PROFITS?

14.12.1 Peak Load Servicing

The philosophy of optimizing business costs leads to operating the most efficient and, therefore, the least costly plants during normal hours and more costly plants during peak hours. Since weather conditions produce typical peak periods for a few hours and for only a few days, one cannot increase the outlay on the initial capacity of the system over these. For a utility, it is best to buy power from a DG to cover these peak periods while operating their low-cost thermal production round the clock. For a DG, this is the best opportunity to earn additional revenue. Should it enter into an annual contract, it achieves marketing security. On the other hand, the utility has achieved supply security.

14.12.2 Selling Contingency Security Reserves to a Utility

Normally, the DG capacities are outside of utility control. As such, these cannot be included in the security reserves required by a utility. However, a DG can package its spare capacity and keep it under the utility area control. Then it becomes a part of utility's security reserve. For an annual fee, the utility gets required reserves without capital investment and the DG gets regular extra income.

14.13 SCOPE FOR GAS-BASED DGS

Gas-based generators have the highest potential of the DGs. Their locations are governed by new and fast-developing load centers. They are economically lucrative, as explained earlier. The gas-based generators fall under the category of "good quality" electric supply.

Generally, gas-based generators will be used within a complex network rather than at the end of a radial line. As such, they raise lots of technical and operating problems. For example, revised load flows will have to be charted out to ensure that there is no overloading of existing facilities. Transmission and distribution costs worked out by regulators will have to be revised.

The spread of gas generators into an EPS will be limited by the fluctuating and rising prices of gas supplies. They are not inhibited by the requirements of energy storage facilities.

14.14 DIESEL GENERATORS

Diesel engine connected electricity generators are widely used by various categories of consumers. If a grid supply fails with or without notice, these sets take over and continue to supply electricity. They come in different sizes and can be activated in very little time, manually or automatically. Their conversion efficiencies are very poor but these efficiencies improve with the size. Generally, the electricity that they produce is rather unaffordable for continuous use. Their voltages can be controlled but the frequencies generally cannot be accurately controlled so as to permit synchronization with the grid frequency. As such, they are not allowed to be switched into a grid. A throw-over switch, with which the load is connected to grid in one position and to the diesel set on the other, is a must.

The cumulative generating capacity locked up in stand-by DGs runs into a very sizeable figure: 60,000 DG sets with a total capacity of 100,000 MVA in the United States [10]. Of these, 70% lie in the 10–200 KVA range.

14.15 EVALUATION OF SERVICE RENDERED BY STAND-BY DGS

See Table 14-5 for DG data. Incremental equipment costs of stand-by generators may be based on costs of supply, costs of preventing losses, and costs of rolling blackouts. If the stand-by reserves were to be included in the initial planning cost, the equipment cost could be low. Today's automated production systems entail a costly loss of production upon the cutoff of electric supply. The customers are exposed to a second transient when the supply comes back. Sensitive equipment must have a tie-up with a regular supply system.

14.16 RELIABILITY COST FOR A DG SET

We base our reliability cost of a DG set on the cost of losses that we suffer. It must be remembered that the hour's loss has to include the hours for which the feeder stayed disconnected, in addition to the hours it took to connect into the customer's entry mains.

The reliability indices such as SAIDI and SAIFI are based on the yearly averages

Table 14-5. Standard sizes and emission norms for diesel generators

Capacity of diesel engines	Date of implementation	Emission limits (g/kW-hr) for NO$_x$	HC	CO	PM	Smoke limit (light absorption coefficient, m – 1) (at full load)	Test cycle Torque %	Weighting factors
Up to 19 kW	1/7/2003	9.2	1.3	5.0	0.6	0.7	100	0.05
	1/7/2004	9.2	1.3	3.5	0.3	0.7	75	0.25
>19 kW up to 50 kW	1/7/2003	9.2	1.3	5.0	0.5	0.7	50	0.30
	1/7/2004	9.2	1.3	3.5	0.3	0.7	10	0.10
>50 kW up to 260 kW	1/7/2003	9.2	1.3	3.5	0.3	0.7		
>260 kW up to 600 kW	1/7/2004	9.2	1.3	3.5	0.3	0.7		

Source: Kirloskar Oil Engines Ltd., Pune, India.

and on the yearly number of loss-of-power instances. These indices help a utility to earn a bonus from a regulator. They do not lead us to the choice of size of a DG. This choice of a DG will naturally depend on the hourly loss that a customer suffers.

14.17 MAINTENANCE AND PROTECTION OF DIESEL GENERATORS

14.17.1 Noise Limit for Diesel Generator Sets (up to 100 KVA)

The maximum permissible sound pressure level for new diesel generator sets with rated capacity up to 100 KVA is 75 dB at 1 m from the enclosure surface. The diesel generator sets are provided without integral acoustic enclosure when manufactured.

The noise limit for diesel generator sets not covered by the above restriction is controlled by providing an acoustic enclosure or by treating the room acoustically at the user's end. The acoustic enclosure or an acoustic treatment of the room should be designed for minimum 25 dB (A) insertion loss or for meeting the ambient noise standards, whichever is on the higher side. Average of measurements at 0.5 m from an acoustic enclosure or room may be made for insertion loss.

The manufacturer must offer the user a standard acoustic enclosure of 25 dB (A) insertion losses and also a suitable exhaust muffler with insertion loss of 25 dB (A). The user must make efforts to bring down the noise levels due to DG sets outside his premises, within the ambient noise requirement of the site and according to control measures. DG sets should be properly maintained so as to prevent deterioration of noise level limits with use.

Every manufacturer must have a valid "type approval" and a valid "conformity of production" certificate for all product models being manufactured by him. These certificates are to be obtained from notified nodal agencies.

14.17.2 Emission Limits for New Diesel Engines (up to 800 kW) for Generator Set Applications

Emission limits and fuel specifications for diesel engines are the same as for commercial HSD (high-speed diesel) vehicles in use in the area. DGs based on renewable energy have different allowable penetration limits into an EPS. They are mostly induction-type generators with an inverter interface. They are dealt in detail in respective chapters on wind and photovoltaic energy.

14.17.3 Pune Pattern of Energy Supply from Stand-By Sets to a Utility

During 2005–2006, Pune, India was in the grip of severe shortage in peak-hour electric power supply. Some major industry groups having substantial stand-by capacity in DGs offered to help. Their generators could not be connected directly into the utility distribution system. There were problems in load matching, staying within constraints of voltage and frequency, and in synchronizing on a grid in which the rate of change in frequency was rather high due to overload.

These industries were major power drawers. Under their supply contract with the industry, they were entitled to and drew power from the utility. This power came at a low cost. They offered to forgo this deal and agreed to generate and use their own power from the stand-by sets during peak hours. This power was costly.

Under the guidance of State Electricity Regulatory Commission, the industries were compensated for the difference between the cost of in-house power production and the cost of the supply from the utility. The utility put in a claim for higher losses arising between supply to bulk consumers at a medium voltage distribution and liberated capacity being supplied to spread-out LT consumers. This claim was accepted and the utility was allowed to put a surcharge in their bills to consumers as "auxiliary service charges" for customers drawing energy above 300 units per month. This arrangement has worked successfully for several years.

With open marketing of electricity, traffic on transmission lines has multiplied and has become multidirectional. Congestion at nodal points has become a burning issue as it hinders growth in electricity. The service charges and the demand-side management charges at nodal points during peak periods should in the near feature promote growth of DGs, supported by energy stores, viable economically and physically. The important players could be wind turbine generators and gas turbine generators.

14.18 UK POLICY ON GENERATION OF LOW-CARBON ELECTRICITY

Distributed generation has been advantageous for providing both power and voltage support to distribution systems. It also accommodates spare capacity from stand-by reserves in generation. It looks like it is going to play an important role in future, in supporting low-carbon electricity production. Let us consider a specific case of the UK electricity system, which has set a policy aiming at 20% low-carbon electricity production by 2020. It has sizeable wind energy in the north to supply sizeable loads in the south. The unpredictably varying wind energy will tie up both generation and transmission capacities against security provisions, particularly during peak-load periods. An ideal solution would have been availability of huge energy reservoirs, hydro or any other viable means, but they are nowhere in sight now.

The solution that the UK system has benchmarked is to promote DG in crowded localities. The DGs will provide the cushion against peak-period demands as well as unpredicted/predicted wind variations. Hundreds of small generators will operate in a free market from hundreds of locations, producing a large number of distributed energy marketplaces. Load balancing can be tough but manageable with strategic location of smart grids composed of electricity, information technology, and communications. This shift to small DGs from continued investments in megasized conventional generation is expected to be a self-perpetuating under a hybrid system of regulation and free markets. This also leads to small grids and smart grids. It can set the pace for others for larger and larger integration of renewable energies into electricity power generation in future. After all, the energy costs in electricity power generation, which are a big chunk, are next to nil with wind.

The DGs for the present are expected to be gas-based turbo generators, operating when it pays and as directed. For the United Kingdom, abundant gas is available from the North Seas [11].

REFERENCES

1. T. Donnelly, Impact of DG on Bulk Transmission with Respect to Transients, *IEEE Transactions on Power Systems,* May 1996, pp. 741–746.
2. R. C. Dugan, T. E. McDermott, and G. J. Ball, Planning for Distributed Generation, *IEEE Transactions on Power Systems,* Volume 7, Issue 2, March–April 2001, pp. 80–88.
3. W. El-Khattam, Y. G. Hegazy, and M. M. A. Salama, An Integrated Distributed Generation Optimization Model for Distribution System Planning, *IEEE Transactions on Power Systems,* Issue 2, May 2005, pp. 1158–1165.
4. I.-S. Bae, J.-O. Kim, J.-C. Kim, and C. Singh, Optimal Operating Strategy for Distributed Generation Considering Hourly Reliability Worth, *IEEE Transactions on Power Systems,* Volume 19, Issue 1, February 2004, pp. 287- 292.
5. M. A. Kashem and G. Ledwich, Distributed Generation as Voltage Support for Single Wire Earth Return Systems, *IEEE Transactions on Power Systems,* Volume 19, Issue 3, July 2004, pp. 1002–1011.
6. J. C. Gomez and M. M. Morcos, Specific Energy Concept Applied to the Voltage Sag Ride-Through Capability of Sensitive Equipment in DG Embedded Systems, *IEEE Transactions on Power Delivery,* Volume 18, Issue 4, October 2003, pp. 1590–1591.
7. I. J. Ramirez-Rosado, L. A. Fernandez-Jimenez, C. Monteiro, V. Miranda, E. Garcia-Arrido, and P. J. Zorzano-Santamaria, Powerful Planning Tools, *IEEE Transactions on Power Systems,* Volume 3, Issue 2, March–April 2005, pp. 56–63.
8. W. Freitas, J. C. M. Vieira, A. Morelato, W. Xu, Influence of Excitation System Control Modes on the Allowable Penetration Level of Distributed Synchronous Generators, *IEEE Transactions on Power Systems,* Volume 20, Issue 2, June 2005, pp. 474–480.
9. M. A. Kashem and G. Ledwich, Multiple Distributed Generators for Distribution Feeder Voltage Support, *IEEE Transactions on Power Systems,* Volume 20, Issue 3, September 2005, pp. 676–684
10. H. B. Puttgen, P. R. MacGregor, F. C. Lambert, Distributed Generation: Semantic Hype or the Dawn of a New Era? *IEEE Transactions on Power Systems,* Volume 1, Issue 1, January– February 2003, pp. 22–29.
11. G. Strbac, C. Ramsay, and R. Moreno This Sustainable Isle, *IEEE Power & Energy Magazine,* September/October 2009, pp. 44–52.

BIBLIOGRAPHY

Baran, M. E., Fault Analysis on Distribution Resources may Contain Many DGs, *IEEE Transactions on Power Systems,* Volume 20, Issue 4, November 2005, pp. 1757-1764.

Bhowmik, A., A. Maitra, S. M. Halpin, and J. E. Schatz, Determination of Allowable Penetration Levels of Distributed Generation Resources Based on Harmonic Limit Considerations, *IEEE Transactions on Power Systems,* Volume 18, Issue 2, April 2003, pp. 619–624.

Blaabjerg, F., A. Consoli, J. A. Ferreira, and J. D. van Wyk, The Future of Electronic Power Processing and Conversion, *IEEE Transactions on Industry Applications,* Volume 41, Issue 1, Jan.-February 2005, pp. 3–8.

Bollen, M. J. and A. Sannino, Voltage Control with Inverter-Based Distributed Generation, *IEEE Transactions on Power Delivery,* Volume 20, Issue 1, January 2005, pp. 519–520.

Borgholm, H. E., A New Heat Recovery and Desulphurization Plant for 4 Wet Kilns in Aalborg Portland, pp. 395–409.

Caisheng, W., M. H. Nehrir, Analytical Approaches for Optimal Placement of Distributed Generation Sources in Power Systems, *IEEE Transactions on Power Systems,* Volume 19, Issue 4, November 2004, pp. 2068–2076.

Celli, G., E. Ghiani, S. Mocci, and F. Pilo, A Multiobjective Evolutionary Algorithm for the Sizing and Siting of Distributed Generation, *IEEE Transactions on Power Systems,* Volume 20, Issue 2, May 2005, pp. 750–757.

Chowdhury, A. A., S. K. Agarwal, and D. O. Koval, Reliability Modeling of Distributed Generation in Conventional Distribution Systems Planning and Analysis, *IEEE Transactions on Power Systems,* Volume 39, Issue 5, September–October 2003, pp. 1493–1498.

Dugan, R. C. and T. E. McDermott, Distributed Generation, *IEEE Transactions on Power Systems,* Volume 8, Issue 2, March-April 2002, pp. 19–25.

Freitas, W. and W. Xu, False Operation of Vector Surge Relays, *IEEE Transactions on Power Delivery,* Volume 19, Issue 1, January 2004, pp. 436–438.

Freitas, W., A. Morelato, W. Xu, and F. Sato, Impacts of AC Generators and DSTATCOM Devices on the Dynamic Performance of Distribution Systems, *IEEE Transactions on Power Systems,* Volume 20, Issue 2, Part 2, April 2005, pp. 1493–1501.

Freitas, W., W. Xu, C. M. Affonso, and Z. Huang, Comparative Analysis Between ROCOF and Vector Surge Relays for Distributed Generation Applications, *IEEE Transactions on Power Delivery,* Volume 20, Issue 2, Part 2, April 2005, pp. 1315–1324.

Freitas, W., Z. Huang, and W. Xu, A Practical Method for Assessing the Effectiveness of Vector Surge Relays for Distributed Generation Applications, *IEEE Transactions on Power Delivery,* Volume 20, Issue 1, January 2005, pp. 57–63.

Gomez, J. C. and M. M. Morcos, Coordination of Voltage Sag and Overcurrent Protection in DG Systems, *IEEE Transactions on Power Delivery,* Volume 20, Issue 1, January 2005, pp. 214–218.

Hegazy, Y. G., M. M. A. Salama, and A. Y. Chikhani, Adequacy Assessment of Distributed Generation Systems using Monte Carlo Simulation, *IEEE Transactions on Power Systems,* Volume 18, Issue 1, February 2003, pp. 48–52.

Hernandez-Aramburo, C. A., T. C. Green, and N. Mugniot, Fuel Consumption Minimization of a Microgrid, *IEEE Transactions on Industry Applications,* Volume 41, Issue 3, May–June 2005, pp. 673–681.

IEEE Standard for Interconnecting Distributed Resources with Electric Power Systems, IEEE Std. 1547, 2003.

Jang, S. I. and K.-H. Kim, An Islanding Detection Method for Distributed Generations Using Voltage Unbalance and Total Harmonic Distortion of Current, *IEEE Transactions on Power Delivery,* Volume 19, Issue 2, April 2004, pp. 745–752.

Jewell, W. T. and T. D. Unruh, Limits on Cloud-Induced Fluctuation in Photovoltaic Generation, *IEEE Transactions on Industry Applications,* Volume 5, Issue 1, March 1990, pp. 8–14.

Katiraei, F., M. R. Iravani, and P. W. Lehn, Micro-Grid Autonomous Operation During and Subsequent to Islanding Process, *IEEE Transactions on Power Delivery,* Volume 20, Issue 1, January 2005, pp. 248–257.

Lin, Y. and G. Gross, Production Cost Analysis of Dispersed Generation Options in a Transmission-Constrained Load Pocket of an Interconnected System, *IEEE Transactions on Power Delivery,* Volume 19, Issue 1, March 2004, pp. 151–156.

Macken, K. P. J., M. H. J. Bollen, and R. J. M. Belmans, Mitigation of Voltage Dips Through Distributed Generation Systems, *IEEE Transactions on Industry Applications,* Volume 40, Issue 6, November–December 2004, pp. 1686–1693.

Mao, Y. and K. N. Miu, Switch Placement to Improve System Reliability for Radial Distribution Systems with Distributed Generation, *IEEE Transactions on Power Systems,* Volume 18, Issue 4, November 2003, pp. 1346–1352.

Marei, M. I., T. K. Abdel-Galil, E. F. El-Saadany, and M. M. A. Salama, Hilbert Transform Based Control Algorithm of the DG Interface for Voltage Flicker Mitigation, *IEEE Transactions on Power Delivery,* Volume 20, Issue 2, Part 1, April 2005, pp. 1129–1133.

McDermott, T. E. and R. C. Dugan, PQ, Reliability and DG, *IEEE Transactions on Power Systems,* Volume 9, Issue 5, Sept.-Oct. 2003, pp. 17–23.

McQueen, D. H. O., P. R. Hyland, and S. J. Watson, Application of a Monte Carlo Simulation Method for Predicting Voltage Regulation on Low-Voltage Networks, *IEEE Transactions on Power Systems,* Volume 20, Issue 1, February 2005, pp. 279–285.

Pipattanasomporn, M., M. Willingham, and S. Rahman, Implications of On-Site Distributed Generation for Commercial/Industrial Facilities, *IEEE Transactions on Power Systems,* Volume 20, Issue 1, February 2005, pp. 206–212.

Pregelj, A., M. Begovic, and A. Rohatgi, Quantitative Techniques for Analysis of Large Data Sets in Renewable Distributed Generation, *IEEE Transactions on Power Systems,* Volume 19, Issue 3, August 2004, pp. 1277–1285.

Wongsaichua, W.,W.-J. Lee, S. Oraintara, C. Kwan, and F. Zhang, Integrated High-Speed Intelligent Utility Tie Unit for Disbursed/Renewable Generation Facilities, *IEEE Transactions on Industry Applications,* Volume 41, Issue 2, March–April 2005, pp. 507–513.

15

INTERCONNECTING DISTRIBUTED RESOURCES WITH ELECTRIC POWER SYSTEMS

15.1 SCOPE

A distributed resource of energy (DR) and an electric power system (EPS) get connected at a point of common coupling (PCC). The nature of operations and the involved duties on both sides of the PCC differ and impact one another. Mutually agreed upon protocols are governed mainly by primary concerns of power quality, reliability, and safety. Abnormal conditions might force separation of a local EPS containing a DR from the main EPS, thus forming an island. Islanding creates many concerns and requires fast action. Reconnection after islanding has been corrected and reestablishing normal conditions in the main EPS area also requires careful scrutiny.

The interconnecting system is strained by both sides across a PCC. Standards and testing procedures applying to normal distribution systems will not hold. These will have to be strengthened.

The DRs have a multiplicity of energy sources such as PV systems, wind-turbine generators and fuel-cell-based systems. These systems require in-depth studies and consideration, particularly under fault conditions, and the ability of generators of verifying capacity to remain integrated even in islanded systems must be determined.

Copyright © 2011 the Institute of Electrical and Electronics Engineers, Inc.

15.2 DEFINITIONS PER IEEE STD 1547-2003

Island. A condition in which a portion of an area EPS is energized solely by one or more Local EPSs through the associated PCCs while that portion of the area EPS is electrically separated from the rest of the area EPS, maintaining the load generation balance.

Interconnection system. The collection of all interconnection equipment and functions, taken as a group, used to interconnect a DR unit or units to an area EPS.

Clearing time. This is defined as the time between the start of the abnormal condition and the DR ceasing to energize the area EPS. It is the sum of the detection time, any adjustable time relay, the operating time for any interposing device (if used), and the operating time for the interrupting device (used to interconnect the DR with EPS such as a circuit breaker).

15.3 DR CEASES TO ENERGIZE THE AREA EPS

This does not mean that it is electrically or mechanically disconnected at the PCC. In fact, it is ready for reconnection when the situation improves and reverts back of to ideal frequency and voltage levels essential for synchronizing the system.

15.4 PROTECTIVE DEVICES

A system consisting of a fuse, a relay, and recloser is a backup for the disconnecting switch. Its interrupting capacity should not be less than that of the disconnecting switch (a circuit breaker).

15.5 SCHEMATIC OF AN INTERCONNECTION BETWEEN A DR AND AN AREA EPS

See Figure 15-1.

15.6 RESTRAINTS ON A DR OPERATOR

A DR operator cannot actively regulate the frequency and voltage at the PCC. It does not inject voltage, but rather injects current generally, at unity power factor with respect to the voltage. Injection of current at the PCC, while holding the voltage there within stipulation, could raise voltage at some other feeder point in the EPS area, should the current through this other feeder increase under impedance levels at that site. The DR is not allowed to do that (except under the guidance of the area controller of the EPS).

A deviation in frequency could result in large-magnitude exchange of energy be-

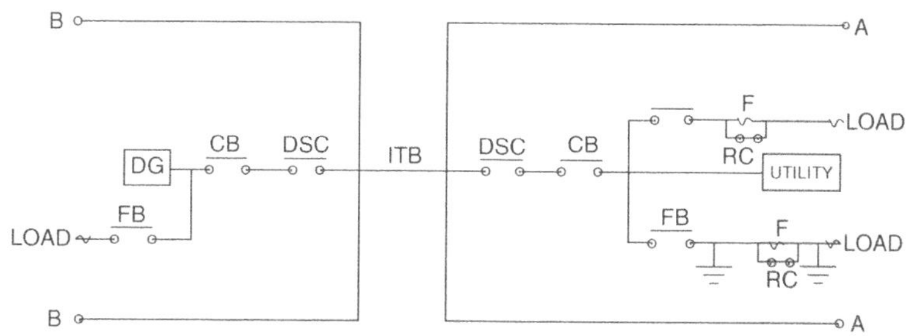

Figure 15-1. An Interconnection between a DG and a utility. A: Area served by EPS, B: Area served by DR, DSC: Disconnecting Isolator, F: Fuse, FB: Fuse Breaker, CB: Circuit Breaker, RC: Reclose.

tween a DR and an EPS. Frequency deviation is not allowed as this could cause the area EPS voltage at the other local EPS to deviate from the standard tolerances beyond a very small deviation.

Some DRs (single-phase inverters, for example) have considerable harmonic outputs that can enter into the EPS at the PCC. The EP system might be having voltage distortions at the PCC on its own account. The DR is allowed a small permitted total harmonic distortion at the PCC (see CBIP publication standards, Appendix 15-1, where the total harmonic distortion is given).

Both active and reactive powers are required at the PCC for maintaining the system voltage. The area operator in the EPS might have a request for reactive power, which a DR has to supply. He will be compensated for that. This is possible with all DRs except those based on induction-type generators. DRs with induction-type generators have to provide capacitors on their own.

Operation of a DR with respect to the capacity to supply extra KVAR under abnormal conditions has to be paid attention to. A DR operating at near unity power factor can quickly change over to maximum. KVAR output by cutting down on the KWs. However, it must have sufficient built-in storage capacity to back up its inverters. Excitation control on the wind-driven synchronous generators will limit control on delivering RKVAR.

The grounding system of the DR has to be properly designed with respect to the grounding of the EPS so that it does not cause overvoltage on the latter and disrupt the coordination of ground fault detection within the area EPS.

15.7 RESPONSIBILITIES AND LIABILITIES OF EPS AREA OPERATORS

Area operators have to design, build, operate, and maintain the EPS. The EPS is bound to maintain power quality on voltages and frequency within declared tolerance values in its area of supply with or without interconnected DRs. They are liable to damage claims by customers, should these limits be transgressed. With today's sensitive and costly microprocessor based equipment, these claims could be crippling. They are equally responsible for safety of personnel, especially for their own employees working on their side of interconnections.

The EPS area operator can guide the transactions at the PCC. However, he has no direct control on the operation of DRs. He is also not liable for damages to the DR's equipment.

EPS are also bound on power security. As such, they have to continuously watch that power balances with load. This requires a flow of large amounts of communication data over the system, taking into consideration impact of the DR [1].

15.8 POWER QUALITY WINDOWS

Voltage levels and their limits are governed by standards. Thus, ANSI C84.1-1995, Range A, for 120 Volts specifies a range of 106–132 V (i.e., 88–110% of the nominal voltage). Trip points to deal with abnormal voltages levels beyond this range are 105 V and 133 V for 60 Hz systems in the United States. During an abnormal time, say, with a short on the feeder or with tripping of a generator, these voltages could go out of range and are required to be regulated through preventive devices.

A DR must automatically cease to energize the system at the PCC so that it does not aggregate the abnormal situation (or hurt itself). It should trip within the periods given in Table 15-1.

The IEEE 1547-2000 standard pertains to inverter-based DRs of the photovoltaic (PV) type, fuel cell type, and so on. The trip control for ceasing to energize is at the inverter location, whereas the reference voltages are taken at the PCC from the EPS side. There could be a voltage drop in the tie system between the inverter and the PCC. This voltage drop has to be taken into account.

The trip times of 120 cycles allowed on both sides of normal voltages take into consideration the ride-through period. A DR itself helps to ride through short time dips in voltages. Please refer to the earlier chapters. Similarly, some time has to be allowed on high voltages for the transients to die down.

If a DR is tripped, it ceases to energize. For its reconnection, it has to be revved up so that it is within stipulated margins of the nominal voltage and nominal frequency.

Table 15-1. Response to abnormal voltages

Voltage (at PCC) time	Maximum trip
$V < 60$ ($V < 50\%$)	6 cycles
$60 \leq V < 106$ ($50\% \leq V < 88\%$)	120 cycles
$106 \leq V \leq 132$ ($88\% \leq V \leq 110\%$)	Normal operation
$132 < V < 165$ ($110\% < V < 137\%$)	120 cycles
$165 \leq V$ ($137\% \leq V$)	2 cycles

Source: [2].

The phase connections have to be ascertained. There is a waiting period of 5 min before it can be reconnected. With these considerations, nuisance tripping is avoided.

Table 15-1 applies for DRs with power rating of 30 kW and below. Low-rated DRs are nonislanding types. For intermediate and higher ratings of DR, the trip points are field adjustable through set points.

For general categories of DRs, IEEE Std 1547 prescribes the same as in Table 15-1, except that it covers the fault clearing time instead of the maximum clearing time.

Very fast acting relays such as ROCOF and VS type as explained in earlier chapter protect the DRs. The ROCOF has trip points rating of 0.10 to 1.20 Hz/Sec.

In case of a short circuit on either side of a PCC, the voltages will dip below 50%. The fault will be cleared and a recloser will automatically reestablish the tie to the system. In such a case, a low-voltage relay operates first and blocks the operation of the ROCOF and VS relays to prevent a nuisance operation.

15.8.1 Frequency

The allowances for frequency to stray away from rated value are very narrow.

15.8.2 Harmonics

Since DRs mainly inject currents, current harmonics have to be measured and restricted. However, the EPS area might be having voltage distortions due to its own loads, prior to the connection by a DR. These voltage distortions result in current distortions again prior to introduction of the DR. The current distortions from all power generation equipment are limited as per Table 15-3. Even harmonics are limited to 25% of the odd harmonic limits. Please refer to Appendix 15-1 for CBIP standards.

15.8.3 Allowable Voltage Distortion Limits for Power Generating Equipment

Wye-connected generator neutrals are grounded through reactors and the output terminals are connected to Wye-connected step-up transformers whose neutrals are ground-

Table 15-2. Interconnection system response to abnormal frequencies

DR size	Frequency range (Hz)	Clearing time(s)
$\leq$ 30 kW	> 60.5	0.16
	< 59.3	0.16
> 30 kW	> 60.5	0.16
	< 59.8–57.0 (adjustable set point)	Adjustable 0.16 to 300
	< 57.0	0.16

Source: [2].
Note: DR $\leq$ 30 kW maximum clearing times; DR > 30 kW fault clearing times. In all the cases where adjustable set points are provided on the trip relays, (a) trip settings shall be coordinated with area EPS operator; (b) 0.16 corresponds to 9.6 Hz or roughly 10 cycles.

Table 15-3. Maximum harmonic current distortion in percent of current for all power generation equipment

Individual harmonic order h (odd harmonics)[b]	$h < 11$	$11 \leq h < 17$	$18 \leq h < 23$	$23 \leq h < 35$	$35 \leq h$	Total demand distortion (TDD)
Percent (%)	4.0	2.0	1.3	0.6	0.3	5.0

Source: IEEE Standard 519-1992.

ed solidly or through earth. The generator voltages always contain zero-sequence harmonic voltages. Under adverse condition of inherent/reflected capacitances and inductances, these voltage harmonics can get amplified and distort measurements. A study for a possible closed-loop formation has to be made and measures taken to break up the loop. A possible combination with delta connections can help.

15.8.4 Maximum Harmonic Voltage Distortions at PCC at Voltages up to 69 kV

These are: for individual voltage distortion, 3.0%; total voltage distortion, 5.0%. For more details, see IEEE Standard 519-1992.

15.9 LIMITATION OF DC INJECTION

The DR and its interconnection system shall not inject dc current greater than 0.5% of the full rated output current at the point of DR connection. In the case of inverter-based DRs, this is achieved through a blocking transformer. An alternative is to monitor continuously the DC input at PCC across a shunt and shut down the inverter in case this exceeds the stipulated value.

15.10 ISLANDING OF A LOCAL-AREA EPS THAT INCLUDES A DR

Under some circumstances, it is likely that a local-area EPS containing a DR will get cut off from the main EPS and the DR will continue to energize this island area. This is dangerous and has to be avoided.

The DR holds onto its voltage and frequency with reference to these parameters on the main system. Once this reference is lost, controls in the DR regulating V and Hz become ineffective.

It is very seldom that the load requirement of the island section and the generation of the DR will match and keep on matching. If the load demand for active power is high, the DR will slow down and lose voltage and frequency rapidly. If the demand is light, the DR will accelerate, increasing voltage and frequency.

If the consumer's critical loads are affected, liability for damages is still with the EPS and it cannot allow islanded DRs to add energy at the PCC. The DR must cease energizing at the PCC almost instantaneously.

A contingency that creates islanding will in all probability demand excessive and instant KVAR, which is beyond the capability of most of the DRs. An increased KVAR load on a DR does not slow it down, since it does not involve exchange of active energy. However, sudden changes in KVAR, particularly across inductive loads, have to be watched out for changed terminal voltages and transients.

Rate of degradation of voltage and frequency under islanding are much steeper than those under overload/underload conditions. Special relays, ROCOF and VS, help to put the DR out of energy supplying mode.

Situations could arise in which the detection of islanding conditions is too slow and the DR protective relays might not operate. The detection times are governed by active power imbalance and excess of active or reactive power, and are specific. Freitas [3] gives an illustration on this for the two types of relays normally used, RCOF and VS types (see Figures 15-2 and 15-3).

Should the supply be resumed while the DR is still energizing the islanded section, there will be mismatches in V, Hz and phases. There will be damages on both sides.

DRs are to be speedily and definitely disconnected under islanding conditions. Islanding could be intentional and preplanned by an EPS operator or islanding could be unintentional.

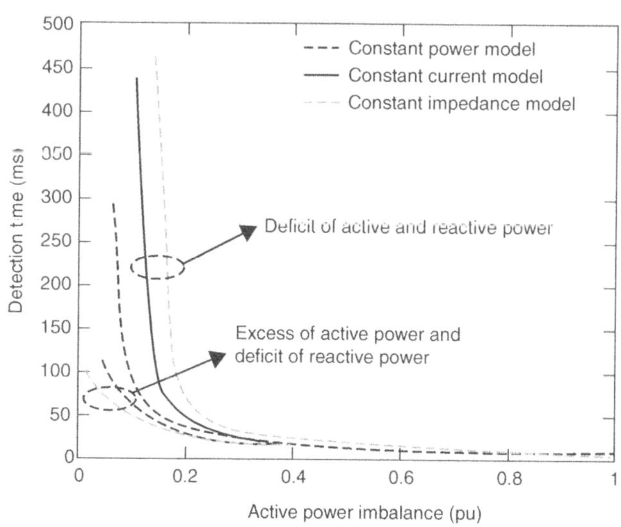

Figure 15-2. ROCOF performance curve. (From [3], © IEEE 2004.)

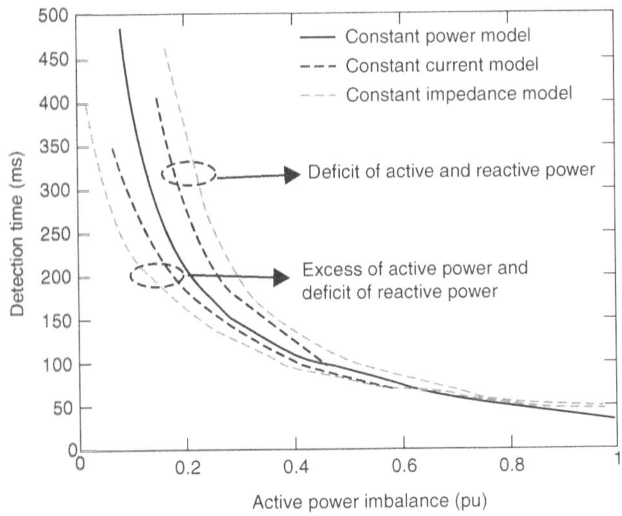

Figure 15-3. VS performance curve. (From [3], © IEEE 2004.)

Small DRs (capacity rated < 30 KWs) are designed to be non-islanding DRs. They have to undergo nonislanding tests.

15.11 RECONNECTION

This has been commented on in Section 15-4. Table 15-4 gives the parameter limits on synchronizing during reconnection/starting.

A DR will parallel with the EPS area without causing voltage fluctuations beyond ±5%, objectionable flicker, cycling of network protections, or overload of the general detection system (fault, etc.).

Check-synchronizing relays are available to prevent operating hazards due to attempted synchronization.

DRs with an aggregate capacity >250 KVA should have a panel monitoring/indicating KW, KVAR, and voltage at the PCC.

Table 15-4. Synchronization parameter limits for reconnection

Aggregate rating of DR units (kVA)	Frequency difference (Δf, Hz)	Voltage difference (ΔV, %)	Phase angle difference ($\Delta\Phi$, °)
0–500	0.3	10	20
> 500–1,500	0.2	5	15
> 1500–10,000	0.1	3	10

Source: [2].

15.12 SAFETY ASPECTS

For personnel:

1. Safe work area clearances have to be maintained when utility workers have to work in close proximity to a live EPS. Only an authorized individual, bearing identification, may be allowed to operate critical equipment such as a disconnect switch. Tags should be attached to live parts.
2. An isolating device on the DR must be readily accessibly, lockable, and of the visible break isolation type.
3. There should be an isolating device on each source of energy. Safe and sound working practices developed over years by the EPS are to be followed. There should be protective grounding. Proper work permits should be issued and cancelled before resumption.

Integrity of interconnections (reliable operation)

1. The interconnection system protect against EMI interference to prevent maloperation of the interconnecting system, which should not misread under EMI influence.
2. The equipment should be rated to withstand surge voltages of up to $2\sqrt{2}$ E across open contacts of circuit breakers or isolators, as per IEEE Standard C62.41.2-2002 or C37.90.1-2002, as applicable. There are different requirements for different environments.
3. Paralleling devices should withstand the system rated voltage of $2\sqrt{2}$ E across open contacts.
4. The interconnection should stay within limits for DC injection and harmonics, as stated earlier.

15.13 TESTING OF INTERCONNECTING EQUIPMENT

IEEE Standard 1547-2005 gives elaborate details on sequence and procedures for testing these. The reader is also referred to IEEE Standards 1547-2005 and 519-1992.

15.14 INTERCONNECTIONS WILL BE IMPORTANT IN TOMORROW'S SCENARIO

Under the free market economy for electricity, more and more DRs are expected to enter the electric supply field. Interconnections will be in demand not only in quantity, but also will spread all over. Developments in energy storage systems are also expected to spur this movement. Interconnections will assume a great importance in the near future.

APPENDIX 15-1 CBIP STANDARD RECOMMENDATION, EXTRACTS FROM PUBLICATION 2517, JULY 1996 [4]

RECOMMENDATIONS

The committee has made the following recommendations.

Target Compatibility Levels

The target compatibility levels (criteria for voltage quality as per Section 5.3 of [4]) of voltage harmonics at point of common coupling are as follows:

Individual Harmonic Voltage Distortion **(D_n)**

Odd harmonics (nonmultiple of 3)				Odd harmonics (multiple of 3)				Even harmonics			
	Harmonic voltage (%)				Harmonic voltage (%)				Harmonic voltage (%)		
Order n	$0.4 \leq U \leq$ 45 kV	45 KV $< U \leq$ 220 KV	$U >$ 220 kV	Order n	$0.4 \leq U \leq$ 45 kV	45 KV $< U \leq$ 220 KV	$U >$ 220 kV	Order n	$0.4 \leq U \leq$ 45 kV	45 KV $< U \leq$ 220 KV	$U >$ 220 kV
5	8.0	3.0	1.0	3	5.5	2.5	1.0	2	2.0	1.8	1.0
7	3.0	2.0	1.0	9	2.5	2.0	1.0	4	1.0	1.0	1.0
11	3.5	1.5	1.0	15	1.0	1.0	1.0	6	0.7	0.7	1.0
13	3.0	1.5	1.0	21	0.3	0.3	1.0	8	08	0.5	1.0
17	2.0	1.0	1.0	>21	0.3	0.3	1.0	10	0.8	0.5	1.0
19	1.5	1.0	1.0					12	0.8	0.5	1.0
23	1.5	0.7	1.0					>12	0.8	0.5	1.0
25	1.5	0.7	1.0								
>25	0.2 + 1.3 × 25/m	0.2 + 0.5 ×25/m									

Total harmonic distortion (THD)/(D_{eff}):

$= 9\%$ in $0.4 \leq U \leq 45$ kV

4% in 45 kV $\leq U \leq 220$ kV

3% in $U > 220$ kV

Generator continuous harmonic current, $\sqrt{I_g = \sum_{m+1}^{m+9} [1.424(m)^{0.87} \times (I_{ga} + I_{gb})^2] + I_{neg}^2}$

Limits of I_g and corresponding duration (T):

	Limits
T $\geq$ 1 hr	0.08 pu
1 min $\geq$ T $<$ 1 hr	0.30 pu
10 sec $\geq$ T $<$ 1 min	0.77 pu

Maximum individual harmonic current values (arithmetic sum) at harmonic:

$3m + 1$ and $3m - 1$	0.015 pu
Single harmonic current	0.010 pu

While sanctioning the new connections to the nonlinear loads, utilities shall assess the harmonics generation from such loads and ensure that connection of such loads does not exceed the harmonic limits at the point of common coupling. Appendix I of [4] gives examples of two nonlinear load cases as follows, in which the specific limits have been maintained by designing suitable filters.

Case 1: Harmonic filters for Lloyd Steel Industries: Third, Fifth, Seventh, Eleventh, Thirteenth Harmonic Filters 3 × 1 ph at 132 kV, 3 ph, 50 Hz.

Case 2: Filters for Rihand and Dadri 1500 MW + 500 kV HVDC Bipole Multiple Filters across 400 kV, 50 Hz, AC system

Utilities shall provide a standard format accompanying the application for seeking new connections. The format shall contain details like harmonic contribution from the equipment/loads to the system and a daily loading schedule of the equipment/load. A recommended "Questionnaire/Format for Data/Information" to be furnished by the consumer or by his authorized installation engineer electricity supplier is given in Appendixes III and IV of [4].

Utilities shall also lay down the guidelines for approving the connections under:

Stage I: Acceptance depending on the network fault level

Stage II: Acceptance depending on the total potentially disturbing power of the network

Stage III: Acceptance on an exceptional and precarious basis

Utilities shall place a sufficient number of harmonic measuring instruments at regular intervals near the sources of harmonics generation.

REFERENCES

1. N. D. Kaushika and N. K. Gautam, Energy Yield Simulations of Interconnected Solar PV Arrays, *IEEE Transactions on Energy Conversion,* March 2003, pp. 127–134.
2. IEEE Standard for Interconnecting Distributed Resources with Electric Power Systems, IEEE Std. 1547-2003.
3. W. Freitas and W. Xu, False Operation of Vector Surge Relays, *IEEE Transactions on Power Delivery,* Volume 19, Issue 1, January 2004, pp. 436–438.
4. Central Board of Irrigation and Power, Guide for Limiting Voltage Harmonics, Publication 251, New Delhi, India.

BIBLIOGRAPHY

Aiello, M., A. Cataliotti, S. Favuzza, and G. Graditi, Theoretical and Experimental Comparison of Total Harmonic Distortion Factors for the Evaluation of Harmonic and Interharmonic Pollution of Grid-Connected Photovoltaic Systems, *IEEE Transactions on Power Delivery,* July 2006, pp. 1390–1397.

Aki, H., S. Yamamoto, Y. Ishikawa, J. Kondoh, T. Maeda, H. Yamaguchi, A. Murata, and I. Ishii, Operational Strategies of Networked Fuel Cells in Residential Homes, *IEEE Transactions on Power Systems,* August 2006, pp. 1405–1414.

Bagnasco, A., B. Delfino, G. B. Denegri, and S. Massucco, Management and Dynamic Performances of Combined Cycle Power Plants During Parallel and Islanding Operation, *IEEE Transactions on Energy Conversion,* Volume 13, Issue 2, June 1998, 194–201.

Baran, M. E. and I. El-Markaby, Fault Analysis on Distribution Feeders with Distributed Generators, *IEEE Transactions on Power Systems,* November 2005, pp. 1757–1764.

Barton, J. P. and D. G. Infield, Energy Storage and its Use with Intermittent Renewable Energy, *IEEE Transactions on Energy Conversion,* June 2004, pp. 441–448.

Castro, R. M. G. and L. A. F. M. Ferreira, A Comparison Between Chronological and Probabilistic Methods to Estimate Wind Power Capacity Credit, *IEEE Transactions on Industry Applications,* Volume 16, Issue 4, November 2001, pp. 904–909.

Ed, M., L. Tostes, U. H. Bezerra, R. D. S. Silva, J. A. L. Valente, C. C. M. de Moura, T. M. M. Branco, Impacts Over the Distribution Grid from the Adoption of Distributed Harmonic Filters on Low-Voltage Customers, *IEEE Transactions on Power Delivery,* January 2005, pp. 384–389.

El-Arroudi, K., G. Joos, I. Kamwa, and D. T. McGillis, Intelligent-Based Approach to Islanding Detection in Distributed Generation, *IEEE Transactions on Power Delivery,* Volume 22, Issue 2, April 2007, pp. 828–835.

Hatziadoniu, C. J., E. N. Nikolov, and F. Pourboghrat, Power Conditioner Control and Protection for Distributed Generators and Storage, *IEEE Transactions on Power Systems,* February 2003, pp. 83–90.

Hernandez-Gonzalez, G. and R. Iravani, Current Injection for Active Islanding Detection of Electronically-Interfaced Distributed Resources, *IEEE Transactions on Power Delivery,* Volume 21, Issue 3, July 2006, pp. 1698–1705.

Huang, S-J. and F.-S. Pai, A New Approach to Islanding Detection of Dispersed Generators with Self-Commutated Static Power Converters, *IEEE Transactions on Power Delivery,* Volume 15, Issue 2, April 2000, pp. 500–507.

IEEE Recommended Practice for Utility Interface of Photovoltaic (PV) Systems, IEEE Std. 929-2000.

IEEE Standard Conformance Test Procedures for Equipment Interconnecting Distributed Resources with Electric Power Systems, IEEE Std. 1547-2005.

Jang, S-I. and K.-H. Kim, An Islanding Detection Method for Distributed Generations Using Voltage Unbalance and Total Harmonic Distortion of Current, *IEEE Transactions on Power Delivery,* April 2004, pp. 745–752.

Katiraei, F., M. R. Iravani, P. W. Lehn, Micro-Grid Autonomous Operation During and Subsequent to Islanding Process, *IEEE Transactions on Power Delivery,* June 2005, pp. 248–257.

Lin, Y. and G. Gross, Production Cost Analysis of Dispersed Generation Options in a Transmission-Constrained Load Pocket of an Interconnected System, *IEEE Transactions on Energy Conversion,* March 2004, pp. 151- 156

Lopes, J. A. P., C. L. Moreira, and A. G. Madureira, Defining Control Strategies for Microgrids Islanded Operation, *IEEE Transactions on Power Systems,* May 2006, pp. 916–924.

Oliva, A. R. and J. C. Balda, A PV Dispersed Generator: A Power Quality Analysis within the IEEE 519, *IEEE Transactions on Power Delivery,* April 2003, pp. 525–530.

Papathanassiou, S. A. and F. Santjer, Power-Quality Measurements in an Autonomous Island Grid with High Wind Penetration, *IEEE Transactions on Power Delivery,* June 2006, pp. 218–224.

Pipattanasomporn, M., M. Willingham, and S. Rahman, Implications of On-Site Distributed Generation for Commercial/Industrial Facilities, *IEEE Transactions on Power Systems,* February 2005, pp. 206–212.

Rifaat, R. M., Critical Considerations for Utility/Cogeneration Inter-Tie Protection Scheme Configuration, *IEEE Transactions on Industry Applications,* Volume 31, Issue 5, September–October 1995, pp. 973–977.

Vieira, J. C. M., W. Freitas, W. Xu and A. Morelato, Efficient Coordination of ROCOF and Frequency Relays for Distributed Generation Protection by Using the Application Region, *IEEE Transactions on Power Delivery,* Volume 21, Issue 4, October 2006, pp. 1878–1884.

Vieira, J. C. M., W. Freitas, W. Xu, and A. Morelato, Performance of Frequency Relays for Distributed Generation Protection, *IEEE Transactions on Power Delivery,* Volume 21, Issue 3, July 2006, pp. 1120–1127.

Woyte, A.,V. Van Thong, R. Belmans, and J. Nijs, Voltage Fluctuations on Distribution Level Introduced by Photovoltaic Systems, *IEEE Transactions on Energy Conversion,* pp. 202–209

You, H., V. Vittal, and W. Xiaoming, Slow Coherency-Based Islanding, *IEEE Transactions on Power Systems,* February 2004, pp. 483–491.

You, H., V. Vittal, and Z. Yang, Self-Healing in Power Systems: An Approach Using Islanding and Rate of Frequency Decline-

Based Load Shedding, *IEEE Transactions on Power Systems,* February 2003, pp. 174–181.

Zeineldin, H. H., E. F. El-Saadany, and M. M. A. Salama, Impact of DG Interface Control on Islanding Detection and

Nondetection Zones, *IEEE Transactions on Power Delivery,* Volume 21, Issue 3, July 2006, pp. 1515–1523.

16

ENERGY STORAGE—POWER STORAGE SUPER CAPACITORS

16.1 ENERGY STORAGE AND THE FUTURE FOR RENEWABLE ENERGY SOURCES

Two types of reserves have to be held for DGs based on renewable energy:

1. Renewable energy sources have a fluctuating nature. A system reserve has to be held to accommodate the fluctuations.
2. Normal spinning reserves have to be held for load balancing, particularly at peak demand times.

Both these reserves have to be held, in basic generation capacity as well as in the distribution capacity of the EPS. Without these reserves, the power quality and reliability are impacted. The investment in locked-up capital, if these reserves are held in main generators, will be nonremunerative. Alternatively, the EPS penetration by renewable energy sources will remain limited.

16.2 ADVANTAGES OF ENERGY STORAGE

An energy store theoretically will absorb the energy when the generation is high and the load demand is low. When the demand goes up and exceeds the generating capaci-

Copyright © 2011 the Institute of Electrical and Electronics Engineers, Inc.

ty of the DGs, the store will release the required quantum of energy. This has multiple advantages:

1. It supports power quality on the voltage and frequency, two basic important parameters.
2. It improves the reliability, upgrading the power quality indices such as SAIFI and SAIDI.
3. It enables a higher penetration of the EPS by renewable energy sources. This is an important aspect, particularly when all over the world strong efforts are being made toward going away from the fossil fuels to renewable energies.
4. With a market-based structure, it helps to utilize optimally electricity production, whether from conventional sources or from nonconventional sources. Take pumped storage system, where we can use wind-based energy, whenever it is available, for pumping up water, instead of thermal energy normally used in times of low demand. The pumped storage will deliver electricity, preferably during peak hours when the price of electricity supply is high.
5. Take again the case of the wind turbine generator. When the wind is blowing, the local demand may be moderate and electricity may be available at competitive prices from normal conventional sources. Selling wind-generated electricity on the market is nonremunerative. The alternative is to dump this generation.
6. Stores help us on the emission front by increasing gains in carbon credits for reducing objectionable gases, thus reducing use of fossil-based fuels.

16.3 FACTORS FOR CHOOSING TYPE AND RATING OF A STORAGE SYSTEM

These include available energy source, network parameters, connection and cycling costs, emissions, fuel bills, capacity, market prices of electricity, and load density.

16.3.1 Network Parameters

A weak network (with low short-circuit capacity) needs a store for electricity. A stronger network with a high short-circuit level itself acts as a store. An extra storage device could probably be put on delay in this case.

16.3.2 Connection and Cycling Costs

These could be high in the absence of a store. Running a thermal generator up and down affects its efficiency and increases its fuel consumption. Voltage fluctuations are wider and faster. A short-capacity store with rapid access and response would be ideal here.

16.4 NATURE OF SUPPORT BY ELECTRICITY STORAGE SYSTEMS

Stores support the EPS during short-time fluctuations in energy supply, as happens when clouds appear in the sky blanking out PV arrays. A store with high storage capacity and a feature of quick delivery over a short time is ideal. A supercapacitor can perform this function.

Stores support the EPS over long-duration fluctuations. As a basic rule, we use electricity at the instant it is produced. A long gap in wind energy supply is made up by stores. In this case, we are trying to separate the instant of use from the instant of generation.

There are seasonal variations, predictable by metrological data. These are best tackled by load management, as detailed in Chapter 3.

Table 16-1 gives an idea of the costing of effective energy, effective power, duration over which these types of stores can be utilized, and their round-trip efficiencies. The following points may be noted:

1. Hydrogen, which is looked upon as a future savior in energy storage and utilization, both for automobiles and for electricity generation, has a low round-trip efficiency.
2. Sodium sulfur technology has been adopted successfully by Japan for support to electrical power systems.
3. Lead–acid is still the universal leader, with the advantage of established presence.
4. Supercapacitors with high round-trip efficiencies are power suppliers mainly and not energy suppliers.

Table 16-1. Properties of storage technologies in the electrical power system

Technology	Energy cost, ¢/kW-hr/year	Power cost, ¢/kW-hr/year	Independent?	Characteristic time, hours	Round-trip efficiency, %
Hydrogen	6.8	128	Y	12+	32
Vanadium redox flow cells	12.5	144	Y	2 to 12	75
Regenesys redox flow cells	9.4	70	Y	6 to 12	75
Nickel metal hydride	24	96	N	1 to 4	64
Sodium suplphur	34	132	N	4 to 8	87
Nickel cadmium	117	351	N	1 to 10	72
Zinc bromide	79	158	N	2 to 5	63
Lead–acid	55	73	N	1 to 5	63
High-speed flywheel	77	0.88	N	10^{-4} to 10^{-2}	89
Supercapacitor	570	4.8	N	10^{-4} to 10^{-2}	86
SMES	3×10^5	47	N	10^{-3} to 10^{-4}	21

Source: [1].

5. The characteristic time scale in Table 16-1 is the typical time taken to discharge the store. It does not indicate charging time.

Figure 16-1 gives the costs per kW/hr of storage for various systems. In the following, we will go into some more detail on storage aspects for PV systems and for wind energy systems.

16.5 LOAD DENSITY, SHORT-CIRCUIT CAPACITY, AND STORAGE OF ENERGY

Let us take an area with a high load density. This has a high grid capacity as well. Demand can be satisfied with a short-term storage capacity. Against a dense load, voltage rise due to injection from stored energy will be modest. Since the load demand also is high, the prices obtainable for stored energy are high. Long-term storage will pay if the stores deliver electricity during peak hours. Again, energy storage will not pay in low-load-density and low-SC-level areas, since we cannot inject too much energy there. Voltages will fluctuate widely. Low-price opportunities might probably be the cause for the low load density as well.

16.6 PHOTOVOLTAIC ENERGY—PV ENERGY IN RESIDENTIAL APPLICATIONS

Except for blanking by clouds, the PV source of energy gives a steady and predicted supply of energy. It maintains performance in tune with statistical metrological data. PV application is expanding fast, particularly within the residential load sector. This sector is supplied by radial feeders starting from a service station. The voltage at the far end has to stay within limits of international grid standards such as IEC-60038. The PVs are connected at low voltages of the distribution system (120 V in United States, 230 V in United Kingdom, Europe, and India). The interconnections are through inverters. They

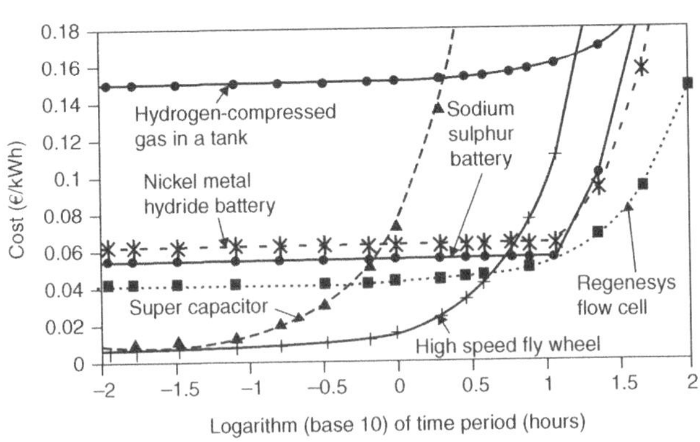

Figure 16-1. Costs of energy storage systems. (From [1], © IEEE 2004.)

exchange energy generally at unity power factor, meaning that they do not inject KVAR into the system. Injection of KVAR alters the voltages at the PCC, which they are not authorized to do.

Let us take a case in which the substation voltage is maintained at 1.0 pu. The voltage at the far end of the feeders rises to 1.03 pu to 1.07 pu, depending on how much current the PV is injecting in a reverse-flow mode. Note that the PV-injected current reduces the feeder current as well as the voltage drop. Not only this, negative voltage drop (voltage rise) is caused by the PV current flowing in the opposite direction.

A cloud suddenly blanks out the PV installation, causing the irradiance to fall to as low as 30% of its average energy output. In a typical case cited, this could cause a voltage variation of 0.03 pu to 0.04 pu [2]. This voltage variation is superimposed on the voltage variation due to normal load fluctuation.

German power system operators limit the permissible voltage increase in low-level distribution systems to 0.02 pu. Obviously, a variation in voltage to the level of 0.03 to 0.04 pu is not acceptable there. An energy storage device that regulates the voltage by absorbing and giving out fluctuating quantum of PV energy is essential there. As the PV intrusion into the grid increases, storage of energy will assume greater importance.

16.7 MAXIMUM PV PENETRATION AND MAXIMUM ALLOWABLE STORAGE GO HAND IN HAND

Let us consider another case in which a PV source is installed in a weak section of a grid, such as a section with low short-circuit capacity. An energy imbalance coupled with intermittent energy exchange is likely to cause a system imbalance. Energy will flow back and forth between the weak grid and the strong store under these conditions. This is not permissible.

In the case of a strong grid with a higher short-circuit level, impact of a distributed PV system on voltage levels will be negligible. This is because the strong grid itself is a very strong storage device for fluctuating energy. Losses involved in storing and retrieving energy from the grid. that is. from the distribution, are quite small, say 5% to 8%. When a storage device accepts and returns energy, the round-trip losses, depending on the type of storage device, are much higher (see Table 16-2). Strictly from the viewpoint of low losses and higher stability, strengthening the grid itself as a storage device appears to be a better proposal than going in for storage. However, it could be a costly proposal. To arrive at a proper conclusion, one must study what factors are involved and the relative costs of planning a proper storage device.

16.8 PLANNING THE SIZE OF A STORE FOR PV INCLUSION IN A DISTRIBUTION SYSTEM

Some indices involved in planning are:

1. Clearness index K_T. This is an instantaneous index defined as solar irradiance normalized (due to unclear skies) on extraterrestrial irradiance. It is a measure

for atmospheric transparency, independent of seasonal or daytime variations. Thus, a low K_T indicates an overcast sky and a higher K_T an almost clear sky. K_T is determined statistically from data drawn over years and is location specific.

2. Persistence is the duration of a cloud cover.
3. Periodogram. This is a measure of the power content of signals over a time scale such as is caused in the PV output when a cloud passes over.

Figure 16-2 shows a typical two-hour node voltage signal and its wavelet periodogram for a number of significant time scales. Note that fluctuations are caused by a dip in irradiance or a loss in PV output. The top portion of Figure 16-2 represents dips

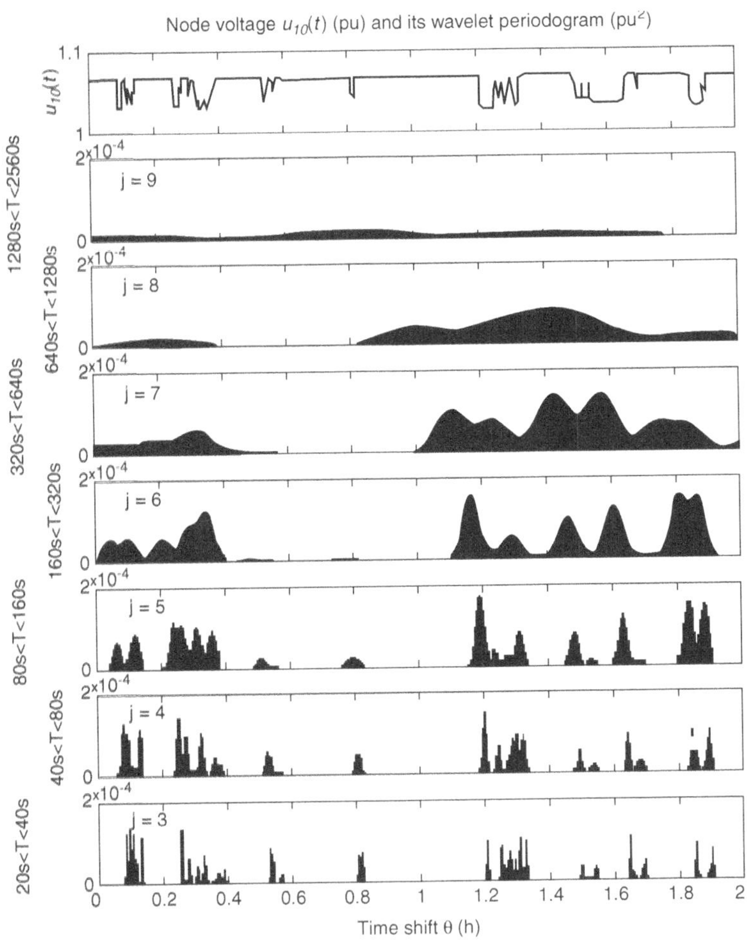

Figure 16-2. Wavelet periodograms of a 2-h time series of the node voltage at node 10. (From [2], © IEEE 2006.)

in voltages over the period. The energy content is calculated for these random wavelets, not necessarily sinusoidal, with the help of a "stationary" wavelet transform.

A fluctuation power index represents mean values of all periodograms for a high number of sample days on each time scale (say, 40 seconds between 2 PM to 2.40 PM).

A hypothetical parallel linear load is assumed to be subjected to these voltage variations. The demand by this load over a given persistence may be called the fluctuation power index. This index, when integrated over that period, gives a fluctuating energy index for that sampling period. It leads us to size up the rating of the storage device in Joules.

Note that the fluctuation power index varies with persistence of fluctuation and its clearness factor K_T. Figure 16-3 shows a typical graph. The index is shown for different K_T values as given in the left-hand corner.

In the above case, both low and high values of K_T represent low-fluctuation power indices. These indices are sizeable for K_T lying between 0.2 and 0.6. Again, in this particular instance, there is a small peak between persistence values of 300 to 700, which could form a base for designing the rating of the storage power.

16.9 TYPES OF STORAGE DEVICES FOR PV SYSTEMS

Widely used lead–acid batteries might attract the first choice. However, these are reported to be unsuitable for a high number of charge–discharge cycles. Supercapacitors and supermagnetic energy storage devices have low bridging times, up to 1 to 2 minutes. These will not be suitable for longer persistences. The necessary buffer capacity for mitigating up to 1 hour persistence has been determined at 80 kJ/kW PV peak pow-

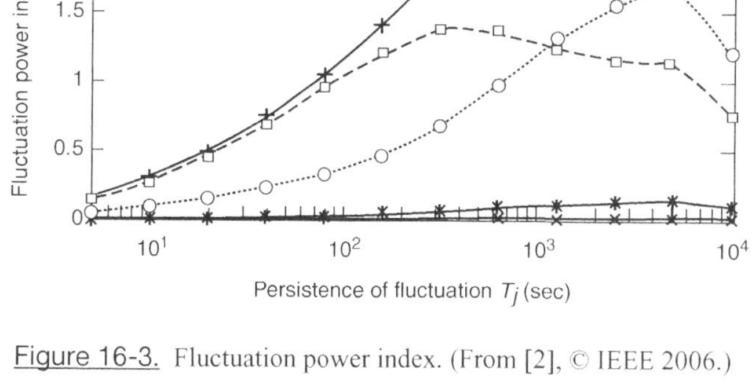

Figure 16-3. Fluctuation power index. (From [2], © IEEE 2006.)

er for distribution feeders and 257 kJ/kW PV peak power for hybrid systems. On persistence of less than 1 hour, the amplitude of voltage fluctuation is reduced from 9.9% to 6.4%.

At 2003 prices in Belgium, the costs of installing suitable supercapacitors were worked out at 18% of average cost of grid-connected PV systems. This increased package cost has to be considered in relation to the site meteorology, penetration factor of PV, grid SC level, and so on, and also against a high power quality demanded by progressively discriminating microbased devices owned by the consumers.

16.10 WIND ENERGY

We will restrict ourselves to wind turbine generators (WTGs) in individual groups rather than regular wind farms. Wind farm sites are preinvestigated and have wind availability for larger periods.

Wind energy differs from PV energy in the following aspects:

1. Daily PV irradiance is steady and constant, except for the cloud effect, whereas daily wind turbulence can vary over a wide range.
2. PV schemes are connected into LV distribution networks. The PV replaces energy supply from grid energy; it does not sell energy into a grid. Conversely, wind energy sells energy into a grid. It is connected into a medium- or high-voltage section of the EPS.
3. PV fluctuations cause a dip in a steady distribution voltage. Wind energy causes a rise in voltage and its energy input into the grid is limited by the maximum voltage allowed at the PCC.
4. Voltage tolerances in medium voltages are more liberal (6%) than those in LV (0.02 pu in Germany).

The storage aspect for wind energy is governed by the following considerations:

1. Storage for a ride-through of wind energy fluctuations. An ideal location will be an EPS with a high load demand. Voltage drops due to high loads (at a distance from service station) are compensated by voltage rises due to injection of power by WTGs. To that extent, a high penetration of the grid system by wind energy is welcome. Short-term storage of electricity in this instance will increase the reliability at the same time it will be helping us on the voltage regulation.
2. Change the time of delivery/price availability. Peak-time electricity prices are high. A high capacity storage system will help to absorb wind energy when it is available and sell it for maximum benefit at peak demand time.

It may be noted that several cost factors are involved in the sale of wind energy. There is a cost of energy production. To this we add cost of storage. Cost of losses fall into two categories:

1. Round-trip losses during energy storage, say into a battery and while taking it out.
2. Transmission loss, which is involved when a WTG without storage is directly delivering energy at a PCC into the grid.

16.11 STORAGE TECHNOLOGIES

Undulations in electric power systems arising at various levels are generally taken care of by controls through use of reserved generation capacities and capacitors. Energy stores can effectively help us in this and spare the reserved generation capacity more economically. Table 16-2 gives different types of storage capacities, duration of the stored output, and level in the EPS where these can be utilized.

16.12 DETERMINING THE SIZE OF STORAGE FOR WIND POWER

The following factors enter into consideration for determining the size of storage:

1. The stores will be location specific.
2. Long-term as well as a daily data on wind velocities and turbulence, on an hourly basis, will have to be obtained.
3. This data will be coupled with the power output of a turbine wind generator (TWG) at this location.
4. The load demand pattern by the EPS, at the point of common coupling will have to be obtained.
5. Data in (3) and (4) above will tell us the possible power flows between the TWG, the proposed stores and the grid, on an hourly basis.

We take data given by Barton and Infield [1] as an illustration. Their data is location specific, the location being Cottesmore in the United Kingdom. They assume a TWG of 1 MW capacity and stores of 300 kW capacities. Figure 16-4 below shows wind speed IEEE spectrum with turbulence added. In this particular case, turbulence IEEE spectrum does not affect the design considerations. He has calculated an hourly spread datasheet for the electricity flows likely to take place between the three system components, namely, the grid, the TWG, and the stores. Table 16-3 gives this spread datasheet.

16.13 CONTROL MODES FOR STORES AND WTG

Barton and Infield [1] list three possible control modes for the stores and TWG:

1. A power curtailment mode. Power available from the TWG at a given instant exceeds the combined load requirement and the storage capacity. This excess power has to be curtailed.

Table 16-2. Storage technologies and applications in the electrical power system

Full power duration of storage	Applications of storage and possible replacement of conventional electricity system controls	Biomass	Hydrogen, electrolysis plus fuel cell	Large hydro	Compressed air energy storage (CAES)	Heat or cold storage plus heat pump	Pumped hydro	Redox flow cells	New and old battery technologies	Flywheel	Super-conducting magnetic energy storage (SMES)	Super-capacitor	Conventional capacitor or inductor
4 Months	Annual smoothing of loads, PV, wind and small hydro	✓	✓	✓									
3 Weeks	Smoothing weather effects: load, PV, wind, small hydro	✓	✓	✓									
3 Days	Weekly smoothing of loads and most weather variations	✓	✓	✓	✓	✓	✓	✓					
8 Hours	Daily load cycle, PV, wind, transmission line repair	✓	✓	✓	✓	✓	✓	✓	✓				
2 Hours	Peak load lopping, standing reserve, wind power smoothing. Minimization of NETA or similar trading penalties	✓	✓	✓	✓	✓	✓	✓	✓				
20 Minutes	Spinning reserve, wind power smoothing clouds on PV		✓	✓	✓	✓	✓	✓	✓	✓			
3 Minutes	Spinning reserve, wind power smoothing of gusts		✓				✓	✓	✓	✓			
20 Seconds	Line or local faults, voltage and frequency control, governor-controlled generation							✓	✓	✓	✓	✓	✓

Source: [1].

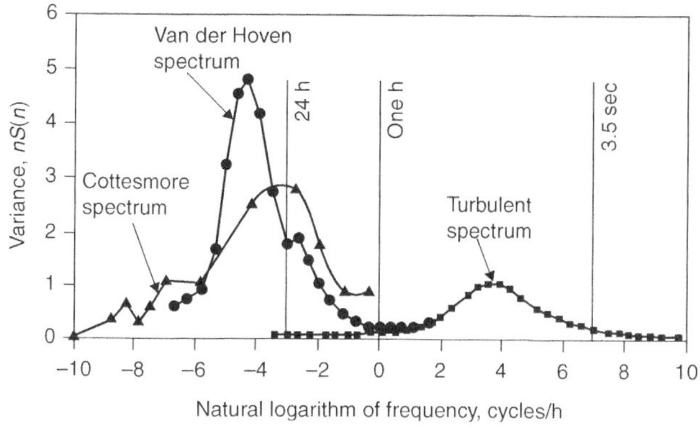

Figure 16-4. Wind speed variation spectra adjusted to 8 m/s mean wind speed and 24% turbulence intensity. (From [1], © IEEE 2004.)

Table 16-3. Illustration of an hourly spreadsheet method

Wind turbine power capacity = 1000 kW	Mean – 1 std. dev.	Mean	Mean + 1 std. dev.	Probability weighted average
Probabilities	0.31	0.38	0.31	
Row 1: U_1 = 4 m/s, probability = 0.43				
$U_1 + U_2$	3.0	4.0	5.0	4.0
Wind power, kW	0	24	67	30
Excess pwer, kW	0	0	0	0
Power stored, kW	0	0	0	0
Extra discharge capacity, kW		300	300	300
Extra power to network, kW				0
Row 2: U_1 = 8 m/s, probability = 0.36				
$U_1 + U_2$	6.0	8.0	10.0	8.0
Wind power, kW	144	360	641	380
Excess power, kW	0	0	39	12
Power stored, kW	0	0	39	12
Extra discharge capacity, kW	300	241	0	185
Extra power to network, kW				11
Row 3: U_1 = 12 m/s, probability = 0.21 $U_1 + U_2$	9.1	12.0	14.9	12.0
Wind power, kW	510	915	1000	816
Excess power, kW	0	313	398	243
Power stored, kW	0	300	300	207
Extra discharge capacity, kW	92	0	0	28
Extra power to network, kW				28

Source: [1].

2. Maximum energy mode. Maximum energy is exported by the store when the EPS capacity to receive energy reaches the upper limit set for overvoltage limitation, by the sum of TWG output and store output at that instant.
3. Power leveling mode. Instead of going by the overvoltage limit, we set as criteria that (a) the grid voltage is held constant and (b) transmission losses in the grid are minimized.

There is a corollary to the above criteria. Let us take a hypothetical contribution of 100 kW by the stores to maintain the above condition 3. The round-trip efficiency of the store is 80%, meaning that it incurs a loss of 8 kW on this supply. The grid is fairly strong and addition of 100 kW reduces its transmission losses by just 1 kW. It is not worthwhile to withdraw costly stores for the grid. On the storage side, it is not operating in a 10 min or 1 hour slab in this type of power leveling mode.

Barton [7] shows that power rating of a store on a 24 hour basis is only about 30% more than that derived on a 10 min or 1 hour basis. He also shows power ratings calculated for 90% benefits come down to 50% of these calculated for 100% benefits. This applies to strong grids; see Table 16-4. On the other hand, he claims that in weak grids, incorporating "energy" storage over a 24 hour basis and energy curtailment [see (1) above] can allow up to three times the wind energy to be absorbed by these grids.

Figure 16-5 illustrates annual energy exported to the grid in different control modes, which were explained earlier. Figure 16-6 illustrates revenue earned. Both these figures illustrate that revenues lost under power curtailments are not very high since curtailments takes place when the load demand is low and electricity prices are also low. On the other hand, power leveling taking place at peak loads with higher electricity prices brings in better revenues, as power is brought in at low price levels and exported at peak levels.

The following points may be noted with respect to the analytical illustration above:

1. The storage capacity at 300 kW is 30% of the rated capacity of 1 MW of the TWG.
2. Benefits may be defined ultimately as the maximum annual revenue returns rather than as the maximum energy exported to the grid. For determining these benefits, spread datasheets on a 24 hour basis with a column added for hourly prices of electricity will be necessary.
3. Costs of a TWG project include costs of:
   - TWG and its installation
   - Shunt capacitors in case of induction generators coupled to the turbine. Generally, induction generators are common.

Table 16-4. Power rating of a store

	10 min or 1 hour	1 day
Power rating for maximum benefit	678 kW	739 kW
Power rating for 90% benefit	300 kW	400 kW

Source: [1].

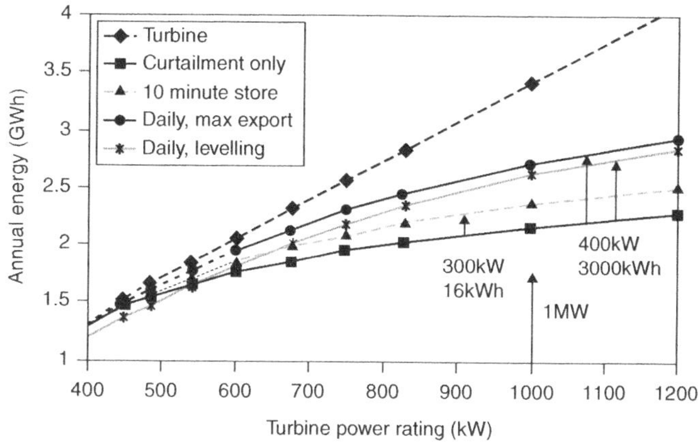

Figure 16-5. Annual energy exported to the grid. (From [1], © IEEE 2004.)

- Transmission systems and controls up to and at the point of common coupling
- Cost of stores, which are considerable. If the stores are storage batteries, the running costs are high.

4. Earning benefits and costs are balanced against environmental benefits offered by a state agency. An example in India is illustrative. Large power consuming industries that invest in wind farms at designated locations are allowed to draw out as much power at their point of consumption as they put in at the point of wind energy generation. They are charged transmission costs. This has promoted a rapid rise of wind farms.
5. There is a rising point of social discontent that cannot be wished away. Agriculturists, on whose lands TWGs are erected, entered long-term contracts for rentals. They could carry on their agriculture normally without hindrance. Now

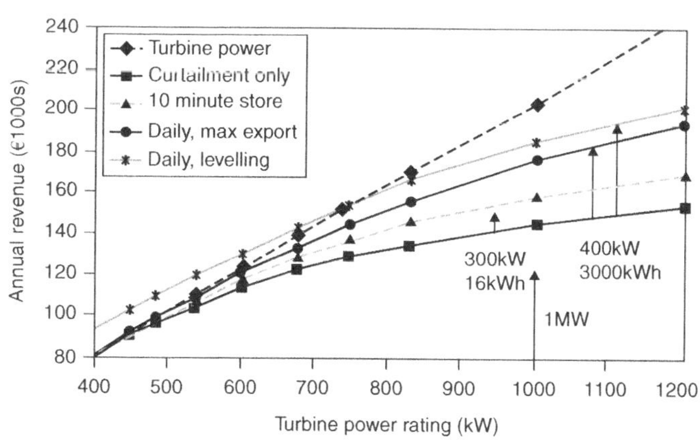

Figure 16-6. Annual revenue earned from electricity. (From [1], © IEEE 2004.)

that they see wealth being created out of wind blowing across their lands, they want a share of this wealth.

16.14 ENERGY RATING OF STORES

The energy rating is derived from spurts of power over different periods on the same basis as we have calculated for loss of energy from PV cells for spurts of blanking out. However, there are differences. Reference energy levels are steady for PV cells whereas they vary for wind energy. Barton and Infield [1] give the following formula for energy rating:

$$\text{Energy capacity in kW-hr} = (\text{power KW} \times \text{time in hours})/\pi$$

Here, time is measured for one cycle and within this time the reservoir fills and empties in a sinusoidal manner.

Continued R and D on different types of energy reserves is changing relative positions of different types of energy storage. Figures 16-7 and 16-8 illustrate this.

Compressed air energy storage is making a strong debut on the energy storage scene. In a small-spot type of reserve, compressing air during lean demand periods and storing it for injection into the starting stage of combined cycle power plants increases their operating efficiencies by something line 40% [3].

On a larger scale, trials are going underway for storing compressed air in large sized caverns and used-up limestone quarries. The up and down efficiencies of compressed air are quite alluring at 70%.

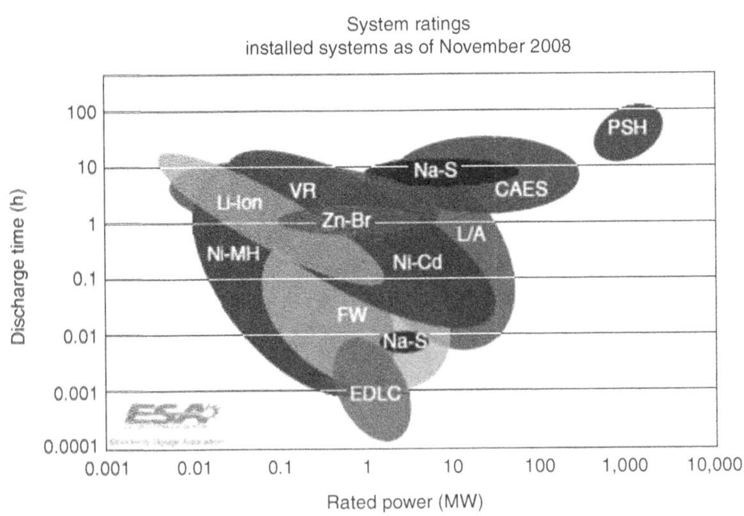

Figure 16-7. Electricity storage by technology. CAES: Compressed Air, EDLC: Double Layer Capacitor, PSH: Pumped Hydro, FW: Flywheel, LA: Lead Acid, Zn-Br: Zinc-Bromide, NaS: Sodium-Sulfur, Li-ion: Lithium Ion. (From [3], © IEEE 2009.)

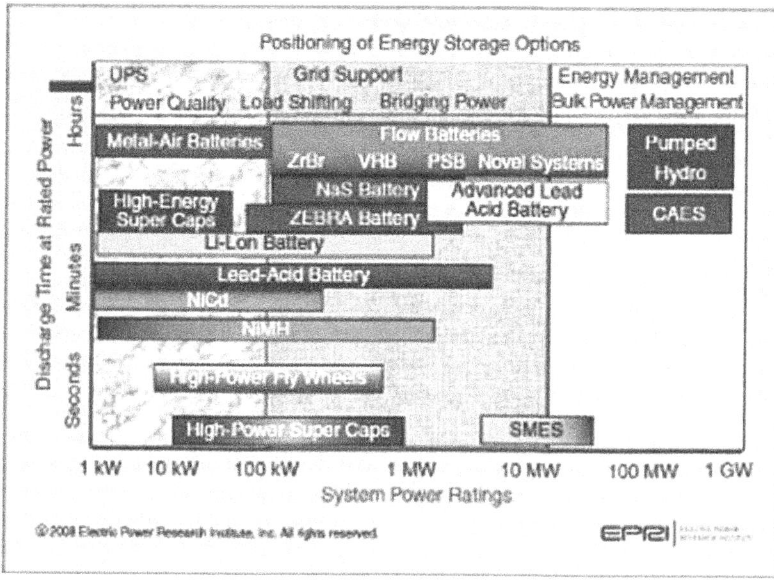

Figure 16-8. Storage technology application comparison. (From [3], © IEEE 2009.)

16.15 CATEGORIES OF ENERGY STORAGES

We can categorize the requirements of energy storage for electricity into three categories:

1. Distributed energy storage compatible with distributed generation by PV cells and isolated wind energy systems. This is currently served by storage batteries. The field of research has shifted to storage batteries of higher round-trip efficiencies and to lowering their costs. Hydrogen stores, when they become viable, will be a great help
2. Medium-sized localized energy stores to take care of medium-sized wind farms and also to support distribution systems. Small localized pumped-water storage could be viable here.
3. Large sized centralized energy storage to help the grid even out demand on generation and ease out electricity prices against longer life on existing generators and deferred investments. Ideally, this calls for long-term planned, large pumped stores. Hydrogen stores will help to absorb, an otherwise wasted wind energy. They might support automobiles also. Their entry into large-scale electricity production will have to be researched both technically and financially. As it is, use of hydrogen directly in automobile engines requires engine modifications [4].

Sometime in future, one may expect breakthroughs for scalable energy storage, either via chemistry or via physics, which will be epoch-making.

APPENDIX 16-1 A SUPERCAPACITOR

In an ordinary capacitor, there is a dielectric medium that consists of layers of polypropylene film between two (conducting) aluminum films. The capacitance is given by

$$C = (\text{DK} \times \text{area})/\text{thickness of dielectric}$$

where DK, the dielectric constant, is a property of the material.

Typically a 400 KVAR, 14 kV, 1 Ph, 50 Hz capacitor has a capacitance of 6.50 microfarads. In a supercapacitor, the capacitance is obtained due to an ionic charge accumulating just over the surface of an electrode. Activated and porous carbon, used as an electrode, has numerous crevices on its surface, giving huge areas over which this charge builds up. One gram of this material has as much as 1000 m^2 of active area. Again, the thickness over which this charge builds up is in electrolytic molecular dimensions, at as low as 10^{-10} m. Together, this gives a very large capacitance for a small volume and small weight.

This phenomenon exists also in ordinary storage batteries, which have the highest capacitance value available across their surface. The difficulty is in positioning a collection plate at a distance of 10^{-10} m from the battery plate.

In a supercapacitor, a paste of activated and porous carbon is applied on both sides of an aluminum film, which acts as a collector. The carbon acts as an electrode, which, when charged, has an ionic cloud around it. Strips of conductors are attached to the aluminum for taking out the current. There is a porous separating plate between two such pasted aluminum collectors. These are then wound in a coil. The wound coil is impregnated with an electrolyte of high electronic activity.

The very small distance across the two faces of a charge precludes applying voltages higher than, say 3 V, for a given electrolyte. This gives us a high-capacitance, low-voltage capacitor cell. Many such cells are connected in parallel to form a group and the groups are balanced and connected in series to obtain the desired capacitance and voltage.

Note that just opposite, say, a positive charge on one carbon paste, occurs if there is a negative charge across the 10^{-10} m distance. A similar negative charge develops across the carbon paste on the other aluminum collector, with a balancing positive charge at 10^{-10} m distance. When we connect two aluminum plates to a DC source, the capacitance, obtained is a result of two capacitances in series across the electrolyte. Hence these capacitors are also called double layered capacitors.

Typical values of such supercapacitors are:

1. 10 F × 2.9 V for each element. Equivalent series resistance (ESR) 60 mΩ. 12 elements in series form an experimental unit. Array voltage 35.2 V, 0.833 F, and ESR = 720 mΩ
2. 1800 F × 2.5 V unit. Three units in parallel per group × 40 groups in series form an array. Array design voltage of 40 × 2.9 V is scaled down to 100 V, operating

voltage. Capacitance = 1800 × 3/80 = 135 F (compare this against 6.50×10^{-6} F for large power capacitors). Stored energy = ½ CV^2 = ½ × 135 × $100^2 \times 10^{-3}$ kJ = 675 kJ

For optimally utilizing this array, we have to remember its internal resistance, ESR. An exponential charging/discharging will result in high current through its ESR, higher losses, and lower efficiencies. Discharging it across a constant load (preferably with a resistance equal to the ESR of the array) gives high storage efficiencies; see Figure 16-9.

This is best obtained with a DC–DC converter as an interface to the supercapacitors. Here again we have to consider the switching losses in the converter as well. One of the methods reported is as follows.

On the DC side (supply-to-load side) a voltage source inverter is used to convert DC into AC. While converting, the commutation takes place at the instant the voltage passes though zero. This involves minimum switching losses. While charging the supercapacitor or returning the energy, a zero-current switching converter, operating at medium frequency, is used. Special thyristors called IGBTs (integrated gate bipolar transistors) have very little stored energy at the switching instant. Together, this combination gives a high performance efficiency. Figure 16-10 gives a schematic for this arrangement.

It may be noted that the current-sourced inverter operates at a medium frequency and the voltage gets transformed through a MF transformer as shown. The above arrangement was used in an experimental elevator with the following details [5]:

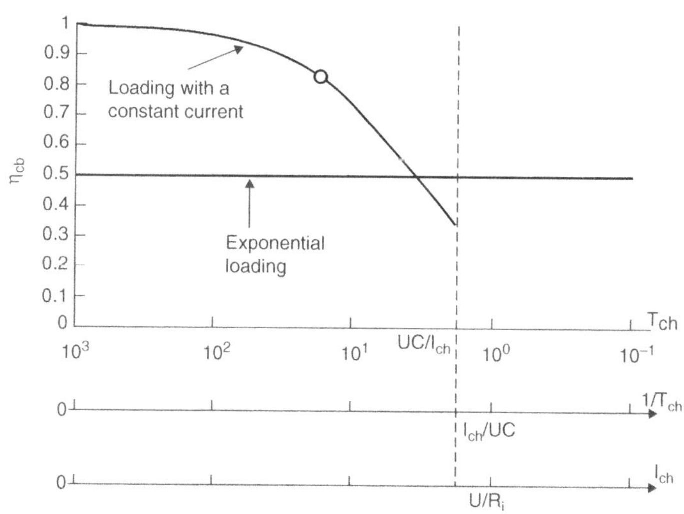

Figure 16-9. Efficiency by exponential and constant current charging. (From [5], © IEEE 2002.)

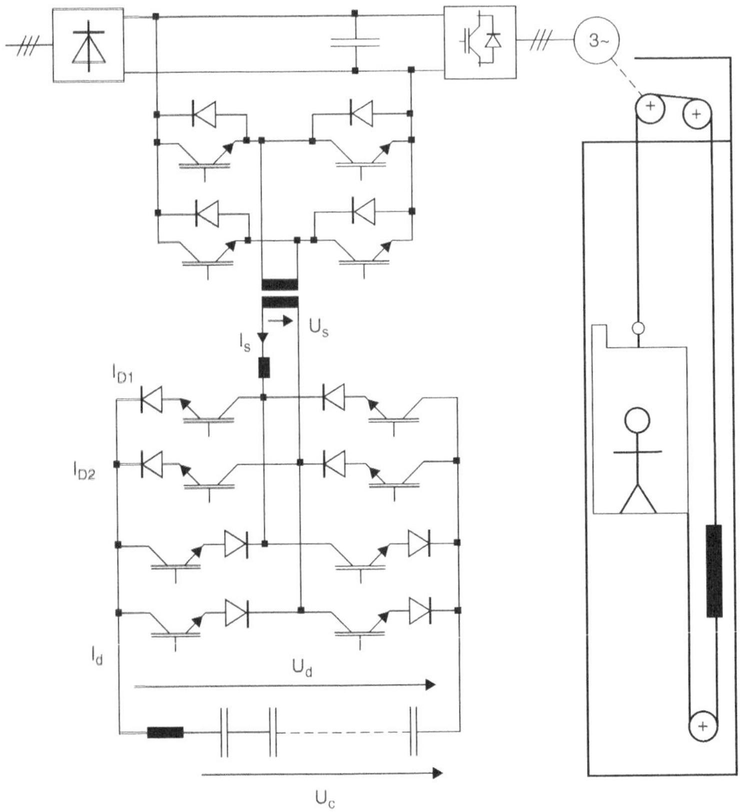

Figure 16-10. General scheme of the elevator drive with supercapacitor storage. (From [5], © IEEE 2002.)

Car weight of elevator—720 kg

Balancing counterweight—1440 kg

Run up energy required—220 kJ or 62 W-hr

The maximum power demand is approximately 33 kW without an energy store.

With supercapacitors as energy storage, the voltage during run-up resulted in a voltage drop down to 82 V on capacitors and power demand on the supply side was only 2.5 kWs. Figure 16-11 shows the experimental results.

There is a great potential for reducing power demand through use of supercapacitors on fluctuating loads. It is expected that electrically driven vehicles of the hybrid type can replace fuel-driven cars with further developments in energy stores. Presently, supercapacitors are used on a large scale as a backup for memory.

Research is going on in using carbon nanotubes as energy storage materials. These tubes are arranged at right angles to the electrolytic surface. These tubes have designed

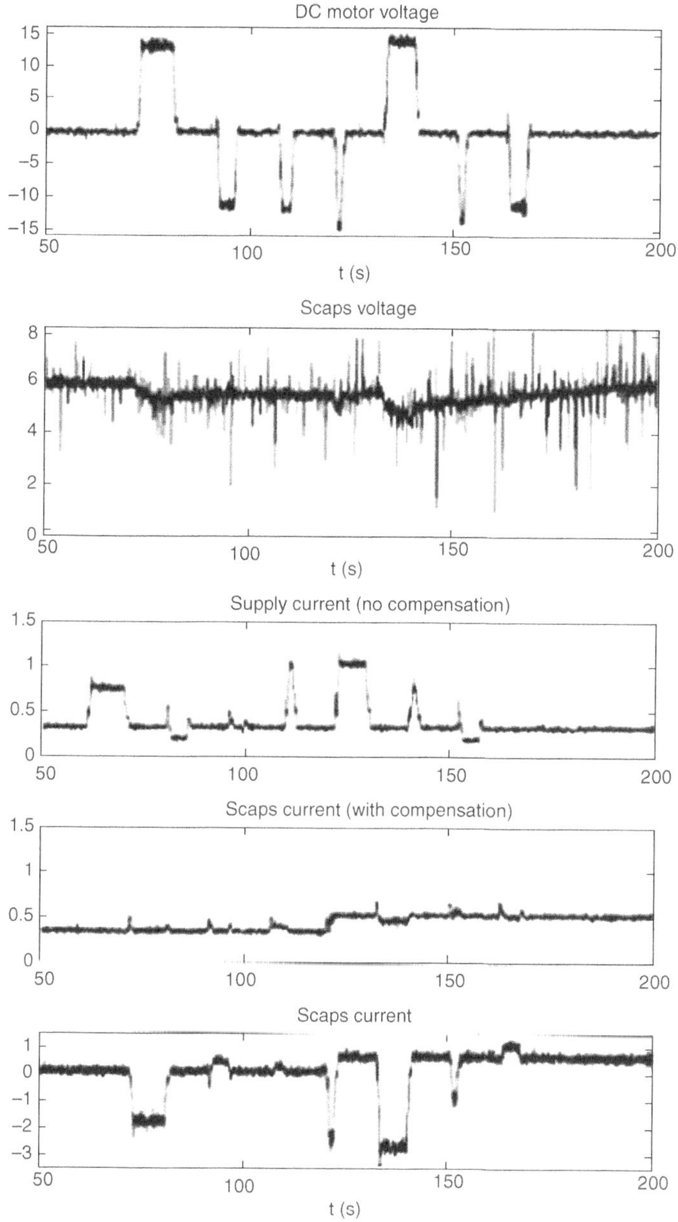

Figure 16-11. Experimental results with/without compensation with supercapacitors. (From [5], © IEEE 2002.)

separation in between and a geometry so that they build up maximum capacitance aimed to be 100 times larger than the present figure of 1500 F/g obtained with activated carbon paste.

Needless to say, these developments will give a big boost to penetration of DGs into power systems and to changing over to renewable energy sources from fossil fuels. They point to a path to the future.

REFERENCES

1. J. P. Barton and D. G. Infield, Energy Storage and its Use with Intermittent Renewable Energy, *IEEE Transactions on Energy Conversion,* Volume 19, Issue 2, June 2004, pp. 441–448.
2. A. Woyte, V. Van Thong, R. Belmans, and J. Nijs, Voltage Fluctuations on Distribution Level Introduced by Photovoltaic Systems, *IEEE Transactions on Energy Conversion,* March 2006, 202–209.
3. B. Roberts, Capturing Grid Power, *IEEE Transactions on Power and Energy,* Volume 7, Issue 4, July–August 2009,pp. 32–41.
4. S. R. Sing, Auto Monitor, Volume 9, No. 14 (Indian Magazine) IOC Ready to supply hydrogen as fuel for automobiles.
5. E. J. van Dijk, J. A. Ferreira, and P. van Gelder, Current Sharing Between Diode Clamps of Polar Capacitor Bank, *IEEE Transactions on Industry Applications,* Volume 38, Issue 6, Nov.-December 2002, pp. 1600–1605.

BIBLIOGRAPHY

Aeloiza, E. C., P. N. Enjeti, L. A. Moran, O.C. Montero-Hernandez, and S. Kim, Analysis and Design of a New Voltage Sag Compensator for Critical Loads in Electrical Power Distribution Systems, *IEEE Transactions on Industry Applications,* Volume 39, Issue 4, July–August 2003, pp. 1143–1150.

Bagen, I. and R. Billinton, Incorporating Well-Being Considerations in Generating Systems Using Energy Storage, *IEEE Transactions on Energy Conversion,* Volume 20, Issue 1, Mar 2005, pp. 225–230.

Bagen, I. and R. Billinton, Incorporating Well-Being Considerations in Generating Systems Using Energy Storage, *IEEE Transactions on Energy Conversion,* Volume 20, Issue 1, March 2005, pp. 225–230.

Ball, G. J., G. Corey, and B. L. Norris, Government, Industry, and Utility Development and Evaluation of a Modular Utility Battery Energy Storage System, *IEEE Transactions on Energy Conversion,* Volume 10, Issue 3, September 1995, pp. 549–554.

Banerjee, S., J. K. Chatterjee, and S. C. Tripathy, Application of Magnetic Energy Storage Unit as Continuous Var Controller, *IEEE Transactions on Energy Conversion,* Volume 5, Issue 1, March 1990, pp. 39–45.

Borowy, B. S. and Z. M. Salameh, Dynamic Response of a Stand-Alone Wind Energy Conversion System with Battery Energy Storage to a Wind Gust, *IEEE Transactions on Energy Conversion,* Volume 12, Issue 1, March 1997, pp. 73–78.

Choi, J.-H. and J.-C. Kim, Advanced Voltage Regulation Method at the Power Distribution Systems Interconnected with Dispersed Storage and Generation Systems, *IEEE Transactions on Power Delivery,* Volume 15, Issue 2, April 2000, pp. 691–696.

Coles, L. R., S. W. Chapel, and J. J. Iamucci, Valuation of Modular Generation, Storage, and Targeted Demand-Side Management, *IEEE Transactions on Energy Conversion,* Volume 10, Issue 1, March 1995, pp. 182–187.

Figueiredo, F. C. and P. C. Flynn, Using Diurnal Power Price to Configure Pumped Storage, *IEEE Transactions on Energy Conversion,* Volume 21, Issue 3, Sept. 2006, pp. 804–809.

Fioravanti, R., K. Vu and W. Stadlin, Large-Scale Solutions, *IEEE Transactions on Power and Energy,* Volume 7, Issue 4, July–August 2009,pp. 48–57.

Gertmar, L., L. Liljestrand, and H. Lendenmann, Wind Energy Powers-That-Be Successor Generation in Globalization, *IEEE Transactions on Energy Conversion,* Volume 22, Issue 1, March 2007, pp. 13–28.

Hopkins, D. C., C. R. Mosling, and S. T. Hung, Dynamic Equalization During Charging of Serial Energy Storage Elements, *IEEE Transactions on Industry Applications,* Volume 29, Issue 2, March–April 1993, pp. 363–368.

Kenny, B. H., P. E. Kascak, R. Jansen, T. Dever, and W. Santiago, Control of a High-Speed Flywheel System for Energy Storage in Space Applications, *IEEE Transactions on Industry Applications,* Volume 41, Issue 4, July–August 2005, pp. 1029–1038.

Kinjo, T., T. Senjyu, N. Urasaki, and H. Fujita, Output Levelling of Renewable Energy by Electric Double-Layer Capacitor Applied for Energy Storage System, *IEEE Transactions on Energy Conversion,* Volume 21, Issue 1, March 2006, pp. 221–227.

Korpas, M. and A. T. Holen, Operation Planning of Hydrogen Storage Connected to Wind Power Operating in a Power Market, *IEEE Transactions on Energy Conversion,* Volume 21, Issue 3, Sept. 2006, pp. 742–749.

Lee, T.-Y. and N. Chen, The Effect of Pumped Storage and Battery Energy Storage Systems

Lewis, R. J., Interfaces are the Dominant Feature of Dielectrics at the Nanometric Level, *IEEE Transactions on Dielectrics and Electrical Insulation,* October 2004, pp. 739–753.

Linzen, D., S. Buller, E. Karden, and R. W. De Doncker, Analysis and Evaluation of Charge-Balancing Circuits on Performance, Reliability, and Lifetime of Supercapacitor Systems, *IEEE Transactions on Industry Applications,* Volume 41, Issue 5, September–October 2005, pp. 1135–1141.

Malesani, L., L. Rossetto, P. Tenti, and P. Tomasin, AC/DC/AC PWM Converter with Reduced Energy Storage in the DC Link, *IEEE Transactions on Industry Applications,* Volume 31, Issue 2, March–April 1995, pp. 287–292.

Monai, T., I. Takano, H. Nishikawa, and Y. Sawada, A Collaborative Operation Method Between New Energy-Type Dispersed Power Supply and EDLC, *IEEE Transactions on Energy Conversion,* Volume 19, Issue 3, September 2004, pp. 590–598.

Nourai, A. and C. Schafer, Changing the Electricity Game, *IEEE Transactions on Power and Energy,* Volume 7, Issue 4, July–August 2009,pp. 42–47.

On Hydrothermal Generation Coordination, *IEEE Transactions on Energy Conversion,* Volume 7, Issue 4, December 1992, pp. 631–637.

Poonpun, P. and W. T. Jewell, Analysis of the Cost per Kilowatt Hour to Store Electricity, *IEEE Transactions on Energy Conversion,* Volume 23, Issue 2, June 2008, pp. 529–534.

Rahman, M. D. H. and S. Yamashiro, Novel Distributed Power Generating System of PVECaSS Using Solar Energy Estimation, *IEEE Transactions on Energy Conversion,* Volume 22, Issue 2, June 2007, pp. 358-367

Rao, P., M. L. Crow, and Z. Yang, STATCOM Control for Power System Voltage Control Applications, *IEEE Transactions on Power Delivery*, Volume 15, Issue 4, October 2000, pp. 1311–1317.

Rau, N. S. and B. Tayor, A Central Inventory of Storage and Other Technologies to Defer Distribution Upgrades-Optimization and Economics, *IEEE Transactions on Power Delivery*, Volume 13, Issue 1, January 1998, pp. 194–202.

Rau, N. S. and W. D. Short, Opportunities for the Integration of Intermittent Renewable Resources into Networks Using Existing Storage, *IEEE Transactions on Energy Conversion*, Volume 11, Issue 1, March 1996, pp. 181–187.

Rufer, A. and P. Barrade, A Supercapacitor-Based Energy-Storage System for Elevators with Soft Commutated Interface, *IEEE Transactions on Industry Applications*, Volume 38, Issue 5, September–October 2002, pp. 1151–1159.

Rufer, A., D. Hotellier, and P. Barrade, A Supercapacitor-Based Energy Storage Substation for Voltage Compensation in Weak Transportation Networks, *IEEE Transactions on Power Delivery*, Volume 19, Issue 2, April 2004, pp. 629–636.

Schoenung, S. M. and C. Burns, Utility Energy Storage Applications Studies, *IEEE Transactions on Energy Conversion*, Volume 11, Issue 3, September 1996, pp. 658–665.

Schoenung, S. M., W. R. Meier, and R. L. Bieri, Small SMES Technology and Cost Reduction Estimates, *IEEE Transactions on Energy Conversion*, Volume 9, Issue 2, June 1994, pp. 231–237.

Spyker, R. L. and R. M. Nelms, Optimization of Double-Layer Capacitor Arrays, *IEEE Transactions on Industry Applications*, Volume 36, Issue 1, January–February 2000, pp. 194–198.

Swider, D. J., Compressed Air Energy Storage in an Electricity System With Significant Wind Power Generation, *IEEE Transactions on Energy Conversion*, Volume 22, Issue 1, March 2007, pp. 95–102.

Thomas, R. J., Putting an Action Plan in Place, *IEEE Transactions on Power and Energy* ,Volume 7, Issue 4, July–August 2009, pp. 26–31.

Uzunoglu, M. and M. S. Alam, Dynamic Modeling, Design, and Simulation of a Combined PEM Fuel Cell and Ultracapacitor System for Stand-Alone Residential Applications, *IEEE Transactions on Energy Conversion*, September 2006, 767–775.

Vilathgamuwa, D. M., H. M. Wijekoon, and S. S. Choi, Interline Dynamic Voltage Restorer: A Novel and Economical Approach for Multiline Power Quality Compensation, *IEEE Transactions on Industry Applications*, Volume 40, Issue 6, November–December 2004, pp. 1678–1685.

Weissbach, R. S., G. G. Karady, and R. G. Farmer, A Combined Uninterruptible Power Supply and Dynamic Voltage Compensator Using a Flywheel Energy Storage System, *IEEE Transactions on Power Delivery*, Volume 16, Issue 2, April 2001, pp. 265–270.

Weissbach, R. S., G. G. Karady, and R. G. Farmer, Dynamic Voltage Compensation on Distribution Feeders Using Flywheel Energy Storage, *IEEE Transactions on Power Delivery*, Volume 14, Issue 2, April 1999, pp. 465–471.

Zubieta, L. and R. Bonert, Characterization of Double-Layer Capacitors for Power Electronics Applications, *IEEE Transactions on Industry Applications*, Tr-IA, January 2000, pp. 199–205.

17

HYDROGEN ERA

17.1 FOSSIL-BASED FUELS

There are three reasons to seek replacements for fossil fuels. First, fossil fuels, which are the major energy providers today, have limited proven reserves on our planet. At the increasing rate at which we are consuming these, we might run into a shortage not too far off in the future. We should find a replacement for them before that day of reckoning. Second, the fossil fuel reserves are concentrated geographically into a few locations, the governments of which can control the supply as well as the prices. Third, there is rising concern about the contamination of our environments. The more we use the fossil fuels, the major contaminants (Table 17-1), the greater is the damage. Strict discipline is being planned and exercised to prevent this. Hydrogen as a fuel does not contaminate.

17.2 HYDROGEN PROPERTIES

Hydrogen is intrinsically a better provider of energy than the best of fossil fuel derivatives (Table 17-2). One kg of hydrogen (H_2) gives 2.8 times the energy given by 1 kg of gasoline. Put differently, hydrogen yields 12 million BTU/$, whereas liquefied nat-

Electricity Power Generation. By Digambar M. Tagare

Copyright © 2011 the Institute of Electrical and Electronics Engineers, Inc.

Table 17-1. Origin and extent of atmospheric pollution

	Long-lived greenhouse gases				Short-lived air pollutions				
Type of pollutant	CO_2	CH4	N_2O	CFCs	CO	NMHC	NO_2	SO_2	Soot
Sources in 1990s (teragrams per year of carbon, nitrogen, or sulphur oxide)									
Industrial and fossil-fuel related	5500	125	1.3	0.65 (1990)	125	120	25	70	7
Tropical agricultural and biomass burning	1600	275	3	—	400	70	10	2	6
Volume mixing ratios									
Preindustrial (about 1850)	280 ppm	750 ppb	270 ppb	0	Unknown				
Present (1990s)	360 ppm	1730 ppb	310 ppb	0.8 ppb	50–500 ppb	2–10 ppb	0.05–5 ppb	0.05–5 Ppb	1–103 ngC/m^3
Lifetime and rates of increase									
Atmospheric lifetime	50-200 years	8 years	120 years	50–100 years	2–3 months	Hours to weeks	1–2 days	A few days	Days to weeks
Annual rate of increase in 1990s	0.5%	0.5%	0.2%	0.5%	Regionally variable				

Source: [1].

ural gas yields up to 4 million BTU/$. In terms of specific energy, H_2 has a specific energy of 150 MJ/kg versus 50 MJ/kg for gasoline.

17.3 HYDROGEN ADVANTAGES

Hydrogen has a very fast and controllable response to energy conversion versus fossil fuels or renewable energy sources. It becomes an ideal fuel for use in automotive transport. It also becomes an ideal energy source to take care of transients and impulses in electrical supply systems.

Hydrogen has a tremendous potential for generating electrical energy. Electrical energy must be utilized at the instant when it is generated. An excess of electrical energy that cannot be used at that instant can be converted into H_2 energy and stored. Hydrogen energy, in turn, can be reconverted when the demand for electric supply exceeds the supply.

Hydrogen with the help of fuel cells, is directly and instantly convertible into electrical energy. The mechanical means of energy conversion of today, such as reciprocating gasoline and diesel engines, steam engines, and rotating machinery such as turbines and generators, are both capital intensive and highly inefficient as energy converters. Hydrogen promises to do away with most of these mechanical means.

Table 17-2. Properties of hydrogen

Energy Content for 1 kg (2.2 lb) of hydrogen = 424 standard cubic feet (Reacting with oxygen to form water)	
Higher heating value	Lower heating value
134,200 BTU	113,400 BTU
39.3 kW-hr	33.2 kW-h
141,600 kJ	119,600 rkJ
33,800 kCal	28,560 kCal

English units			Normal boiling point (1 atm)		Gas phase properties @ 32°F and @ 1 atm		
Substance	Chemical symbol	Mol. Weight	Temperature, °F	Latent heat of vaporization, BTU/lb	Specific gravity, air = 1	Specific heat (Cp), BTU/lb, °F	Density, ft³
Hydrogen	H_2	2.02	–423	191.7	0.06998	3.425	0.005611

English Units			Liquid phase properties @ BP and @ 1 atm		Triple point		Critical point		
Substance	Chemical symbol	Mol. weight	Specific gravity, = 1	Specific heat (Cp), water °F	Temperature BTU/lb °F	Pressure, psia	Temperature, °F	Pressure, psia	Density, ft³
Hydrogen	H_2	2.02	0.071	2.309	–434.6	1.045	–399.93	190.8	1.88

Metric Units			Boiling point @ 101.325 kPa		Gas phase properties @ 0° C and @ 101.325 kPa			Liquid phase properties @ B.P. and @ 101.325 kPa	
Substance	Chemical symbol	Mol. weight	Temperature, °C	Latent heat of vaporization kJ/kg	Specific gravity, air = 1	Specific heat (Cp), kJ/kg, °C	Density, kg/m³	Specific gravity, water = 1	Specific heat (Cp), kJ/kg °C
Hydrogen	H_2	2.02	–252.8	446.0	0.06998	14.34	0.08988	0.071	9.668

Metric units			Triple point		Critical point		
Substance	Chemical symbol	Mol. weight	Temperature, °C	Pressure, kPa abs	Temperature, °C	Pressure, kPa abs	Density, kg/m³
Hydrogen	H_2	2.02	–259.2	7.205	–239.96	1315	30.12

Source: Universal Industrial Gases, Inc., http://www.uigi.com/hydrogen.html.

17.4 PRODUCTION OF HYDROGEN

17.4.1 Presently Developed Processes for Production of H_2

Electrolysis of Water. This is the simplest method and can be adopted to suit any scale of production. The drawbacks are:

- Cost of production is high
- Use of an electrolyte which is environmentally harmful
- Electrodes are costly

Earlier, it was mentioned that intermittent energies from renewable resources can be gainfully utilized via the hydrogen route, for example, via PV fuel cells, windmill fuel cells, and grid-supporting fuel cells in which hydrogen acts as a convenient energy storage media compared to storage batteries.

Chemical Plants. There are large-scale chloralkali plants that split NaCl under electrolysis. Hydrogen is a generated as a by-product and is rather a nuisance.

Petroleum Refineries and LNG. LNG is a major source of methane. Methane, when acted upon by steam, yields a good supply of hydrogen under a process called steam methane reforming (SMR). This process is used in the manufacture of fertilizers using natural gas. Natural gas contains 60% methane. There are two sources of methane presently being worked upon on a production basis: LNG from gas fields and from refineries. It is expected that H_2 production on a commercial scale will be coordinated with oil refineries.

Methane from City Wastes. Methane from anaerobic digestion of biomass. The Port Authority of New York has reported installation of eight waste disposal stations in which methane is generated from city solid wastes. Together, these stations produce 1.8 MW of electric power [6].

Methane from Agricultural Products. A chain of sugar factories in India actually produce methane as a biofuel for mixing with gasoline and diesel. India has many opportunities to utilize hydrogen (see Appendix 15-1).

See Table 17-3 for a summary of processes.

17.4.2 Processes under Development for Bulk Production of H_2—Coal Gasification

Supplies of coal are vastly spread all over the world. Presently, they are the main sources for energy. However the environmental contamination mainly by CO_2, SO_2, NO_x, and CO_x, is not acceptable. Coal washeries remove some of the objectionable matter. Coal gasification is being seriously experimented upon. It yields methane.

Table 17-3. Costs and performance characteristics of various hydrogen production processes

Hydrogen production process	Energy required ($kW\text{-}hr/Nm^3$ of H_2) Ideal	Practical	Status of technology	Efficiency (%)	Costs relative to SMR	% of total production	Need for CO_2 sequestration
Steam methane reforming	0.78	2–2.5	Mature	70–80	1	48	Y
Methane/NG pyrolysis			RandD to mature	72–54	0.9		N
H_2S methane reforming	1.5	—	RandD	50	<1	—	N
Landfill gas dry reformation			RandD	47–58	~1	—	Y
Partial oxidation of heavy oil	0.94	4.9	Mature	70	1.8	30	Y
Naphta reforming			Mature				
Steam reforming of waste oils			RandD	75	<1	—	Y
Coal gasification (Texaco)	1.01	8.6	Mature	60	1.4–2.6	—	Y
Partial oxidation of coal			Mature	55		18	Y
Steam–iron process			RandD	46	1.9		Y
Chloralkali electrolysis			Mature		by-product	4	N
Grid electrolysis of water	3.54	4.9	RandD	27	3–10		Y
Solar and PV electrolysis of water			RandD to mature	10	>3		N
High-temperature electrolysis of water			RandD	48	2.2		N
Thermochemical water splitting cycles			Early RandD	35–45	6		N
Biomass gasification			RandD	45–50	2.0–2.4	—	N
Photobiological			Early RandD	<1			N
Photolysis of water			Early RandD	<10			N
Photoelectrochemical decomposition of water			Early RandD				N
Photocatalytic decomposition of water			Early RandD				N

Source: [2].

17.4.3 Processes under Laboratory/Scientific Exploration—Thermochemical Water Splitting

When steam is heated above 800–850°C, the H_2O molecule splits into H_2 and O. With a high-temperature membrane, H_2 can be separated. Actually, this amounts to running a fuel cell in reverse [3]. However, if we spend much more energy in splitting the water molecule than we get back by recombining it, where is the merit in this exercise?

We expect to fall back on thermonuclear power as a main source for energy. Presently, the route of utilizing thermonuclear energy is via steam-driven mechanical–electrical processes. H_2 generation and conversion into electrical energy process can be made more economic.

It has some merits. Thermonuclear energy does not contaminate the environment and will be available profusely.

17.5 POTENTIAL MARKET SEGMENTS FOR HYDROGEN

The automobile market with its electrical vehicles is the first and fastest growing market. Presently, of all the energy uses in the world, use by automobiles account for 42%. This market is vast.

Next, we can shift our vision to residential consumers, who presently use gas cylinders (LNG) on a large scale. A switch over to H_2 cylinders in place of LNG could be swift. Home appliances have created a dominant market for electrical energy.

The biggest uses of energy are presently electrical utilities. This market is undergoing convulsions, as we have seen in earlier chapters. In the vision of a hydrogen era, we can see this market switching over to distributed generation; it could be based on renewable resources or it could be direct-hydrogen based.

17.6 PRESENT ROADBLOCKS TO USE OF HYDROGEN

17.6.1 Costs of H_2 Are High

The cost of H_2 must be brought down to the level or slightly below the cost of gasoline, for use in automobiles. Generating H_2 by electrolysis costs \$7 to \$8 per kg. Generating it from LNG is expected to bring this cost to between \$4 and \$5 per kg. As against this, a liter of gasoline today costs slightly over \$1. One has to add the costs of well-head-to-consumer transport, storage, distribution, marketing, and so on to this. We have a long way to go and we come back to our basic “chicken first or egg first” problem, with nil production of H_2 as of today. Large-scale H_2 is not available today, so there is no infrastructure to distribute it. Since there is no infrastructure, H_2 is not produced on a large scale.

17.6.2 Basic Infrastructure Does Not Exist

The basic tools—technical means for production, storage, transport, delivery, and consumption—have to be researched on. These must be field tested. Standards and codes will have to be developed and brought into use. All this will be a battle against huge established spreads in the well-established, existing petroleum-product-based supply systems.

17.6.3 Petroleum Products Are Well Established

Petroleum products involve extensive investments in storage, pipelines, depots, service stations, pumps, marketing outfits, and so on. These employ millions of people in various categories. Can an upstart like hydrogen compete with and replace them in the foreseeable future? As one pessimist put, "It is an idea whose time will never come." Not all people agree. Almost all the important government and industry bodies in many countries believe in the hydrogen vision. What happens then to the present electrical industries in operating, transmitting, and distributing?

17.7 GOVERNMENTS ENVISION A HYDROGEN ERA

The government of India has instituted a hydrogen board under the Ministry of Energy. The board has drawn a road map for development of hydrogen production and use. This road map is similar to that briefly sketched in Appendix 17-1.

17.8 AN EXAMPLE TO CONSIDER

I would like to present an example of an energy meter manufacturing organization in India. They were producing millions of electromechanical measuring meters, mainly single phase, per year. Over a period of 30 years, they had fine tuned their production and quality control splendidly. They had production bases spread all over the country and employed thousands of people.

Then came the IC chips, competing just in some basic parameter readings to begin with. Within a span of 3–4 years, the ICs advanced to a stage where they could read out, store, and transmit data in about 20–25 parameters. The meters with ICs were tamper-proof. They replaced the old meters en-block.

The old organization sold out all their presses, dies, tools, jigs, and stocks of their springs and jeweled bearings as scrap and took to IC-based meters. They retrenched the old staff and closed some of the old manufacturing centers.

With scores of talented brains backed by national and international organizations such as this one, one may expect the hydrogen era to come any time, and none too soon.

APPENDIX 17-1 PROCEEDINGS OF THE NATIONAL HYDROGEN ENERGY ROAD MAP WORKSHOP ARRANGED BY U.S. DOE

Basic purposes—Breakouts required

Barriers—What are the scientific, engineering, environmental, institutional, economic, and market barriers (to . . . .)

Needs—What are the most important needs including research and development, demonstrations, analysis, policy, codes and standards, to address the barriers.

Top needs and entities—What are the top needs and which entities will address these needs?

Next steps—What are the top priority needs, including their descriptions, key milestones, dates, primary funding entities, and next steps?

Workshop sessions:

1. Hydrogen production breakout session
   Hydrogen delivery breakout session
   Hydrogen storage breakout session
   Hydrogen energy conversion breakout session
   Hydrogen application breakout session
   Hydrogen outreach and educational breakout session
   Hydrogen system integration breakout session
   Striking points in various sessions
2. Hydrogen production breakout
   Limited available of source material. Renewables, few. Nonrenewable, plenty.
   Safe sites for H_2 production.
   Codes and standards to be developed and finished.
   Carbon sequestration and self-disposal problem.
   Catalysts, membranes, refractory, adsorbents, and so on to be further evolved.
   Low efficiencies and high cost of production of H_2.
   Seawater use (extensively) problem to be solved. Mass desalinization methods.
   Nuclear off-peak thermochemical process development.
   Coal gasification to be concentrated upon.
   Integration with refinery process.
   Advanced research into utilizing nanotechnology.
   Develop lower costs and longer life materials for FCs as well as for semiconductors.
   Genetic engineering of biological-based photolytic processes.
   Membrane for present off-peak nuclear application. 900°C electrolyte also.
   Eliminate precious metals.
   Develop H_2 detection technology.
   Couple with cogeneration.
   New methods for low concentration of H_2 capture.
   Validation of performance.
   Establish venues for government and industries to work together.
   Establish academic programs to train next-generation researchers. We have the vision. We need the mission.
   Work on biomass feedstocks. Identify organisms for doing conversion.
   Distributed production using small-scale reformers.
   Reduce costs, improve reliability and safety, develop codes and standards for this, fuel flexibility, materials development, system integration (with existing systems).

3. Hydrogen delivery breakouts (required)

   Code and permits for

   a) Access to roads
   b) Safety perception
   c) H_2 venting during transportation

   National and international harmonization on codes and standards.

   Costs to liquefy H_2. Costs to store and costs to transport.

   Trailer—weight and capacity at present. Compressed H_2 has low energy density, poor economics for transport over long distance.

   Dispensing to customers—costs, safety, performance.

   Filter for delivery to fuel cells.

   ISO/SAE codes development (drafts) for tanks, connectors, refilling stations.

   Qualifying the vision.

   Develop high-capacity, light-weight, low-temperature, fast-charging/discharging hydrides for transposition.

4. Hydrogen storage breakouts (required)

   How to prevent H_2 embritlement? Stress corrosion and cracking?

   Lifetime of existing metal hydrides.

   Storage media for high-density, reversible, low-temperature H_2 and CH_4.

   Inadequate volumetric density in nanotubes.

   Safety aspects of storage materials plus H_2.

   OEM and Infrastructure must support high pressures—350 bar and 700 bar.

   Define "fast charging."

   Standards for storage of H_2 in metal hydrides.

   Education on cryogenic characteristics of gaseous energy systems.

   "Well-head-to-wheels" type energy analysis for H_2 systems.

   Stability of hydrides under cyclic loading.

   Kinetics of large H_2 uptake and discharge.

5. Hydrogen energy conversion breakouts

   Unsolved sealing, joining and interconnecting materials in SOFC.

   Safety implications of H_2 burning.

   All fuel cells are

   a) Costly
   b) Durability and reliability in performance unproved
   c) Safety issues to be addressed
   d) Conversion issues in MCFC
   e) Performance under different environments and geographic locations to be laid down

Flames in turbines

a) Emission
b) Efficiency
c) Safety
d) Flame management in turbine

Business model for widespread distributed energy market.

Lack of confidence by financial markets. No "patient" capital available. Tax credits not enough.

Software tools to simulate vehicle collision for H_2 fuels. H_2 vehicles safety standards.

Accelerated testing.

Existing building codes to be amended to include

a) Fuel cells, turbines, and engines in building
b) H_2 plumbing and storage

6. Hydrogen application barriers

There are no national or international standards. Similarly, safety codes and standards are not uniform. Fuel costs are high.

Customer may be risk averse. Zeppelin mishap, almost a century earlier, has left a deep scar.

Fuel cell life is low, costs are high, and cost of storage is high.

Technologies are still immature. Reformers are also costly.

Infrastructure for distribution and delivery is still to be developed.

There is not enough public awareness of the whole subject.

Requires RandD on high-pressure hydrides, nanotechnology, codes, and standards.

7. Hydrogen outreach and break out session

a) Target audiences—college students, policy makers, foundations, RandD communities.
b) Cultural barriers:

Not enough patience to hold an for a long term vision.

Too many immature ideas pursing one another, no compulsion to change.

Curriculum not widely developed.

c) Financial barriers:

Not enough incentives.

Weak public private partnership and policy.

8. Hydrogen system integration. Following need attention:

a) Cross cutting R&D on catalysis, materials, manufacturability, and mass production required.
b) Technology on carbon sequestration refueling facilities.

Analysis of life cycle costs.
Uneven development of subsystems.

A lot of the needs and barriers are getting mixed up in the above. However, there is much need for research concentration and funding. Both are being highlighted and attended to.

APPENDIX 17-2 HTGR KNOWLEDGE BASE

IAEA-TECDOC—1085: Hydrogen as an Energy Carrier and its Production by Nuclear Power*

Abstract

The impact of power generation on environment is becoming an ever increasing concern in decision making when considering the energy options and power systems required by a country in order to sustain its economic growth and development. Hydrogen is a strong emerging candidate with a significant role as a clean, environmentally benign and safe to handle major energy carrier in the future. Its enhanced utilization in distributed power generation as well as in propulsion systems for mobile applications will help to significantly mitigate the strong negative effects on the environment. It is also the nuclear power that will be of utmost importance in the energy supply of many countries over the next decades. The development of new, innovative reactor concepts utilizing passive safety features for process heat and electricity generation are considered by many to play a substantial role in the world's energy future in helping to reduce greenhouse gas emissions. This report produced by IAEA documents past and current activities in Member States in the development of hydrogen production as an energy carrier and its corresponding production through the use of nuclear power. It provides an introduction to nuclear technology as a means of producing hydrogen or other upgraded fuels and to the energy carrier hydrogen and its main fields of application. Emphasis is placed on high-temperature reactor technology which can achieve the simultaneous generation of electricity and the production of high-temperature process heat.

REFERENCES

1. G. Zorpette, Hawaii's Geothermal Program, *IEEE Spectrum*, Volume 29, Issue 2, February 1992, pp. 49–50.
2. A. T-Raissi, and D. L. Block, Hydrogen: Automotive Fuel of the Future, Power and Energy Magazine, Volume 2, Issue 6, November–December 2004, pp. 40–45.

*International Atomic Energy Agency, Vienna (Austria). http://www.iaea.or.at/inis/aws/htgr/abstracts/abst_30027279.html.

3. P. P. Predd, Cheaper Hydrogen Beckons [Hydrogen Energy], *IEEE Spectrum,* Volume 42, Issue 3, March 2005, p. 16.

BIBLIOGRAPHY

Agbossou, K., M. Kolhe, J. Hamelin, and T. K. Bose, Performance of a Stand-Alone Renewable Energy System Based on Energy Storage as Hydrogen, *IEEE Transactions on Energy Conversion,* Volume 19, Issue 3, Sept. 2004, pp. 633–640.

Andrews, C. J. and S. A. Weiner, Visions of a Hydrogen Future, *Power and Energy Magazine,* Volume 2, Issue 2, March–April 2004, pp. 26–34.

Boenig, H. J., J. B. Schillig, H. E. Konkel, P. L. Klingner, T. L. Petersen, and J. D. Rogers, Design Installation And Commissioning of the Los Alamos National Laboratory Pulsed Power Generator, *IEEE Transactions on Energy Conversion,* Volume 7, Issue 2, June 1992, pp. 260–266.

Bretz, A. A. and W. Sweet, Addressing the Coal Conundrum, *IEEE Spectrum,* Volume 36, Issue 12, December 1999, pp. 40–41

Energy Magazine, Volume 2, Issue 2, March–April 2004, pp. 4–6.

Fairley, P., The Greening of GE [Alternative Energy], *IEEE Spectrum,* Volume 42, Issue 7, July 2005, pp. 28–33.

Gurney, J. H., Building a Case for the Hydrogen Economy, *Power and Energy Magazine,* Volume 2, Issue 2, March–April 2004, pp. 35–39.

Jain, P., U. C. Bhakta, and S. K. Sanyal, Hydrogen Leakage Due to Stress Corrosion Cracking of Stator Conductors in 210 MW Generator: A Case Study, *IEEE Transactions on Energy Conversion,* Volume 15, Issue 1, March 2000, pp. 57–60.

Jones, W. D., Take This Car and Plug it [Plug-in Hybrid Vehicles], *IEEE Spectrum,* Volume 42, Issue 7, July 2005, pp. 10–13.

Kishinevsky, Y. and S. Zelingher, Coming Clean with Fuel Cells, *Power and Energy Magazine,* pp. 20–25.

Kolhe, M., J. C. Joshi, and D. P. Kothari, Performance Analysis of a Directly Coupled Photovoltaic Water-Pumping System, *IEEE Transactions on Energy Conversion,* Volume 19, Issue 3, Sept. 2004, pp. 613–618

Lahey Jr., R. T., R. P. Taleyarkhan, and R. I. Nigmatulin, Bubble Power [Other Sources of Nuclear Energy], *IEEE Spectrum,* Volume 42, Issue 5, May 2005, pp. 38–43.

Lelieveld, J., V. Ramanathan, and P. J. Crutzen, The Global Effects of Asian Haze [Air Pollution], *IEEE Spectrum,* pp. 50–54.

Li, Y. H., S. S. Choi, and S. Rajakaruna, An Analysis of the Control and Operation of a Solid Oxide Fuel-Cell Power Plant in an Isolated System, *IEEE Transactions on Energy Conversion,* Volume 20, Issue 2, June 2005, pp. 381–387.

Olken, M., Hydrogen Economics this Issue Brought to You by the Letter "H", Power and

Perry, T. S., General Motors on the HY-Wire, *IEEE Spectrum,* Volume 41, Issue 1, January 2004, pp. 64–65.

Roberts, B., The promise of Hydrogen is it the Super Power of the Future? *Power and Energy Magazine,* Volume 2, Issue 2, March–April 2004, pp. 21–24.

Senjyu, T., T. Nakaji, K. Uezato, and T. Funabashi, A Hybrid Power System Using Alternative Energy Facilities in Isolated Island, *IEEE Transactions on Energy Conversion,* Volume 20, Issue 2, June 2005, pp. 406–414.

Voelcker, J. Top 10 Tech Cars, *IEEE Spectrum,* Volume 41, Issue 3, March 2004, pp. 28–35.

18

BASIC STRUCTURE OF POWER MARKETING

18.1 RECONSTRUCTION OF THE ELECTRICITY BUSINESS

The reconstruction has been proceeding stage-wise as below:

Stage 1. Old vertically integrated electricity business had four sectors that operated in a follow-up sequence. These were generation → transmission → distribution → customers. The main feature was a central office that covered every operation from planning to bill collections. The returns were fixed on the basis of reasonable returns on investment by an independent regulator.

Stage 2. Competition was introduced in generation. Independent power producers (IPPs) were allowed to generate. There was only one central buyer, the old monopoly.

Stage 3. IPP's were allowed to sell directly to large customers. It was compulsory for the monopoly to allow transmission and distribution over their networks for charges fixed by the regulator.

Stage 4. Competition was opened up in distribution also. Distribution was sectionalized into zones. These zones were either sold or leased to independent distributors. These independents had the freedom to buy from a supplier of their choice. Open access to transmission as well as distribution was compulsory. Large customers traded independently.

Electricity Power Generation. By Digambar M. Tagare

Copyright © 2011 the Institute of Electrical and Electronics Engineers, Inc.

Stage 5. Large customers were, step-wise, along with clusters of retail customers, now free to buy anywhere and use the networks for a charge, subject to operational coordination. The electric supply started from generators and was carried by transmission to the distribution. The distribution had a nodal point from which various entities like DISTCO, retail consumer groups, and large consumers drew their supply. The reconstruction system that we are discussing ends at the nodal point as far as new authorities in reconstruction are concerned.

18.2 UNBUNDLING OF OLD MONOPOLY

In this whole process, the old monopoly was divided into independent identities:

Genco—Covering generation old and new when added
Transco—Covering transmission lines and substations from 33 kV upward
Distco—Distribution company covering activities below 33 kV systems

At step 5 above there were a multitude of possible interactions, as shown in Figure 18-1.

18.3 OPEN ACCESS TO CRITICAL FACILITIES

The key to the successful operation of the whole scheme was an open access. The system was now available to any party who requested for use of transmission and distribution networks for their contracted quantity of electricity at their contracted time.

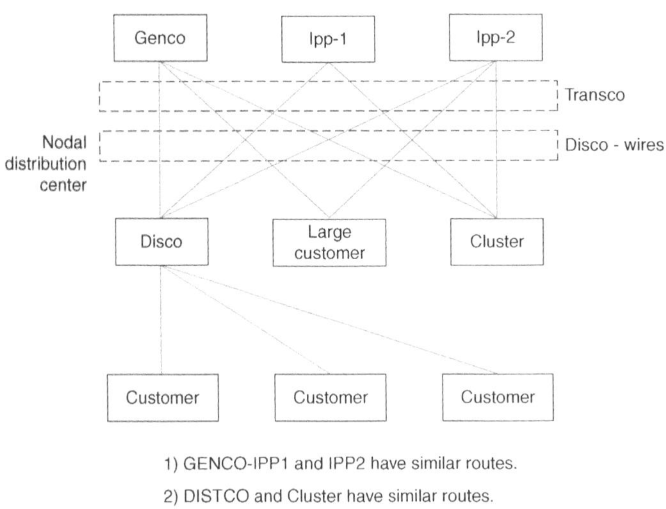

Figure 18-1. Competition at stage 5.

Stage 6. With the open access system in position, the next stage was ripe to treat electricity as a commodity and trade it on a power exchange. At this stage, traders were also allowed in to buy and sell.

Somewhere between stage 5 and stage 6, the electricity business changed from a bilateral trade to a multiparticipant, free/open marketing business. Electricity became a tradable community. Tariffs went away, their position being taken by spot or market prices. Contracts for supply could also be traded. Prices stabilized between a limit where the price of the marginal last suppliers highest bracket (peak time) was accepted in the market and a limit where the lowest bid of the marginal purchases was accepted.

18.4 HOW DOES THE NEW SYSTEM WORK?

We now have a picture with a number of stakeholders. To begin with we have IPPs (Independent Power Producers) and Genco. They have to compete in a free-for-all market to get business through various means, such as prices, quality and reliability of supply, and long-term contracts. They also have an opportunity/temptation to take unfair advantage in a free system.

We have a number of customers including large industries, Distco, load distribution agencies, and even retail customer groups. All of them are looking for a price advantage along with quality and reliability.

Electricity has become a tradable commodity that one can bid to supply or bid to purchase in a marketplace. The marketplaces are electricity exchanges that are regional and centrally controlled. If there is more than one exchange for a region, they are interconnected. This makes the trading and supply–purchase operations uniform, central, and almost instantaneous.

Electricity travels at the speed of light. Even with today's progress in science and technology, it cannot be stored. This means that production and consumption must match in a system on an instant-to-instant basis. With a multiplicity of suppliers and consumers spread over a region, balancing supply and consumption needs a central system operator (technical), who can corelate this requirement on a continuous basis.

What applies to physical coordination, or managing the load flow, also applies to marketing. We have a number of supply bidders and purchase bidders and the level where the stock prices of this bidders match have to be proclaimed on an hourly basis. Here we require another central system operator (financial).

It is imperative that both the above operators should be independent of any influence from any of the market participants. To achieve this, there is a regulator or a regulatory commission that acts as a watchdog entity.

18.5 MARKET PARTICIPANTS AND THEIR FUNCTIONS

The Northern Electricity Regional Council (NERC) of the United States is considered a leader in setting down self-imposed standards for electricity industries. Their norms have long been respected. They have defined market operators as:

1. Transmission companies
2. Generation companie
3. Traders, marketers, brokers, and aggregators
4. Distribution companie
5. Load-serving entities
6. End users

These entities are further divided into two groups as per their functions (see Table 18-1).

18.6 NEW KEY PERSONNEL

The following key personnell play important roles in the new scenario:

A regulator or regulatory commission (regional or central)
A system operator—technical side
A system operator—financial side

18.6.1 Role of a Systems Operator (Technical)

A system operator controls power flow between multiple suppliers and customers over a variety of transmission paths. A system operator should have no connection with or interest in other market players such as generators, distributors, retailers, and traders. His functions are:

- Coordinate generation schedules and load dispatch schedules in advance.
- Arrange real-time power flow from generation to load on an instant-to-instant basis. There could be last-minute variations.
- Call for special outputs bracketed under auxiliary services.
- Hold sufficient reserves in a ready-to-run state.
- Avoid traffic congestion.
- Arrange least loss/least cost transmission routes.

Table 18-1. Market participants and their functions

Service functions	Operating functions
Reliability authority	Generator
Interchange authority	Load serving entity
Balancing authority	Purchase, selling authority
Transmission service provider	Transmission owner
Planning authority	Distribution provider

Source: [1].

18.6.1 Role of a System Operator (Financial)

This is best understood if we study the operation of free electricity markets. The market operator evaluates demand bids against day-ahead generator offers. He produces a set of prices and schedules that are consistent with both the bids and offers. These are posted at regular intervals on the Internet. At the end of the day ahead, next day's schedules are fixed. In fact, a majority of the contracts are finalized in the day-ahead market. The next day in the real-time market, the participants are allowed to trade. Actual transactions are finalized at regular intervals. The results are put finally on the Internet. At the end of the day, balancing of participant's accounts are also put up.

If the bids are accepted in the day-ahead market, a buyer can plan consumption for the next day. If he cannot utilize the bid quantity, it is bought back at real-time prices and sold. If he requires electricity over and above what he has bid, the operator will buy it for him at real-time spot prices.

18.7 ROLE OF A REGULATOR OR REGULATORY COMMISSION

This is an independent role not subject to government intervention. Its functions are:

1. Licensing of power plants
2. Tariff price cap setting with respect to tariffs in the case of old structures or price caps in the case of free markets, with considerations of:
   - Social balance
   - Nonexploitation
   - Profitability and investments prospects. Promoting capital investment depends on this aspect
   - Stability of business
   - Setting service and efficiency models
   - Setting transmission charges
   - Keeping a watch on and preventing misuse of market power
3. Addressing environmental and social issues
4. Act as a public redress system

18.8 TOOLS FOR THE SYSTEM OPERATOR

The system operator has the following tools in his hands:

Function	**Tools**
Scheduling	Bids from buyers and offers from suppliers.
Imbalance	Imbalance market.

Function	Tools
Congestion	Congestion market with tradable transmission rights
Market spot prices	Short-term contract trading
Reserves	Reserves market
Transmission rights	Physical or financial rights

18.9 SECONDARY MARKETS

Apart from the wholesale electricity power markets, secondary markets also exist on the commodity stock exchange for the following:

1. Fixed transmission rights (FTRs) to cover congestion charges. Please note that the transmission service provider will ramp up his congestion charges to an extent that will discourage dispensable load and he can then balance the system. These congestion charges could go high.
2. Some generators can hold back a part of their capacity as spinning reserve in the operational reserve and unload it when the demand comes. These reserves are tradable. However, there is a market watch to ensure that this does not lead to unfair market power.
3. What applies to the supply side also applies to the demand side. A portion of supply under a supply contract can be relinquished for a price. For a specific consumer, consumption-based returns are diminishing beyond a certain point. It will be profitable for him to sell and relinquish his hold. His consent is tradable on the secondary market with respect to unserviced contract demand.

The penalties for nonperformance and not honoring demand supply contracts are very heavy. Secondary markets help out the affected parties.

18.10 FREE MARKET OBJECTIVES

The Federal Energy Regulatory Commission (FERC) had issued a working paper inviting comments on standard service and wholesale markets design [2]. The objectives laid down in this third phase of reform are as below.

18.10.1 Objectives for the Transmission Systems [2, 3]

1. In principle, all load should be under a single transmission tariff.
2. Demand response is essential in a competitive market to:

    Assure an efficient interaction between supply and demand

    Check market power

    Provide consumer a choice

3. Network access service charges to replace old transmission service charges under FERC's old order of No. OA-888.
4. Transmission rights to be tradable. Method of their allocation and trading to be transparent.
5. For the wholesale electricity market, operating reserves become ancillary, tradable services.
6. Marketing power is to be monitored and inhibited.

18.10.2 Objectives for the Wholesale Market: A Standard Market Design (SMD)

The objectives of the SMD are laid down as below:

1. It should lead to economically efficient operations, leading to lowered costs of delivered wholesale energy.
2. While maintaining reliability, it should offer increased choices to the participants.
3. Between regions, the standards and regulations should be compatible and, if possible, uniform. This will take care of in-between "seams" issues.
4. Market rules should be fair and transparent. These should be "fuel and technology" neutral.
5. Transmission system operations must be independent of participants.

The system is still evolving.

18.11 SUCCESS OF THE FREE MARKET

This depends on the following

1. There are many buyers and many sellers. There has to be lack of market power on both sides.
2. Demand and supply should be responsive to price. A rise in price should contract the demand. A fall in price should diminish production and supply.
3. The marketplace (exchange) should promote liquidity and efficiency in transactions.
4. There should be equal access to essential facilities—transmission and distribution networks.
5. Subsidies and environmental controls should be so treated that they do not interfere with the working of the market.

In this scenario, a regulator or a regulatory commission—regional and/or central—plays a pivotal role.

18.12 HOW DO ELECTRICITY MARKETS OPERATE?

There are two markets operating simultaneously: a day-ahead market and a real-time market.

There are two separate commodities traded: regular energy in megawatt hours and secondary commodities like transmission reserves and generator reserves.

At the beginning of a day, the generators advise the pool of the availability and schedule of energy delivery for the next day. Transmission advises on the same. A program of supplies and bids are finalized and put on the Internet for the next day's operations.

In the next day-ahead market, bidding starts for both volumes and time slots. They are tradable. However, the bids and supplies are balanced continuously and the results are put on the Internet at regular intervals. There is only trading. At the end of the day, the balanced results are fixed for the next day and put on the Internet.

The next day in the real-time market, the participants are allowed to trade. Actual transactions are finalized at regular intervals. The results are put finally on the Internet. At the end of the day, balancing of participant's accounts are also put up.

18.13 FLOW OF OPERATING FUNDS

The distributor has all the responsibilities beyond the nodal center. From there on up to the customer's premises, it is his responsibility to manage the power flow and also look after the financial side.

On the technical management side, he has to maintain the distribution system in good order, adding lines and service stations as the business expands. His charges on distributing energy are determined by the regulator. His metering accuracy and coverage are dictated by the regulator. On the financial side, he collects the revenue and passes them on to the central financial operator. This operator, in turn, makes out payments to market players such as generators, transmitters, and distributors. The operator also collects fines and balances settlement payments from the market operators.

18.14 EFFECT OF RECONSTRUCTION ON ELECTRICITY BUSINESS—CAPITAL INVESTMENT PROSPECTS

18.14.1 Generation

Prices and trends of prices lead the way to decisions on investment and on the type of generation. Under the old system, it was compulsory for an electricity utility to supply all the electric power that has was asked for. A "golden rule" allowed one day of partial outages in 10 years. Penalties for irregular supply were crippling. The planners had to predict rise in demand, add reserves to it, and start designing. The planning process started from arithmetic or statistical figures.

Under the free market system, prices and price trends as well as their variation during the time of the day start the decision process. An important supporting factor for this is VOLL—value of lost load. A customer loses production/service due to nonsupply of electricity. His cost of loss could be equal to the maximum price he is ready to pay. Thus, VOLL could be a deciding factor in a decision to invest in generation. Information collected on location, duration, and period of day when VOLL is likely, helps an investor to decide on the type of generator he can profitably install his likely price realization, and so on. Thus, the price of energy and its trend become the starting point for planning investment in generation.

18.14.2 Peak-Load Generators and Base-load generators

Let us consider the peak-load period, in which prices are the highest since the demand is overreaching the supply. The upper level that prices can reach could match VOLL for a needy customer. Gas turbine sets, which can be quickly started and having a reasonably low medium capacity, are ideal here. Their cost of electricity production is the highest. The cost of input energy, the gas, is highly volatile. The new regime provides good incentives for investment in this type of generation. They are also ideal for situating at nodal distribution points.

When a region is developing fast, its 24 hour demand for supply is also growing fast. This region will promote investment in new coal-based or nuclear-fuel-based megagenerators. Here again, the starting incentive for generation planning is the price.

There are mid load suppliers, generally old, re-engineered plants, purchased from the old utilities. Although their energy costs are low, maintenance costs are high.

18.14.3 Investment and Costs of Compliance with Emission Control Measures

A pound of coal produces x RWH of electricity along with y g of SO_2, NO_x, and CO_2. Thus, the emissions in tons of polluting gases are tied with megawatt hours of power associated with them. There is an understanding on limiting y and reducing it progressively so that the air we breathe is enjoyable and healthy. Countries have voluntarily accepted caps on these emissions. These are proportionately passed on until they reach the generator stage. Indirectly, the generation is limited unless measures are taken to deal with this cap.

18.14.4 BACT Favored by Regulators

In the old system, a generator invests in the best available (emission) control technology (BACT) approved by a regulator and the regulator allows him to produce more power. The regulator also allows him to pass on this cost to the consumer. The investment in BACT is balanced by the extra power a generator can produce and increase his profits.

Under the free market system, he has still to balance his investment decision on BACT against emission control profits to be earned. Since the prices are determined by the market with a spot pricing system, he cannot pass on the extra costs to the customer. He has to deduct them from his net profits. On the negative side, he has to curtail his production.

18.14.5 Output Limitations

The generator has an option of not investing in BACT but of reducing his production, which is more economical. The supply gets lower and the prices will go up. In this case, the consumer pays environmental costs in addition to curtailed power supply. A regulator will be on the watch to see whether the generator is cutting down production to increase the spot prices. That is market power and is prohibited.

18.14.6 Cap and Trade

The pollution controller has another tool in his bag—cap and trade—to encourage investment in BACT. A generator is allowed so much yearly capacity to produce so many tons of CO_2. This capacity is tradable. He exhausts his cap and then buys it from others to keep up his generation. Alternatively, he uses less of his cap and sells the balance. His investment decision on BACT will be governed by the price and volume of cap he has to buy. This method is being universally adopted.

Some forms of renewable energy such as wind power and small hydro power can earn cap or carbon trading credits in terms of the megawatt hours they produce. They can sell these on the secondary market and earn extra profits, another incentive to invest in generation through renewable energy sources.

18.14.7 Effect on Transmission Systems: Investment Incentives and Responsibilities

Background. Old monopolies planned for transmission systems from their generators to the nodal distribution centers. They had all the design parameters. Under the new scenario, they continue to own the transmission systems but have no control of them. An independent system operator takes charge of power flow arrangements as well as the business settlements. Interregional transmission was created by national agencies. This transmission also goes under independent control.

Revenues for Transmission. The owners of the transmission system now get paid for conveying the electricity. These charges include costs of transmission losses plus costs of congestion changes plus overhead costs. The overhead costs include a reasonable rate of investment (RRI).

PTR and FTR. Over and above this, as explained earlier, they can sectionalize their systems and sell transmission rights over given sections and at scheduled times to market participants who want to hedge themselves against unpredictable congestion changes. These are physical transmission rights without reference to the amount of energy that can be conveyed. There is a proposal [4] for FTRs that includes the amount of energy, say, 100 MW from point A to point B between 6.30 PM to 8.30 PM. This 100 MW can be split into two FTRs of 50 MHW each. The PTR and FTR can be traded on the secondary markets.

Over and above these, a Transco owner can collect connection charges and access charges from a generator.

A location-wise pricing system (LMP) is recommended for universal acceptance. Essentially, it consist of:

1. Access charge—a standard charge to be paid by all users to allow the owner to recover embedded costs of the transmission system.
2. Costs of losses over the pathway from the source to the sink.
3. Costs of congestion across the pathway.

It is expected that all the above constitute sufficient incentives for the existing Transco to invest in expansion or in new regional lines.

A wind energy farm might be interested in participation in laying additional transmission lines or supporting existing ones. Please see Chapter 10 on this.

18.15 NATIONAL GRID TRANSMISSION SYSTEM

National grid transmission lines are looked upon as a national responsibilities, on a par with national highways. In most countries, the government builds and expands these. They are also subject to operation by the ISO and payments through the ISC.

With increase in number of transactions and increase in energy demand, and also with a multiplicity of market operators, transmission congestion is a natural outcome. A great deal of R&D is going onto tackle this subject.

APPENDIX 18-1 A VAST ARRAY OF TOOLS TO SUPPORT TOMORROW'S MARKET PARTICIPANTS

Today's power market may be likened to a multiheaded hydra, where all the heads have to be fed, listened to, and kept active against a raving competition. The system has not finally evolved yet. It requires a vast array of programs. Common requirements of these programs are that they should be very nearly standard, adaptable and interfaceable for whatever changes and new applications evolve. Albuyeh [1] sums up these arrays as below (see Figure 18-2).

A) Integration infrastructure:

This is an important messaging tool that integrates various components of the system. These components consist of 11 basic services, such as configuration, alarming, messaging, time, task, scheduling, database, archiving, logging, calculations, security, and user base. It aims for long and short term developments for its own and third-party use. It aims for reusability of components and for easing in legal systems.

B) Generation management suite:

a) Real-time communication and monitoring, along the lines of old SCADA systems.

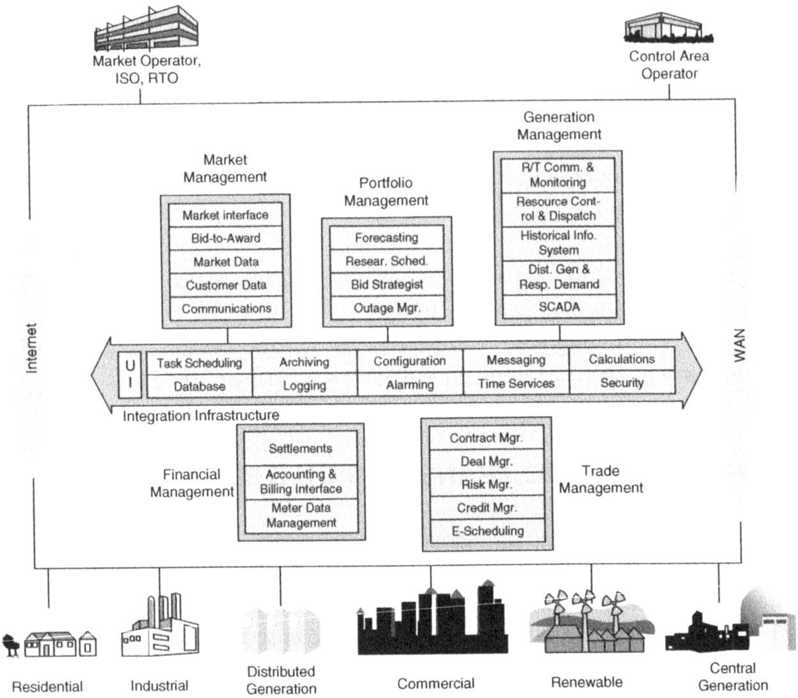

Figure 18-2. Overview of market participant package. (From [1], © IEEE 2003.)

b) Resource control and dispatch, based mainly on the price and cost of dispatch (old-time economic dispatch). Load control and dispatch has to match along with frequency in permitted variation limits.
c) Distributed generation has to be monitored and controlled and be properly fitted in the slots.
d) Responsive demand management—a new version of demand side management (DSM).

C) Market management suite:

a) Market interface manager who looks after energy, ancillary service, trading rules, data exchanges between central market operator and market participants, as also (A) and (B) above.
b) Bid-to-award manager who really fits all the pieces in the continuously changing jigsaw puzzle, tying the bidder to the market operator with the help of a variety of data and tools.
c) Market data manager, a data handler who covers data from central market, congestion data, outages schedules, LMPs, zonal prices, demand forecasts, and so on.

D) Trade management suite:
a) Contract manager
b) Trade manager
c) Risk manager
d) Credit manager
e) Electronic scheduling (open access; all time requirements have to be met)

E) Portfolio management suite:
a) Forecasting
b) Resource scheduling and optimum schedules for generators and loads have to be worked out. A variety of programs are available for this.
c) Bid strategist support [C (c) above].

F) Financial management suite:
a) Settlement statements monthly
b) Standing data input management—integrates C, D, and E above on standing data.
c) Dynamic data input management—integrates C above and meter data interface to handle bids, awards, deployment, and metered values.
d) Settlement calculations
e) Dispute management
f) Report management
g) Accounting and billing interface
h) Meter data management

REFERENCES

1. F. Albuyeh and J. Kumar, Decision Support Tools for Market Participants, *IEEE Transactions on Power Systems,* Issue 2, May 2003, pp. 512–516.
2. Federal Energy Regulatory Commission, Working paper on Transmission Service and Wholesale Electric Market Design.
3. Analysis of "Federal Energy Regulatory Commission, working paper on standard service and wholesale electric market design," The E-cubed Company, LLC, New York, April 2002.
4. S. Hunt, *Making Competition Work in Electricity,* Wiley, New York, 2002.

BIBLIOGRAPHY

Bjorgan, R., C.-C. Liu and J. Lawrence, Financial Risk Management in a Competitive Electricity Market, *IEEE Transactions on Power Systems,* Issue 4, November 1999, pp. 1285–1291.

Bouffard, F., F. D. Galiana, and A. J. Conejo, Market-Clearing with Stochastic Security-Part I: Formulation, *IEEE Transactions on Power Systems,* Issue 4, November 2005, pp. 1818–1826.

Brown, R. E. and J. J. Burke, Managing the Risk of Performance Based Rates, *IEEE Transactions on Power Systems,* Issue 2, May 2000, pp. 893–898.

Chattopadhyay, D., An Energy Brokerage System with Emission Trading and Allocation of Cost Savings, *IEEE Transactions on Power Systems,* Issue 4, November 1995, pp. 1939–1945.

De Tuglie, E. and F. Torelli, Load Following Control Schemes for Deregulated Energy Markets, *IEEE Transactions on Power Systems,* Issue 4, November 2006, pp. 1691–1698.

Dubhash, N. K, A Critical Overview of Recent Global Experience with Electricity Restructuring, *Economic and Political Weekly,* Volume XL, No. 50.

Dubhash, N. K., Global Experience with Electricity Reform, *Economic and Political Weekly,* Volume XL, No. 50

Fahrioglu, M. and F. L. Alvarado, Designing Incentive Compatible Contracts for Effective Demand Management, *IEEE Transactions on Power Systems,* Issue 4, November 2000, pp. 1255–1260.

Hamoud, G. and I. Bradley, Assessment of Transmission Congestion Cost and Locational Marginal Pricing in a Competitive Electricity Market, *IEEE Transactions on Power Systems,* Issue 2, May 2004, pp. 769–775.

Kirschen, D. S., Demand-Side View of Electricity Markets, *IEEE Transactions on Power Systems,* Issue 2, May 2003, pp. 520–527.

Lamoureux, Ul Electrical Energy Policy, *Power Engineering Review,* July 2002, pp. 12–16.

Li, C.-A, A. J. Svoboda, X. Guan, and H. Singh, Revenue Adequate Bidding Strategies in Competitive Electricity Markets, *IEEE Transactions on Power Systems,* Issue 2, May 1999, pp. 492–497.

Plazas, M. A., A. J. Conejo, and F. J. Prieto, Multimarket Optimal Bidding for a Power Producer, *IEEE Transactions on Power Systems,* Issue 4, November 2005, pp. 2041–2050.

Rau, N. S., Issues in the Path Toward an RTO and Standard Markets, *IEEE Transactions on Power Systems,* Issue 2, May 2003, pp. 435–443.

Silva, C., B. F. Wollenberg, and C. Z. Zheng, Application of Mechanism Design to Electric Power Markets (Republished), *IEEE Transactions on Power Systems,* Issue 4, November 2001, pp. 862–869.

Singh, H. and A. Papalexopoulos, Competitive Procurement of Ancillary Services by an Independent System Operator, *IEEE Transactions on Power Systems,* Issue 2, May 1999, pp. 498–504.

19

LOOKING INTO THE FUTURE

Predictions in the rise in electricity demand were touched upon in Chapter 2. In the near future, the automobile industry will be the biggest consumer of oil and gas fuels, as compared to the production of electricity, to the extent of some 80%. It also is important in respect to atmospheric pollution. Emission compliance limits are specified under European and national standards for cars. These standards are continuously upgraded.

Although SO_2 and NO_x can be detained in case of automobiles with the help of catalytic scrubbers, CO_2 emissions cannot be detained. A move to electrically driven cars is being seriously pressed upon the automobile industry.

The electrically driven cars of the near future will be serviced at the equivalent of gas stations along the road. At best, they will be charged up at nighttime at residences and at places of deposit. At some time not too far off in the future, the thermal–mechanical power generation of cars will be added to the total electricity power generation. If anything, the thermal electricity power generation will shoot up, a relief being granted on CO_2 emission limits by transferring the CO_2 allocation from automobiles to electricity. A point to note is that efficiency norms will have to be further increased by thermal electricity power generation with a hope that some innovations will arrive in due time to give relief.

Electricity Power Generation. By Digambar M. Tagare

Copyright © 2011 the Institute of Electrical and Electronics Engineers, Inc.

Let us a take a look at a companion industry engaged in mass serving as is the electricity power industry—the telephone industry. It has more or less the same distribution network and customer spread as electricity. It was a leader in reconstruction. The electricity industry followed its lead and is still facing similar grid congestion problems.

At the customer end, the telephone industry has been continuously innovating. As a result, in almost all the countries, hand-held mobile cell phones outnumber distributed land lines. It serves increasingly more customers.

Can we forecast a similar electricity energy storage package for a customer who can pick it up in a store and take it home or put it in a car and drive away?

INDEX

Copyright © 2011 the Institute of Electrical and Electronics Engineers, Inc.

Printed and bound by CPI Group (UK) Ltd, Croydon, CR0 4YY
16/09/2026
14963461-0003